Developments in Earth Surface Processes
Volume 21

Mountain Ice and Water

Developments in Earth Surface Processes, 21

Series Editor – J.F. Shroder, Jr.

For previous volumes refer
http://www.sciencedirect.com/science/bookseries/09282025

Developments in Earth Surface Processes
Volume 21

Mountain Ice and Water

Investigations of the Hydrologic Cycle in Alpine Environments

Edited by

Gregory B. Greenwood
Mountain Research Initiative, Institute of Geography,
University of Bern, Switzerland

J.F. Shroder, Jr.
Center for Afghanistan Studies and Geography and Geology,
University of Nebraska, Omaha

ELSEVIER

AMSTERDAM • BOSTON • HEIDELBERG • LONDON • NEW YORK • OXFORD
PARIS • SAN DIEGO • SAN FRANCISCO • SINGAPORE • SYDNEY • TOKYO

Elsevier
Radarweg 29, PO Box 211, 1000 AE Amsterdam, Netherlands
The Boulevard, Langford Lane, Kidlington, Oxford OX5 1GB, United Kingdom
50 Hampshire Street, 5th Floor, Cambridge, MA 02139, United States

Library of Congress Cataloging-in-Publication Data
A catalog record for this book is available from the Library of Congress

British Library Cataloguing-in-Publication Data
A catalogue record for this book is available from the British Library

ISBN: 978-0-444-63787-1
ISSN: 0928-2025

Publisher: Candice Janco
Acquisition Editor: Amy Shapiro
Editorial Project Manager: Tasha Frank
Production Project Manager: Anitha Sivaraj
Designer: Greg Harris

Typeset by TNQ Books and Journals

Contents

List of Contributors

N. Barker, School of Plant Sciences, University of Pretoria, South Africa

L. Chernova, Institute of Geography Russian Academy of Sciences, Moscow, Russia

G. Choudhury, University of North Bengal, Siliguri, West Bengal, India

V.R. Clark, Botany Department, Rhodes University, South Africa

A. Claustres, Université de Toulouse; INP, UPS, EcoLab/ENSAT, Castanet Tolosan, France; CNRS, EcoLab, Castanet Tolosan, France

F.A. Engelbrecht, Climate Studies, Modelling and Environmental Health, Natural Resources and the Environment, Council for Scientific and Industrial Research (CSIR), South Africa

J.W.H. Ferguson, Centre for Environmental Studies, University of Pretoria, South Africa

A. Fischer, Institute for Interdisciplinary Mountain Research, Innsbruck, Austria

S.V. Hansson, Université de Toulouse; INP, UPS, EcoLab/ENSAT, Castanet Tolosan, France; CNRS, EcoLab, Castanet Tolosan, France

L. Hartl, Institute for Interdisciplinary Mountain Research, Innsbruck, Austria

K. Helfricht, Institute for Interdisciplinary Mountain Research, Innsbruck, Austria

M.T. Hicks, City University of New York, New York, United States

V. Hreinsson, Reykjavik Academy, Reykjavik, Iceland

Á.D. Júlíusson, Reykjavik Academy and National Museum of Iceland, Reykjavik, Iceland

T. Khromova, Institute of Geography Russian Academy of Sciences, Moscow, Russia

G. Le Roux, Université de Toulouse; INP, UPS, EcoLab/ENSAT, Castanet Tolosan, France; CNRS, EcoLab, Castanet Tolosan, France

A. Muraviev, Institute of Geography Russian Academy of Sciences, Moscow, Russia

S. Nikitin, Institute of Geography Russian Academy of Sciences, Moscow, Russia

G. Nosenko, Institute of Geography Russian Academy of Sciences, Moscow, Russia

A.E.J. Ogilvie, Stefansson Arctic Institute, Akureyri, Iceland and Institute of Arctic and Alpine Research, Boulder, Colorado, United States

F.O. Sarmiento, University of Georgia, Neotropical Montology Collaboratory, Athens, Georgia, United States

R. Seidler, University of Massachusetts Boston, Boston, MA, United States

B. Seiser, Institute for Interdisciplinary Mountain Research, Innsbruck, Austria

G. Sharma, The Mountain Institute-India, Gangtok, Sikkim, India

R. Sigurðardóttir, Reykjavik Academy, Reykjavik, Iceland

M. Stocker-Waldhuber, Institute for Interdisciplinary Mountain Research, Innsbruck, Austria

S. Strachan, University of Nevada, Reno, NV, United States

S.J. Taylor, Centre for Environmental Studies, University of Pretoria, South Africa

Y. Telwala, Global Mechanism, UNCCD, Rome, Italy

S. Van Rensburg, South African Environmental Observation Network, Pretoria, South Africa

H. Wiesenegger, Hydrographical Survey of the Federal Government of Salzburg, Salzburg, Austria

N. Zverkova, Institute of Geography Russian Academy of Sciences, Moscow, Russia

Editorial Foreword

Mountain Ice and Water is a new volume of papers reviewed and edited by John (Jack) Shroder, Emeritus Professor of Geography and Geology at the University of Nebraska at Omaha, USA, and Greg Greenwood, Director of the Mountain Research Initiative from Geneva, Switzerland. The chapters were derived from research papers that were delivered at the Perth III Conference on *Mountains of our Future Earth* in Scotland in October 1915. The conference was established to help develop the knowledge necessary for responding effectively in the coming decades to the risks and opportunities of global environmental change and supporting transformations toward global sustainability. To this end, the conference and the book have investigated the future situation in mountains from three points of view: (1) Dynamic Planet: Observing, explaining, understanding, and projecting Earth, environmental, and societal system trends, drivers, and processes and their interactions to anticipate global thresholds and risks; (2) Global Sustainable Development: Increasing knowledge for sustainable, secure, and fair stewardship of biodiversity, food, water, health, energy, materials, and other ecosystem services; (3) Transformations Toward Sustainability: Understanding transformation processes and options, assessing how these relate to human values, emerging technologies and social and economic development pathways, and evaluating strategies for governing and managing the global environment across sectors and scales.

This three-fold structure has permitted the research community working on global change in mountains to align itself more easily with the broader framework established by the *Future Earth* initiative (http://www.futureearth.org/) to provide knowledge and support to accelerate transformations to a sustainable world. Thus this book has a number of chapters from topics all over our planet in terms of major water and ice issues pertaining to planetary dynamics, sustainable development, and ideas about possible transformations toward sustainability. For example, among the diversity of presentations are glacier changes throughout the whole of Russia, water from the highlands in South Africa, water and ice in northern Iceland, hydropower in the Eastern Himalaya and downstream impacts in the same region, ice, water, and mummies in the Andes and other tropical mountains, along with environmental issues in these regions, public policy and governance in mountains, mountain snows and droughts, and changes in natural alpine riverine landscapes. The resulting volume is a number of themes that present some new thinking on mountain ice and water that will help in planning for an uncertain but somewhat threatening future in the face of the climate change that is threatening ever more people and ecosystems.

Because of the need to accelerate production of papers after the conference, and as a result of our desire to maintain reasonably high quality, we had to maintain pressure on the chapter authors to get their papers finished quickly. We also reviewed all papers and requested revisions on a basis of fairly rapid turnaround, thus eliminating more submissions than we would have cared for, but we trust that we have achieved some measure of diversity and interest that can further at least some of the objectives of the conference on *Mountains of Our Future Earth*. Certainly the dynamics of the planet, some transformations toward sustainability, and more sustainable development on a global basis are in evidence in these chapters so that we can recommend them to the scientific community.

J.F. Shroder, Jr.
Gregory B. Greenwood
1 September 2016

Chapter 1

The Drakensberg Escarpment as the Great Supplier of Water to South Africa

S.J. Taylor,*[,1] J.W.H. Ferguson,* F.A. Engelbrecht,[§] V.R. Clark,[¶] S. Van Rensburg[||] and N. Barker**

**Centre for Environmental Studies, University of Pretoria, South Africa; [§]Climate Studies, Modelling and Environmental Health, Natural Resources and the Environment, Council for Scientific and Industrial Research (CSIR), South Africa; [¶]Botany Department, Rhodes University, South Africa; [||]South African Environmental Observation Network, Pretoria, South Africa; [**]School of Plant Sciences, University of Pretoria, South Africa*

[1]Corresponding author: E-mail: staylor@zoology.up.ac.za

Chapter Outline

Mountain Ice and Water. http://dx.doi.org/10.1016/B978-0-444-63787-1.00001-9

INTRODUCTION

The world's freshwater supplies are increasingly constrained by population growth, urbanization, pollution, eutrophication, changing rainfall patterns, and shrinking cryospheric storage, a situation that could cause conflicts (Jury and Vaux, 2007; Rijsberman, 2006; Ohlsson, 2000). Water security, along with food and energy security, has become a global issue (Ludwig and Roson, 2015). Mountains are vital sources of freshwater for rural and commercial agriculture as well as cities, for national and local economies, and for ecosystems and ecosystem services, often with a trans-boundary dimension. Mountains induce precipitation and are able to store large amounts of water either as snow or ice, or in soil, wetlands, or peat areas, for slow release throughout the year. The Intergovernmental Panel on Climate Change (IPCC) warned in 2007 that mountain regions around the world will be particularly affected by climate change and that hydrological changes can already be observed (Christensen et al., 2007).

South Africa is a water-deficient country with mean annual precipitation (MAP) ranging from around 1250 mm in the east to 50 mm in the west with a mean value of around 500 mm (Tyson, 1991). Half of the country has an annual precipitation less than 400 mm. In addition, potential evaporation increases strongly from around 1200 mm annually in the east to 2750 mm in the west (de Villiers, 1996). For this reason, water scarcity in South Africa is considered a potentially significant constraint for economic development, and it is predicted the demand will outstrip supply in the near future (Blignaut and van Heerden, 2009). In addition to this, the impacts of climate change on regional water resources are uncertain. While there is high confidence in the projections of drastic increases in temperature over southern Africa under low mitigation (Engelbrecht et al., 2015), and while the region is likely to become generally drier (Niang et al., 2014), the rainfall features of eastern South

Africa are uncertain. Some models project drier scenarios while others suggested wetter futures (DEA, 2013, p. 3–10; Niang et al., 2014). Even in wetter scenarios, the increase in temperature may still result in reduced water availability (DEA, 2013, pp. 3–10; Engelbrecht et al., 2015).

In 2000, surface water resources were already over-allocated in 5 of 19 Water Management Areas (WMAs) (DEA, 2013, p. 77). The Department of Environmental Affairs (DEA) estimates that South Africa faces shortages of between 2% and 13% of total water requirements by 2025, but if climate change projections and other uncertainties are included, these shortages could be as high as 19–33% by 2025 (DEA, 2013, p. 21).

While not in the group of the world's 30 most water-stressed countries (Maddocks et al., 2015), South Africa's water resources are considered stressed, bordering on water scarce, with a current water availability of 1100 m^3 per person per annum (StatsSA, 2010, p. 7). The United Nations considers a water availability of less than 1700 m^3 per person per annum as "water stress," with values below 1000 m^3 per person per annum considered as "water scarcity" (StatsSA, 2010, p. 7). The Intergovernmental Panel on Climate Change (IPCC) projected that water availability in South Africa could decline in time to below the 1000 m^3 per capita per annum threshold deemed to be a global standard for human "well-being" because of climate change impacts (Bates et al., 2008). As elsewhere in the world, this decline in South Africa will be driven by increased demand by people, farms, cities, and business rather than by the potential for modified precipitation patterns (Beniston, 2003).

During the past century, South Africa focused on engineered systems (dams, inter-basin transfers, tunnels, and pipelines) to safeguard its water supply. National water security in South Africa in the future is now expected to depend on an ability to plan development compatible with mountain catchment ecological infrastructure (Nel et al., 2013). Using existing infrastructure, since 1994, the South African state has provided access to basic water services for 9 million people, mostly those concentrated in the urban areas. The Free Basic Water policy of 2001 aimed to provide the first 6 kL of water free to all households. In 2013, DEA reported that during 2006, 3.3 million people still lacked access to adequate, clean water supplies, with another 15.3 million being without access to sanitation services at that time (DEA, 2013, p. 21).

In terms of the Great Southern Africa Escarpment (Fig. 1.1), key water resource areas include the Eastern Cape Drakensberg (also called the Cape Midlands Escarpment (CME)), the southern Maloti-Drakensberg, and Northern Drakensberg [including the eKangala Drakensberg and the Mpumalanga Drakensberg (MD)] (Nel et al., 2013).

In South Africa and Lesotho, these mountain catchments have now been designated as strategic water areas (Nel et al., 2013). These strategic water source areas together contribute 50% of the region's water supply, captured from less than 8% of the land surface area. The Drakensberg/Maloti escarpment segment of the Great Escarpment of Southern Africa forms a major catchment that supplies water to large parts of southern Africa including

FIGURE 1.1 Champagne Castle and Cathkin Peak in the central Maloti-Drakensberg, reaching an altitude of 3377 m.

Swaziland, Mozambique, and Namibia (Nel et al., 2013; Tyson, 1986), while the other Drakensberg segments are important locally. The biggest catchment is that of the Orange-Senqu River some 100,000 km^2 in extent and draining from the Drakensberg/Maloti mountains westward to the Atlantic Ocean. In the modern era, long-distance transfer of water to several urban and irrigation farming areas has been possible, particularly from the Senqu-Orange catchment of the Maloti-Drakensberg (de Villiers, 1996).

The entire Drakensberg thus forms a critical resource for southern Africa, not only because it provides so much of the water but also because several international cooperative initiatives secure water from this mountain range.

This chapter focuses on the water dynamics and physical hydrology of the entire Drakensberg, management and uses of the water from the Drakensberg, and a perspective on the way forward for an already-overused water resource in a region that is generally water deficient. The discussion also considers long-term changes in these processes. The aims of the chapter are as follows:

1. To summarize the readily available knowledge about the factors affecting water supply from the Drakensberg. These include climatic, socioeconomic, as well as water governance issues.
2. To identify the critical deficiencies with respect to the sustainable management of water from the Drakensberg escarpment.
3. To propose ways of addressing the science and governance deficiencies affecting water from the Drakensberg in order to promote sustainable management of this resource.

THE GREAT ESCARPMENT OF SOUTHERN AFRICA AND THE DRAKENSBERG

After the breakup of Gondwanaland, the interior of southern Africa weathered inland, leaving a raised inland basin surrounded by a retreating wall of rock, the Great Escarpment, the highest part of which is the eastern section, the Drakensberg/Maloti mountains. During the past 20 million years, significant uplift of this plateau basin also occurred, with a larger effect in the east of South Africa (Truswell, 1977). The Drakensberg is, therefore, a relatively young feature with altitudes up to 2900–3400 m above sea level (asl), the highest mountains south of Kilimanjaro (Nel and Sumner, 2008). Erosion and, consequently, inland movement of the uplifted escarpment gave rise to much of the coastal plains between the Drakensberg and the Indian Ocean (Truswell, 1977). To the north-east, the Manica Highlands, part of the eastern arm of the Great Escarpment, forms the border between Zimbabwe and Mozambique. In the west, elements of the escarpment extend to Namibia (the Schwarzrand and edge of the Khomas Highland in Namibia) and Angola (the Serra da Chela).

The Drakensberg is divided into four geographical parts, the Mpumalanga Drakensberg (MD), the eKangala Drakensberg, the Maloti-Drakensberg and the Cape Midlands Escarpment (CME).

The Mpumalanga Drakensberg

The northern section of the Drakensberg lies in South Africa on the eastern edge of the Bushveld igneous geological complex, from about 23.5 degrees to about 27 degrees south. This part of South Africa is only some 160 km from the Indian Ocean with the foot of the Drakensberg at about 500 m asl and the crown between 2000 m asl and 2200 m asl, an increase in altitude of about 1400–1700 m. Mariepskop (1948 m asl) and Mount Anderson (2200 m asl) are two important peaks. In the south, the escarpment runs into the mountains of Swaziland (1100–1300 m asl), a much older geological phenomenon.

Large rivers do not arise in the MD, but some of the more noticeable smaller rivers include the Letaba River in the north, the Blyde River in the central section, and the Usuthu River in the south. The Olifants River that receives runoff in the central plateau, far behind the mountains, flows through the Drakensberg toward the sea.

Forestry is a major activity along the length of the Mpumalanga escarpment, extending onto the plateau behind the MD. Three large towns, Mashishing, Mbombela, and Phalaborwa, lie close to the mountain. There are large rural populations below the mountain, depending largely on subsistence farming and migrant labor for a livelihood. Commercial farming below the mountain is mostly fruit, nuts, and game farming, while the farmers on the plateau behind the mountain mostly farm with livestock and maize.

Only small parts of the MD falls within protected areas, the most prominent being the Blyde River Nature Reserve (290 km^2).

The eKangala Drakensberg

The eKangala Drakensberg is a relatively low section of the escarpment lying along the north-western borders of Kwazulu-Natal Province with an altitude of around 1700 m asl, dropping to around 1200 m over a 15 km horizontal distance. It gives rise to the Buffalo River, a major tributary of the Tugela River further south, as well as to the Pongola River in the north (joining the Usuthu River to the north). Agriculture in this area consists mostly of stock farming and mixed irrigation farming with maize and citrus. Important towns close to the mountain include Ladysmith, Newcastle, and Vryheid.

The Maloti-Drakensberg

The Maloti-Drakensberg is the highest part of the Drakensberg, forming the eastern border of Lesotho and South Africa, ranging from around 1200–1500 m asl in the foothills to around 3100 m asl in South Africa, and up to 3482 m asl at Thabana Ntlenyana within Lesotho. Thabana Ntlenyana is the highest point in Africa south of Mount Kilimanjaro. The height of the mountains above the plains is around 1700 m. Within Lesotho the mountains are usually referred to as the Maloti Mountains and the larger complex including South African mountains is referred to as the Maloti-Drakensberg.

To the north, the "Amphitheatre" presents a steep, cliff-like escarpment that falls from 3000 m asl to below 2500 m over horizontal distance of less than 500 m. This region is the source of the Tugela River, the second biggest mountain catchment in South Africa. In the central part of this range near Champagne Castle (3100 m asl), the eastern slopes are not as steep as in the north and the foothills form a continuous plateau at around 1900 m asl, the "Little Berg" that falls off to plains at about 1300 m asl. Further to the south, the slopes become even less steep, although the altitudes remain similar. At Sani Pass, the only vehicle access and between the east of Lesotho and South Africa, the altitude drops from 2900 m to 2200 m asl within the horizontal distance of 4 km. This part of the mountain gives rise to the Mkomaas River as well as the Umzimvubu/Umzimkulu River. The indigenous vegetation on the Maloti-Drakensberg escarpment and in Lesotho is grassland with montane forest or thicket in the valleys or ravines.

A large part of this section of the Drakensberg forms a UNESCO World Heritage Site, comprising the Maloti-Drakensberg Park and adjacent areas in Lesotho as well as South Africa. The Maloti-Drakensberg Park includes the Sehlathebe National Park in Lesotho and the uKhahlamba Drakensberg National Park in South Africa. This section of the Drakensberg also forms part of the Maloti-Drakensberg Transfrontier Park, an initiative to promote cross-border development, governance, and management of the mountains. A large part of the South African slope and foothills of the Drakensberg, therefore, falls within conservation areas. The Lesotho Maloti-Drakensberg is the source of a major hydroelectric initiative that supplies both South Africa and Lesotho.

Several areas in South Africa adjacent to the Drakensberg foothills are densely populated with villages within which residents practice subsistence farming or work on nearby farms, in towns, or as migrant workers in more distant cities. Cattle form a very important part of the social system, acting as a traditional indicator of prestige and a means of conserving wealth (ILRI, 1995). Commercial farming of crops and livestock occurs in many areas, with the latter especially at higher altitudes. Tourism is well developed in the Maloti-Drakensberg with hundreds of private-managed hotels, lodges, and resorts. In the conservation areas, the provincial wildlife department operates several tourist facilities (mostly for trekking) including accommodation along the length of the Maloti-Drakensberg escarpment. In Lesotho the tourism facilities are more basic.

The Maloti-Drakensberg does not have nearby cities, the closest ones being Pietermaritzburg (100 km away) and Durban (160 km away). Forestry has always been a very important commercial activity on the lower slopes of the Kwazulu-Natal Drakensberg, especially in the central part below Champagne Castle and Cathedral Peak. Much of the research in the Maloti-Drakensberg originated in response to the problems encountered by the forestry industry.

The Cape Midlands Escarpment

The CME is the most southerly part of the Drakensberg complex. Covering around 31,500 km^2, it comprises the Sneeuberg, Great Winterberg-Amatholes (GWA), and Stormberg (mostly in the Eastern Cape Province, South Africa; Fig. 1.2A), forming the western remnants of the Drakensberg massif (Clark et al., 2011). While now fragmented and dissected by millennia of fluvial erosion, this section of the Great Escarpment is an important catchment area for the arid Cape Midlands area of the Great Karoo and Albany regions of the Eastern Cape (Clark et al., 2014). Since the Eastern Cape water yield for human consumption is approaching catchment discharge, the ecological value of these mountains cannot be overemphasized (Phillipson, 1987).

The CME comprises an archipelago-like mountain range with isolated mesic anomalies (rainfall 700–1000+ mm/annum) surrounded by extensive semi-arid to arid lower altitude plains (300–500 mm/annum). These mountain islands act as "moisture-nets," harvesting precipitation in the form of rain, mist, and snow by orographic effect from the easterly winds blowing off the warm Indian Ocean, by intercepting cold fronts and cut-off lows from the south and west, and inducing convectional precipitation by creating thermal islands (Clark et al., 2009).

With a mean summit plateau altitude of 1800–2100 m (and peaks reaching 2300–2500 m), the Sneeuberg is named for the heavy snowfalls during winter (Clark et al., 2009). The Winterberg-Amathole are not as high (average plateau altitude c.1700 m asl; highest peak 2369 m asl) but are overall much wetter due to their central location in the trajectory of the Indian Ocean south-easterlies. The Stormberg is overall lower in altitude (max. 2207 m asl), but forms an extensive elevated area and has both the lowest recorded and mean minimum temperatures in South Africa.

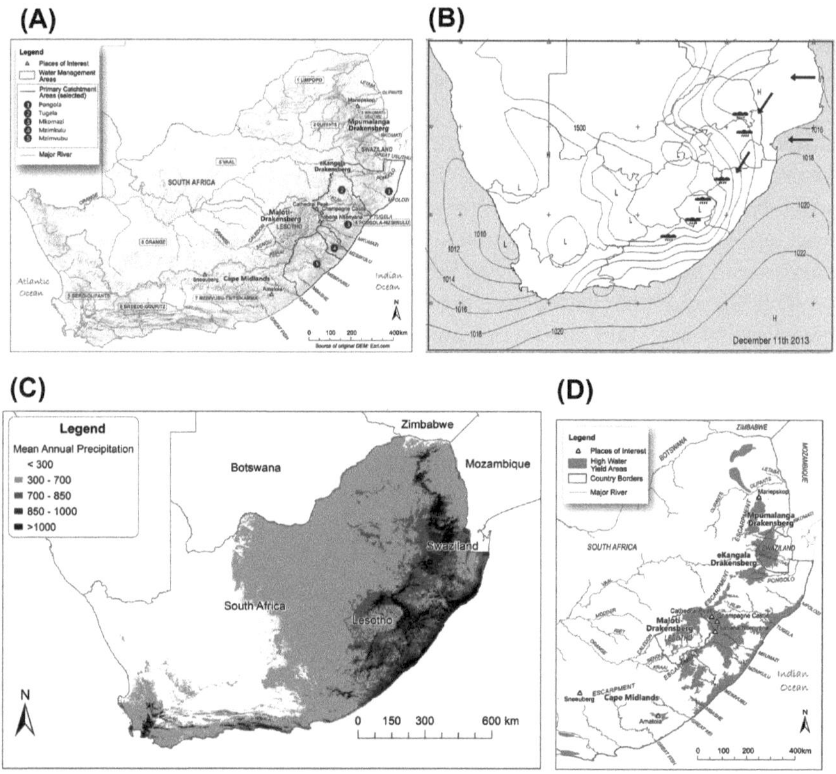

FIGURE 1.2 (A) Map of southern Africa showing the Drakensberg of southern Africa and Lesotho, as well as some of the important catchments (numbered in legend) arising in or close to the Maloti-Drakensberg. (B) Synoptic weather map of southern Africa showing the most common mechanism of precipitation in the Drakensberg. The ridging high-pressure area to the south east (>1000 mB at sea level) advects moist air (indicated by arrows) from the Indian Ocean onto the Drakensberg, bringing rain (indicated by dark clouds). The low-pressure system to the northwest of the mountains facilitates the inflow of sufficient moisture for rain *(Air pressure data courtesy SA Weather Service).* (C) Rainfall map for southern Africa, indicating that the Drakensberg is a regional source of high rainfall. (D) Map of 1 min square geographical units of eastern southern Africa indicating catchment areas with relative water yield greater than 0.5 *(Modified from Nel, J., Colvin, C., Le Maitre, D., Snith, J., Haines, I., 2013. Defining South Africa's Water Source Areas. WWF South Africa, Cape Town, 32 pp. With permission, WWF-SA 2016).*

Important rivers with all or substantial parts of their catchments in these mountains are the Buffalo (supplying East London), Great Fish (supplemented by the Orange-Fish Transfer Scheme and supplying the towns of Cookhouse, Cradock, Bedford, Grahamstown, and extensive irrigation schemes), Swart Kei (supplying water to Tarkastad, Wittlesea/Sada, and Cathcart), Sundays (supplying Port Elizabeth), and numerous smaller systems (collectively termed the Amatola coastal catchments) such as the Keiskamma, Koonap, Kap, and others, which supply water to the smaller towns of Bedford, Adelaide, Fort Beaufort, Alice, Stutterheim, and a large rural population (Clark et al., 2014). The CME is furthermore a continental watershed, draining toward both Atlantic and Indian Oceans (Clark et al., 2009).

In addition to surface water, the CME is an exceptionally important ground water recharge system. The dolerite, which dominates much of these mountains, lends itself to aquifer recharge through an extensive series of fissures and joints (Du Toit, 1920; Agnew, 1958). As a result ground water is often a much more tangible benefit to landowners living in and around the CME: most farms in the Sneeuberg and Stormberg for instance rely on springs on the mountains or in their foothills for domestic and farm use rather than streams. Local landowners report that winters of good snowfalls result in strong spring and ground water yields during the following summer season.

Past land uses have degraded aquatic habitats in the CME catchments. The Department of Environmental Affairs and Tourism (undated report based on the National Spatial Biodiversity Assessment status of river reaches) indicates that all river reaches in the CME vary in status from vulnerable (most of the GWA and Stormberg) to Endangered (parts of the GWA and Stormberg) to critically endangered (most of the Sneeuberg). Although at a visual level (from personal observation throughout the CME) the overall condition of these catchments at higher elevations (i.e., >1500 m) is good, these upper catchments were severely degraded due to livestock overgrazing during the mid-1900s (Keay-Bright and Boardman, 2007). While modern farmers have reduced livestock densities and are implementing sustainable grazing regimes, many of these upper river reaches still show the impacts of catchment abuse.

Past mismanagement is often masked by secondary grassland in which grazing-sensitive climax grasses have been grazed out, numerous naturalized alien species having become integrated into the indigenous flora, and "monoculture" shrublands (e.g., of *Euryops floribundus*) have replaced grasslands in marginal areas. Fortunately, except in communal rural landscapes, there is now little evidence of current detrimental land management practice. The primary concerns on commercial grazing land are wetland and upland riparian disturbance from myriad small earth dams, trampling by livestock, and the numerous 4 × 4 access tracks that provide farmers with access to the upper regions of their farms. The effects are difficult to quantify, but the cumulative impacts could be significant. Perhaps the most concerning is the local ambivalence by landowners to current threats to from montane alien invasive species such as *Nassella trichotoma* and *Rosa rubiginosa* (Clark et al., 2009).

THE DRAKENSBERG AS A SOURCE OF WATER

Weather Systems Bringing Rain to the Drakensberg

Tyson et al. (1976), Preston-Whyte et al. (1991), and Preston-Whyte and Tyson (1997) described a number of weather systems that affect precipitation in the Drakensberg. Preston-Whyte et al. (1991) classified the weather affecting the Drakensberg into a number of distinct systems:

Ridging high: High-pressure cells from the south-east causing moist air from the Indian Ocean to be advected on the eastern aspect of the mountain, causing thunders storms and orographic rain.

Tropical temperate troughs: Moist tropical air advected in a southerly direction by low pressure from the north west, bringing hot, moist tropical air that is forced on the northern and western aspects of the Drakensburg, causing thundershowers or orographic rain, especially in summer.

Mid-latitude cyclones: Cold air associated with Antarctic fronts from the south-west that cool the atmosphere, causing precipitation. This phenomenon frequently causes snow on the Drakensberg.

Cut-off low: A low-pressure cell that is cut off from the westerly winds to the south of the subcontinent, sucking in large quantities of moist tropical air and causing rain when that air cools. These systems are more prevalent during autumn and spring and bring both rain and snow to the Drakensberg

Coastal low: Low-pressure cell that moves along the coast, advecting moist air from the Indian Ocean and causing rain.

Although Preston-Whyte et al. (1991) did not rank these different sources of rain, the above list represents these different events more or less in decreasing frequency of occurrence in the Drakensberg.

Important Water Resource Areas in the Drakensberg

Several approaches exist for measuring the relative importance of water delivery from mountains (Messerli et al., 2004). A useful approach is that of Viviroli et al. (2007) who defined a dimensionless coefficient, the relative water yield (RWY), which measures the amount of water discharge from a mountain as a fraction of the water provided by the whole catchment (including lowlands) within which that mountain is situated. Values of RWY values greater than one indicate high importance of mountain catchments in providing water. The quantities involved in these calculations are normally obtained from models and interpolations that take into account existing databases with information on rainfall and other climatic data, as well as on soil and discharge characteristics. This exercise was performed for the Drakenberg by Nel et al. (2013), based on previously published models of MAP and rainfall–runoff relationships that take into account the anthropogenic alterations to runoff (Middleton and Bailey, 2009). Fig. 1.2D presents a map with the final result from their modeling and depicting the RWY over southern

Africa, indicating RWY larger than 1.0. Eight percent of the southern Africa surface produces more than 50% of the water of the region. A number of points are clear and are as follows:

1. Except for the southern Cape, the Drakensberg is the single most important source of water in southern Africa and supplies regions where the bulk of the population resides.
2. Lesotho and Swaziland, comprising part of the Drakensberg escarpment, are regionally critically important water resource areas.
3. In the eastern part of southern Africa and apart from the Drakensberg, only two coastal areas and two higher lying areas are significant in terms of RWY. Only 3 of the 13 important water resource areas in eastern southern Africa, defined by Nel et al. (2013), are not mountains. Two of these mountain areas (Soutpansberg in the north and Amatole in the south) do not form part of the Drakensberg.

Thus, from a geographic point of view, the Drakensberg is by far the most important water resource area, as evident from RWY data.

Rainfall

For southern Africa, few quantitative studies have been performed to measure precipitation trends in mountain areas (Fig. 1.2B,C). Sene et al. (1998) performed a study of rainfall and discharge in the Maloti-Drakensberg, using data from the weather stations of the Lesotho Weather Service and the South African Weather Service. Their data set comprised 32 weather stations in Lesotho and 31 in South Africa close to the Lesotho frontier. Most of these weather stations had data for over 20 years during the period 1950–90. However, data are lacking for the altitudinal range 2600–3000 m and include data from only one weather station higher than 3000 m. This paucity of data at high altitudes is marked within all the southern African data sets. Sene et al. (1998) divided the Maloti-Drakensberg mountains into the following three rainfall zones:

1. The eastern escarpment (extending from 1100 m asl in South Africa to 3000 m asl in Lesotho) with annual precipitation increasing from some 800 mm to around 1200 mm.
2. The western aspect of the eastern escarpment, assumed to be a rain shadow, with annual precipitation around 600 mm.
3. The Maloti Mountains in the western half of Lesotho with annual rainfall in the order of 800 mm per annum.

For each of the above zones, Sene et al. (1998) calculated a linear positive relationship between altitude and rainfall, these being respectively 51 mm, 27 mm, and 48 mm per 100 m gain in altitude.

Similar studies for the rest of the Drakensberg at smaller spatial scales did not find this linear pattern. Instead they have found that the central

Drakensberg escarpment experiences annual rainfall ranging from 700 mm at altitudes around 1100 m asl up to around 1100 mm at 2000 m asl and a bit higher, but that the highest peaks lie above the mist belt and consequently have lower rainfall. For instance, analyzing a 30-year rainfall data set for the central Drakensberg, Nel & Sumner (2006) found an increase in rainfall with altitude (41 mm/100 m) from 1100 m up to 2100 m asl where the annual precipitation was 1000 mm/year. However, reliable data for the high Drakensberg (at above 3000 m asl) were deficient. In a later study, Nel and Sumner (2008) reported on rainfall over a three-year period along two altitudinal transects (Sentinel Peak and Sani Pass) in the Drakensberg (Fig. 1.3). Each transect included four

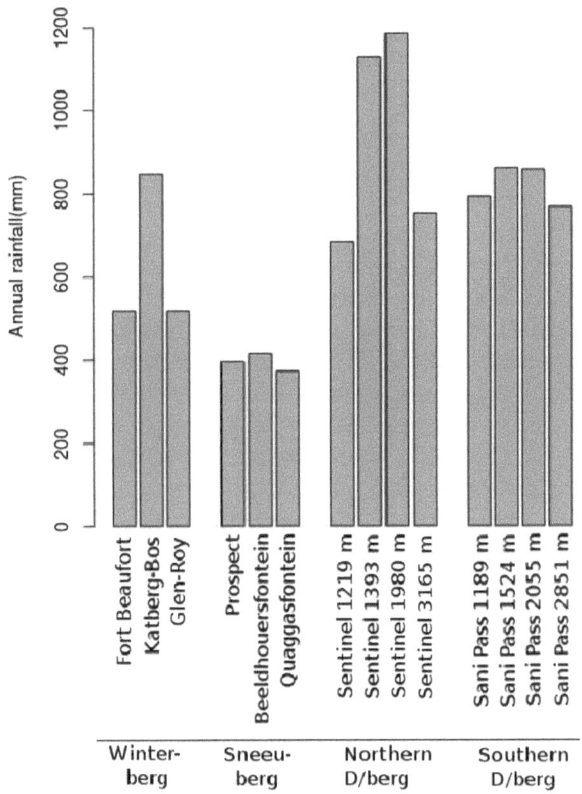

FIGURE 1.3 Mean annual precipitation at selected weather stations at slope bottom, mid-slope, and behind mountain top (from left to right): (A) Great Winterberg-Amatolas, (B) Sneeuberg; data supplied by the South African Weather Service for these six weather stations (1980–2009; *Data from Clark, V.R., Barker, N.P., Mucina, L., 2011. The Boschberg (Somerset East, Eastern Cape) – a floristic cross-roads of the southern great escarpment. South African Journal of Botany 77, 94–104*), and (C) precipitation on Northern Drakensberg (2002–05) and (D) on southern Drakensberg (2002–05, *Data from Nel, W., Sumner, P.D., 2008. Rainfall and temperature attributes on the Lesotho Drakensberg escarpment edge, southern Africa. Geografiska Annaler A, 90, 97–108*; left to right showing a transect from bottom of slope to top of slope).

rainfall gauges, spanning a range from around 1200 m to 3000 m asl. Their results showed a decrease in rainfall above about 2000 m asl indicating the importance of the mist belt (generally lying below 2500 m) for rain, with the crest of the escarpment being above the mist belt. Rainfall at Sani Pass top and at the Sentinel (both around 3000 m) was 10–30% lower than that at 2000 m (Fig. 1.3). Using interpolation between rainfall stations in the Drakensberg, Schulze (1979) modeled that, as one proceeds westward and up the escarpment, rainfall increases from about 800 mm at an altitude of 1500 m asl to about 1300 mm just below the crest at 3000 m asl. Rainfall variability also decreased as one moves up this slope.

Rainfall drops off rapidly as one progresses over the crest into the rain shadow. In Lesotho, the rainfall is mostly lower, around 600–800 mm per annum (Sene et al., 1998). Similarly, rainfall at Mariepskop in the MD, as measured over a four-year period (2012–15) showed broadly similar patterns with annual precipitation of 723 mm at 650 m asl, 1115 mm at 1300 m asl, and 670 mm at 1940 m asl. Here, the vertical gradient is comparable to that of the central Drakensberg, but shifted 1000 m lower in altitude. These observations reveal similarity of the Drakensberg to East African mountains with altitudes around 5000 m asl, which also exhibit extensive dry areas above the mist belt (Hemp, 2006).

In the CME, there is a negative moisture gradient from east to west and south to north; this makes the Amatholes the wettest part of the CME (>1000 mm/annum) and the most northern areas of the Sneeuberg and Stormberg the driest (400–500 mm/annum) (Clark and Barker, 2014) (Fig. 1.3). Rainfall is generally summer dominated, but with significant peaks in spring and autumn (Clark et al., 2011). As a result of the high local precipitation in these mountains compared to the arid lowlands, many of the rivers rising in the CME are exotic (i.e., the mountain catchments provide virtually all of the river flow) or only flow after exceptionally rainfall events.

Heavy Rains and the Effect of Land Management on Discharge and Siltation

Heavy rainfall, poor soil management, and erosion interact negatively to increase discharge and reduce the water retention of soils, causing perennial rivers and dams to become seasonal, affecting the temporal patterns of water supply and base flow within a mountain catchment (Morin and Benjamini, 1977; Schultze, 1979; Hurni et al., 2005). In addition, increased soil transport causes silting of dams and lakes, negatively affecting the water supply from these impoundments (Tamene et al., 2006).

Schulze (1979) studied heavy rainfall events at a mountain site at 2200 m asl and an eastern foothill site, at 1400 m asl, both near Cathedral Peak in the central Drakensberg. At the one extreme, both rainfall sites had a mean of four short, intensive bouts of rainfall (>50 mm/h) per year. However, when it came to extended periods of soft rain (24 h > 1 mm/h), the high site had about twice as many (2.6 events/y) events compared to the low site (1.2events/y). Thus, at both altitudes the probability of floods from short-term heavy rain was about equal, while the higher site tended to have more long bouts of soft, penetrating rain.

Schultze calculated that the low-lying site at Cathedral Peak is exposed to rainfall with significantly more kinetic energy (20–25 J/m^2) than the high-lying site at 2200 m asl (10–15 J/m^2). Soil erosion is thus more likely at the low-lying site, exacerbated by the fact the low-lying site had large areas under intensive cultivation. However, a broader scale analysis by Schulze (1979) covering some 2000 km^2 from Cathedral Peak in the north to Champagne Castle in the south suggested that the situation is much more complex. High-lying areas (above 3000 m asl), which are more extensive close to Champagne Castle than at Schulze's intensive study site at Cathedral Peak, experience more heavy bouts of rain. It, therefore, appears that there are two gradients in kinetic energy from rain in the Drakensberg: (1) from north to south along this section of the escarpment and (2) distance eastward from the escarpment.

Kinetic energy is only one factor that affects soil erodibility. Many other factors are also important, including soil type, slope, slope length, and vegetation cover. Schulze modeled these factors in estimating the soil loss patterns in the aforementioned study area, incorporating climatological factors (e.g., rainfall and radiation), topography, and soils as well as different vegetation characteristics relating to the dominant vegetation type. This resulted in two areas of relatively high soil loss: (1) the high-altitude areas to the south at Champagne Castle (Fig. 1.2A) where high rainfall and steep, long slopes promote soil loss and (2) the foothill areas to the east with very low vegetation cover due to agriculture and other human activity and with intensive rainfall bouts. He estimated that soil loss ranged from >20 metric tons/ha/y in the north-eastern foothills to around 17 tons/ha/y in the southern high mountains to around 1 ton/ha/y in the low-lying flat areas furthest from the escarpment. Together, these results indicate that the high rain intensity is likely to result in increased soil erosion in two mountain landscapes: steep, long slopes at higher elevations and areas with high levels of agriculture or other human activity resulting in low plant cover. More importantly, these calculations demonstrate the effect of human management of the mountain slope. Consistent with the measurements of Schulze (1979), the high-lying slopes of the Maloti-Drakensberg historically (i.e., >50 y ago) contributed to a significant part of the silt in the Senqu-Orange (Compton et al., 2010).

Over the past 50 years, there has been a 100-fold increase in river sediment derived from the Drakensburg, mainly from the cultivated and overgrazed foothills and Maloti regions. The sediments from high-lying areas themselves are miniscule. For example, most of the erosion in the Senqu-Orange catchment occurs in the western foothills of the Maloti-Drakensberg in areas of intense human disturbance (Compton and Maake, 2007; Compton et al., 2010).

Similar to the case of Ethiopia (Hurni et al., 2005), human land management in the Drakensberg plays a major role in determining discharge and downstream water quality.

Long-Term Change in Mountain Rainfall

Several studies have attempted to detect changes in rainfall over time. Different approaches of the studies described in the following sections make conclusive general trends difficult to discern, highlighting the need for a consistent and well-considered approach to change detections studies that are intended to inform policy.

Sene et al. (1998), in their analysis of rainfall in Lesotho, compared rainfall over two periods: 1931–60 and 1961–90. For each rainfall station, they then tested for statistically significant differences between these periods for rainfall and none showed a statistically significant difference.

The Climate Research Unit at the University of East Anglia houses historical climate data for Lesotho. We analyzed data for the period up to 2000 for a number of rainfall stations in Lesotho as well as for a number of close-by stations using rainfall data obtained from the South African Weather Service (Table 1.1). The trends in rainfall were analyzed by means of generalized least-squares ANOVA, taking account of the temporal autocorrelation using ARMA, employing function gls of R V3.1.2 (R Development Team, 2014). A significant decreasing trend was observed only for one rainfall station (Leribe in Lesotho, Table 1.1). With the caveat that rainfall data for Lesotho could not be found after the year 2000; these data suggest that there has been no long-term trend in regional annual rainfall over the Drakensberg.

Kruger (2006) performed a study of long-term rainfall trends (1910–2004) in South Africa. In general there were no marked changes in rainfall with different trends at closely situated weather stations. Over the central Drakensberg, four out of eight rainfall stations showed a significant decrease in rainfall over that period. However, more complete data are available for two of these eight stations and, when these more recent data were added to the original data set, an ARMA generalized mean-squared ANOVA did not reveal a significant trend (Table 1.1).

We conclude that there is no strong evidence for any linear regional change in annual precipitation in the Drakensberg. However, this does not negate changes in the seasonal distribution and shower intensity of rainfall over the mountains, a topic currently with very fragmentary evidence for the Drakensberg.

These points highlight the paucity of long-term data sets at high altitudes, with most of the aforementioned rainfall stations having been discontinued. Our experience is that, due to between-year variation in climate variables, records of about 30 years' duration are often not sufficient to detect reliable long-term climate change.

TABLE 1.1 Mean Annual Precipitation (MAP) at Several Sites in the Drakensberg

Station ID	Coordinates		Altitude (m)	Observatory Period		MAP (mm)	*g/s* Slope	*p*
	Latitude	Longitude		From	Until			
Maloti-Drakensberg, Lesotho								
Quithing	30.42	27.72	1650	1901	1995	814	−0.15	0.43
Mohaleshoek	30.15	27.47	1620	1901	1995	812	−0.02	0.86
Qachasnek	30.10	28.60	1970	1901	1989	1088	−0.33	0.08
Mafeteng	29.82	27.25	1660	1901	1995	740	−0.08	0.51
Maseru	29.45	27.57	1628	1901	1995	748	−0.05	0.64
Teyeteyana	29.15	27.73	1690	1901	1995	830	−0.03	0.78
Leribe	28.88	28.05	1670	1906	1995	907	−0.37	* 0.02
Matsoakeng	28.63	28.87	1844	1908	1983	896	−0.13	0.60
Caledon	28.57	28.65	1981	1914	1982	1127	−0.35	0.28
Buthabuthe	28.25	28.25	1680	1920	1993	883	−0.23	0.15

Maloti-Drakensberg, South Africa								
Barkley East	30.97	27.60	1780	1901	1996	963	0.16	0.24
Lady Gray	30.70	27.22	1632	1901	1994	753	−0.25	0.17
Shaleburn	29.78	29.35	1909	1975	2014	1062	0.53	0.60
Estcourt	29.00	29.88	1160	1901	1996	963	0.16	0.24
Ficksburg	28.87	27.88	1600	1906	1994	850	−0.12	0.41
Royal Natal National Park	28.69	28.95	1398	1960	2014	1272	−0.71	0.48
The Cavern	28.63	28.96	1480	1975	2014	1144	0.07	0.95
Witsieshoek	28.53	28.80	1780	1908	1994	974	−0.16	0.43
Mpumalanga Drakensberg, South Africa								
Hebron	24.66	30.93	980	1950	2014	1541	−0.21	0.95
Salique	24.60	30.90	934	1936	2014	1228	−1.57	0.49
Mariepskop	24.58	30.87	1300	1928	2014	1302	−1.45	0.53

Slope and *p* indicate the results of generalized least-squares ANOVA with ARMA correction, testing the hypotheses that there is a long-term trend in MAP. Asterisk in *p* column indicates statistically significant temporal trend.
Lesotho data provided by University of East Anglia, UK; South African data provided by the South African Weather Service.

Mist

Very little is known of the contribution of mist to total precipitation in the Drakensberg. Schulze (1979) estimated that Cathedral Peak on the central Drakensberg escarpment received a monthly mist precipitation of about 50 mm (around 5% of total precipitation) during the months from October to March. Quantitative measurement of the water equivalence of the Drakensberg mist remains to be published. Current studies are underway in the South African Environmental Observation network (SAEON) Cathedral Peak research area to determine this.

Snow

For most high-altitude mountains, snow also contributes significantly to precipitation. Given that the central Drakensberg is above 3000 m a.s.l. and is situated relatively far from the equator, one would expect that snow would be more important than on tropical mountains such as Kilimanjaro (Cullen et al., 2006). However, despite the higher latitude, snow appears to be of lesser importance for the Drakensberg. Mulder and Grab (2009) identified snowfall events on the Drakensberg from Landsat photography and measured the snow-covered areas. From 1989 to 2004, there were a mean of 2.5 snowfall events per year (Fig. 1.4), substantially less than the eight snowfall event per year mentioned by Tyson et al. (1976). Also using Landsat images, Mulder & Grab (2002) estimated snow cover after a single snowfall in May 1996 and

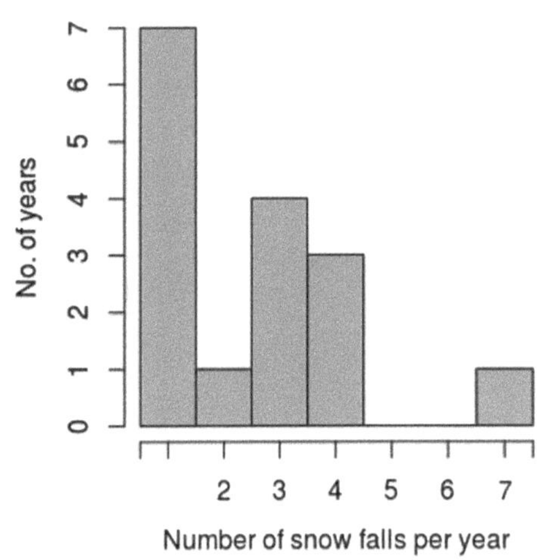

FIGURE 1.4 Distribution of snowfall events in the Maloti-Drakensberg during the period 1989–2004. Most years had four or fewer snow events, but all years had at least one snowfall. *Constructed using data from Mulder, N., Grab, S.W., 2009. Contemporary spatio-temporal patterns of snow cover over the Drakensberg. South African Journal of Science 105, 228–233.*

concluded that the coverage was less than 1% of the mountain above 3000 m asl. However, Mulder & Grab (2009) reported on much heavier snowfalls during the period 1989–2004, when snow on the Drakensberg was encountered over extensive areas. On the Landsat images, with many of the pixels indicating snow, the South African Drakensberg had a snow cover of up to 68% of their study area covering the eastern Maloti-Drakensberg and most of the mountain above 3000 m asl had snow. In this study, the most important weather patterns bringing snow to the Drakensberg were winter cut-off lows and Antarctic cold fronts from the south-west. However, the snow thaws within weeks (Harvey and Marker, 1992) and the Maloti-Drakenberg does not have permanent snow.

Nel and Sumner (2005) argued that the mean annual snowfall in the high Drakensberg is likely to be less than 1 m in depth and that the water yield in heavy snowfall areas is likely to be less than 100 mm (some 10% of total annual precipitation). However, these are qualitative arguments: reliable measurements remain to be published. While the importance of snow in the Drakensberg is, therefore, not well understood, most scientists would expect that snow is much less important for water yield in the Drakensberg than in most northern-hemisphere mountains above 3000 m (Harvey and Marker, 1992).

Evapotranspiration

Evapotranspiration is one of the major processes returning water vapor to the atmosphere. Schulze (1979) estimated mean annual actual evaporation (MAAE) by subtracting annual runoff estimates mean annual runoff (MAR) from annual rainfall mean annual precipitation (MAP). If done over an annual period and on a catchment scale, the interaction with soil water is minimized. For the eastern central Drakensberg escarpment, this resulted in a more or less linear relationship with a slope of 0.26. This means that, although runoff increases with rainfall, the difference between MAP and MAR, that is, actual evapotranspiration, also increases at the catchment scale. For an annual rainfall (MAP) of 1000 mm, the actual evapotranspiration was estimated at 693 mm, illustrating the importance of evapotranspiration in the water balance of the mountain. As rainfall increases and as one ascends the escarpment, actual evapotranspiration also increases and there was a fairly monotonic relationship between MAAE and distance from the crest of the escarpment: Schulze estimated the annual MAAE to range from some 600 mm in the low-lying foothills of the escarpment to about 850 mm near the crest of the escarpment.

Measurements of actual evapotranspiration are relatively crude and more modern estimates (including estimates of potential evapotranspiration) based on mass balance calculations remain to be published. Modeling the effects on the water balance of increased energy input to the surface due to temperature increases under climate change is another future research priority.

Water Discharge

In terms of water supply to humans, water discharge is a critical factor. Runoff is measured as the number of millimeters of water that drains from a point, just as rainfall on that same point is measured in millimeter. On the other hand, discharge is the volume of water that drains from a catchment. Runoff is usually calculated from measurements of discharge. However, discharge is more complex to measure than rainfall (Dingman, 2015), primarily because of the strong spatial dependency of discharge within a particular catchment: discharge along a river is expected to differ at each measurement station. Secondly, expressing a relationship between rainfall and runoff is complex, depending on topography, soil structure, vegetation, and human activities. This is often achieved by the calculation of a runoff coefficient (the fraction of total rainfall that drains from the surface of a fixed-size catchment).

For the CME, Clark et al. (2014) provided a rough estimate of water discharge from the Winterberg-Amatole (332 million m^3/annum, or 15% of the total discharge from the Great Fish, Great Kei, and the smaller catchments). However, there has been no comprehensive hydrological assessment of the CME specifically.

For the Central Drakensberg, Schulze (1979) derived an exponential curve for relating discharge to rainfall for the central eastern Drakensberg escarpment. Roughly, a doubling of rainfall produced a fourfold increase in discharge at the lower levels of precipitation and a six—fold increase at higher levels of precipitation.

Sene et al. (1998) calculated runoff coefficients for Lesotho. Long-term average values of almost 50% were estimated for the high mountain areas in the north, while coefficients for the flatter lower lying areas were around 20%. These average values mask a linear relationship between the change in discharge as a function of a change in rainfall with a slope around 1.4: a given absolute annual anomaly in precipitation (e.g., an increase or decrease measured in mm) produced a larger absolute anomaly in discharge. These estimates are lower than those derived by Schulze (1979), indicating the effect of the differences in topography between the Lesotho landscape and that of the Eastern escarpment, the latter characterized by many steep slopes.

Given that Lesotho generally experiences less rainfall than the eastern South African Drakensberg escarpment, why does the westward-flowing Orange-Senqu River have a higher discharge than the eastward-flowing Tugela River that originates at the top of the escarpment? The answer is that the catchment size of the Orange-Senqu is much larger than that of the Tugela (compare lower Tugela with lower Orange in Table 1.2). Even though Fig. 1.2D shows that the highest RWY are on the escarpment and that Lesotho in general has a lower yield, the large area constituting Lesotho with relatively high rainfall feeds a much larger volume of water into the Orange-Senqu River (Table 1.2).

TABLE 1.2 Long-Term Trends in Mean Annual Discharge (Million Cubic Meter Water) of Rivers Emanating From or Close to the Drakensberg

Catchment;	Station	Coordinates		Observatory Period		Catchment	Mean Discharge	g/s	
Station Name	ID	Latitude	Longitude	From	Until	Size (km^2)	(m^3 10^6/y)	Slope	*p*
Blyde river: Driehoek	B6H005	24.51752	30.83174	1958	2015	2204	227.2	−0.07	0.96
Nkomati: Tonga	X1H003	25.68225	31.78185	1940	2016	8785	546.9	−4.5	*0.015
Usuthu: upper: Newstead	W5H036	26.50572	30.62667	1970	2015	531	17.1	−0.25	0.53
Tugela: upper: Tugela Drift	V1H001	28.73592	29.82111	1951	2016	4176	860.6	−7.6	0.25
Tugela: middle: Tugela Ferry	V6H002	28.75003	30.44222	1927	2015	12,862	2455.0	−17.6	0.16
Tugela: lower: Mandini	V5H002	29.14050	31.39258	1959	2015	28,920	2194.0	−52.3	***0.0001
Mkomazi: upper: Lot 93 1821	U1H005	29.74369	29.90494	1960	2015	1744	587.9	−2.7	0.35
Mkomazi: lower: Delos Estate	U1H006	30.16908	30.69717	1962	2007	4349	633.7	−4.8	0.18
Mzimkulu: upper: Fp 1609030	T5H004	29.77714	29.47083	1963	2015	545	222.9	0.255	0.74

Continued

TABLE 1.2 Long-Term Trends in Mean Annual Discharge (Million Cubic Meter Water) of Rivers Emanating From or Close to the Drakensberg—cont'd

Catchment;	Station	Coordinates		Observatory Period		Catchment	Mean Discharge	*gls*	
Station Name	ID	Latitude	Longitude	From	Until	Size (km^2)	(m^3 10^6/y)	Slope	*p*
Mzimvubu: upper: Ku-Makola	T3H007	30.86014	29.07125	1984	2015	6906	715.4	14.7	0.41
Mzimvubu: lower: Kromdraai	T3H008	30.57067	29.15017	1962	2015	2471	204.4	0.66	0.54
Caledon River: Wilgerdraai	D2H022	29.61667	27.06556	1993	2015	12,852	1001.0	−20.6	0.13
Orange: upper: Oranjedraai	D1H009	30.33639	27.35861	1960	2015	24,685	3846.8	−10.1	0.67
Orange: lower: Marksdrift	D3H008	29.16202	23.69594	1962	2015	99,316	4850.9	−34.82	0.55

Months with questionable data were omitted and the annual flow rate estimated from the remaining months. Catchment size is an indication of how far downstream a measuring station is. Trends calculated using generalized least-squares ANOVA (*gls*) with ARMA correction. Asterisk in *p* column indicates statistical significant temporal trend.

Based on the river flow database of the Hydrological Services directorate of the South African Department of Water and Sanitation.

Long-Term Changes in River Discharge

River discharge in South Africa is measured at a number of measuring stations by the South African Department of Water and Sanitation. Discharge measurements were obtained for a number of important rivers originating in or near the Drakensberg (Table 1.2) and analyzed using generalized least-squares ANOVA, taking account of the temporal autocorrelation using ARMA, employing function gls of R V3.1.2 (R Development Team, 2014). Only two rivers (Tugela and Nkomati) showed a significant declining trend in flow rate, both situated in areas with extensive sugarcane planting and dam building (for the Nkomati) since 1960. In addition, the Tugela is subject to inter-basin transfer near the mountain upstream of the discharge measurement points. These data indicate that there has not been any significant linear change in the annual discharge of the rivers originating in the Drakensberg. Although measurement points as close as possible to the Drakensberg were selected, these discharge rates should not be taken as "natural" discharge unaffected by human usage, but should be understood as the interaction between discharge and human water use close to the mountain: the numbers in Table 1.2 are meaningful with respect to water availability, not necessarily with respect to long term-climate change.

HUMAN WATER DEMAND FROM THE DRAKENSBERG

Water Requirements of the South African Economy

The total water demand for the national economy was 12,871 million m^3/y in 2000 (DWA, 2004), with the Drakensberg being of critical importance for six of the South African nine provinces (Table 1.3). At that time, the biggest single sector using water was irrigation farming using 63% of the total water requirement. Water use per sector in 2013 was as follows: agricultural irrigation (60% of total annual yield), domestic household use and drinking water (27%), mining (3%) and industrial processes (3%), electricity generation (coal-fired cooling towers and hydropower), (4%) and "other" (3%) (DWS, 2013); the national increase in water needs from 2000 to 2013 had already surpassed the upper "high" worst-case scenario envisaged in the 2004 assessment (DWA, 2004).

The Economic Value of Water From the Drakensberg

Water has obvious value in the economic activities it allows (e.g., agriculture and industry) and is generally recognized as a critical economic resource (SA Treasury, 2011). An indicative national gross domestic product (GDP) value for South Africa is $350 billion for 2013, and this could be taken as the proxy total value of water in the economy, assuming that no economic activities would be possible without water. Some sectors are more dependent on water

TABLE 1.3 Details of Provinces That Derive Their Water Directly From the Drakensberg Escarpment and/or Lesotho, Either Directly or via Transfer Schemes

Province	Population at 2011 Census	Provincial Gross Product (2013)[a] in ZAR	Links to Major River Systems That Originate in the Drakensberg Escarpment and/or Lesotho	Water Management Areas (WMA) and Inter-Basin Transfer Systems (Bold)
Eastern Cape	6,562,053	R272.7 billion (2013)	Water is piped from the Orange River to Port Elizabeth, East London to augment their supplies	Mzimvubu to Keiskamma WMA **The Orange-Fish transfer scheme**
Northern Cape	1,145,861	R71.1 billion (2013)	The Orange River itself is a major source of water for Northern Cape towns and the irrigated agricultural activities along the Orange River	Lower Vaal WMA Upper Orange WMA Lower Orange WMA **The Orange-Riet transfer scheme**
Free state	2,745,590	R179.8 billion (2013)	Water piped from the Orange River to some small towns (e.g., Philippolis) to augment supplies	Upper Orange WMA Middle Vaal WMA
KwaZulu-Natal	10,267,300	R565.2 billion (2013)	18 significant dams including the Spioenkop dam and Woodstock dams on the Tugela River	Mvoti to Umzimkulu Thukela WMA Usutu to Mhlatuze WMA **Tugela-Vaal transfer scheme**
Gauteng	12,272,263	R1.12 trillion (2013)	The Lesotho highlands water Project Phase I and II pumps water from Lesotho into the Vaal system via the Liebenbergsvlei and Wilge Rivers to augment supplies. Phase III is being planned	Upper Vaal WMA Crocodile (west)—Marico WMA **Lesotho highlands water Project** **Orange-Vaal transfer scheme** **Thukela-Vaal pumped storage scheme**

[a] *StatsSA (2012).*

than others, specifically irrigated agriculture and horticulture, which also has a strategic importance in terms of national food security (Table 1.4). Statistics South Africa values agriculture's direct contribution to the national GDP at 3–4% per year, but because of economic multipliers and the food value chain, the GDP value of agriculture in the South African economy is closer to 25%, which means that the impact of a decline in this sector because of water constraints can have serious economic and job security consequences (StatsSA, 2016).

TABLE 1.4 South African Provinces and Water Management Areas That Derive Most of Their Water From the Greater Drakensberg and the Value (ZAR) of Irrigated Agricultural Production in These Provinces

Provinces That Derive the Bulk of Their Water From Water Management Areas Supplied by the Orange-Senqu River Basin, Lesotho, and the Drakensberg Escarpment	Water Management Areas Within the Orange-Senqu River Basin or Downstream From the Drakensberg Escarpment	Water Utilization Per Province (Million m^3/y 2004)[a]	Value (ZAR) of Irrigated Agricultural Production Per Province[b]
Eastern Cape	Mzimvubu to Keiskamma	190	81.2 million
	Fish to Tsitsikamma	763	349.2 million
Northern Cape	Lower Vaal	525	701 million
	Lower Orange	977	74.8 million
N/Cape and free state	Upper Orange	780	181.6 million
KwaZulu-Natal	Mvoti to Umzimkulu	207	117.4 million
	Thukela	204	55.8 million
	Usutu to Mhlatuze	432	150.5 million
Free state	Middle Vaal	159	51.8 million
Total national value		**7920**	**4.4 billion**

ZAR = South African rand
[a]*Department of Water Affairs (DWA, 2004).*
[b]*Statistical South Africa (StatsSA, 2010).*

Other unusual value additions of water include the value that municipalities derive by buying water from bulk sellers and selling it on to households and businesses and hydroelectric power generation. In 2014, municipalities in South African made a profit of R4 billion by buying and selling water (StatsSA, 2015). Gauteng Province and its municipalities buy the most water at a cost of R2173 million per annum for 2014 (StatsSA, 2015).

The Gariep Dam has a 360 MW hydropower generation capacity and Vanderkloof Dam has a 240 MW capacity, while the Tugela-Vaal pumped storage system delivers 1 GW of hydroelectricity of which a significant proportion is used to pump water during off-peak hours. Eskom provides South Africa with 95% of its electricity, and of this only 5% comes from hydroelectric generation (Eskom, 2011). The Muela Hydropower Station in Lesotho generates 72 MW, used in Lesotho. Low water levels in all these dams is a potential constraint to the generation of this electricity. Water also has value within natural ecosystems, creating value through biodiversity conservation and informal economic activities like artisanal fishing, reed harvesting, and water abstraction in rural areas, as well as value to the species in the landscape.

Utilization of Water From the Drakensberg

The national authoritative document with systematic quantitative estimates of water utilization in different catchments is the National Water Resource Strategy of South Africa (DWA, 2004) and its revision, the NWRS-2 of 2012, which were both developed under the National Water Act (Act 36 of 1998). The first edition, the NWRS-1, was published in 2004 and set out the framework for an integrated water resources management strategy for the first time, including quantitative data. The new NWRS-2 sets out the strategic direction for water resources management in the country over the next 20 years, with a particular focus on priorities and objectives for the period 2013–2017 (DWA, 2012), but lacks any up-to-date data.

The concept of yield, critical in the aforementioned documents, is crucial for understanding integrated water resource management. The discharge from a catchment is not always usable for humans because of the technical ability to extract the water. In fact, assuming that human water needs are approximately constant through a year (as with household and industrial requirements), utilization of discharge is limited by the lowest water levels in rivers. In addition, agricultural water utilization is often seasonal. The ability of a catchment to satisfy human water needs in the long-term determines the yield of the catchment. Several factors determine yield: (1) discharge quantity, (2) seasonality of discharge, (3) ecological flow requirement to maintain river health, (4) dams or lakes that capture peak flows and that can supplement water supply during low flow periods, (5) engineering structures such as inter-basin water transfer, and (6) ground water interactions with surface flows. Yield is, therefore, expected to be significantly less than discharge. In South Africa, inter-basin water transfer is a factor that strongly affects the yield of a catchment in either a positive (for a receiving catchment) or a negative way (for a water providing catchment). Water yield is, therefore, understood to be the maximum amount of water that can be sustainably diverted for human use within a particular catchment. During 2000, except for the Orange-Senqu, the yield of rivers from the Drakensberg was in the order of 20% of the discharge (Table 1.4). Obviously yield could be increased by more engineering structures such as dams or inter-basin transfers (assuming discharge is sufficient), dependent on the magnitude and consistency of discharge that would make it sustainable.

A summary of water yield sufficiency for the larger rivers emanating from the Drakensberg is given in Table 1.5. The importance of the westward-flowing Orange-Senqu River in providing water to South Africa is obvious. It is the only river with a large excess of yield over demand (more than 300 million m^3/y), despite the fact that this river is the source of two major inter-basin transfers as well as a number of smaller ones. This, in turn, emphasizes the importance of the Drakensberg/Maloti mountains in supplying South Africa with water.

The second largest river from the Drakensberg, the Tugela which runs eastward, was supplying water to capacity during 2000 with a deficit that was satisfied with boreholes and other technical means, and seasonal

TABLE 1.5 Water Balance of Several Rivers Originating on or Close to the Drakensberg Escarpment, Using Data Extracted From DWA (2004)

	For the Year 2000					
River Name	Discharge	Yield	Use	Transfer Out	Balance 2000	Balance 2025
Lower Olifants (Blyde River)	698	101	164	0	−63	−68
Nkomati	749	118	65	97	−44	−45
Usuthu	901	202	69	114	19	12
Tugela (upper)	1502	396	94	377	−75	−80
Tugela (total)	3799	737	334	506	−103	−150
Orange-Senqu	4765	4449	968	3148	335	−45
Umkomaas	1080	31	98	1	−68	−70
Umzimkulu	1373	16	40	10	−34	−35

The balance estimates for the year 2000 are given, as well as an estimate of the yield balance in the year 2025. The numbers indicate millions of m^3/year.

non-sustainable extraction possibly consuming some of the ecological reserve of the river. The Tugela is the source of a major inter-basin water transfer and some smaller transfers which are the major cause of the balance deficit (Table 1.4) and which consumes most of the upper Tugela river yield. However, the lower Tugela has large human water demands that amplifies the deficit originating in the upper Tugela (Table 1.4), a factor demonstrated by the significant flow reduction for the lower Tugela (Table 1.2).

Most of the rivers flowing from the Drakensberg share a similar situation with the Tugela. In the southern rivers (e.g., Umkomaas and Umzimkulu), the deficit is actually larger than the yield estimate (Table 1.5). Apart from the Orange-Senqu River, the only river with a surplus of yield is the Usuthu, a small river compared to the Tugela. However, the Usutu is heavily utilized and, although no statistics are available for Swaziland, it is expected that this river incurs a deficit while flowing through Swaziland. Clearly the rivers from the Drakensberg are under strong human pressure and the current utilization patterns are not sustainable given the present combination of discharge and infrastructure.

Projections of Sectoral Water Use From the Drakensberg

DWA (2004) performed projections of water demands in 2025 using a so-called "base scenario" (assuming low population growth, around 0.5%/y) and a "high scenario", assuming more rapid population growth (around 1%/y). Currently, 12 years after that report, the South African population has grown to 55 million and an annual rate of increase of 1.6% (StatsSA, 2015), significantly higher than the "high scenario" envisaged in that report. In the following discussion, the "high scenario" is therefore adopted. More importantly, a larger proportion of the South African population is expected to live in cities where dependency on formal water provisioning infrastructure is larger.

The Orange-Senqu River is currently the largest single water source that brings about a positive yield balance for the country as a whole. It is foreseen that a significant increase in inter-basin transfer will occur before 2025 (DWA, 2004). In addition, the largest current use of non-transferred water in the Orange River is irrigation agriculture which is also foreseen to increase in future, bringing the Orange River from a large positive yield balance into a deficit (i.e., negative yield balance) in 2025 (Table 1.5). This is likely to strongly affect water management policy in South Africa.

The Tugela River, strongly affected by irrigation agriculture (mostly sugarcane with about 2.5 times the magnitude of other water demands), is expected to grow moderately in water demands, in line with general population growth. Larger inter-basin transfers from the Tugela are not envisioned and, while some dam construction is anticipated (DWA, 2004), the negative yield balance is foreseen to increase by almost 50% from 103 million m^3/y in 2004 to 150 million m^3/y by 2025 (Table 1.5).

For the other rivers emanating from the Drakensberg, a slight worsening of the existing deficit is expected, not as marked as for the Orange and Tugela. The relatively small Usutu River is expected to be the only one with a positive yield balance (albeit with diminished magnitude) in 2025 (Table 1.5). These estimates underscore the dependency of the largest part of South Africa on water from the Drakensberg and the need to manage this resource efficiently. The frequent occurrence of negative yield balances in Table 1.5 indicates the need to increase yield where it could be done in a sustainable way.

SOUTH AFRICAN WATER GOVERNANCE AND MANAGEMENT

South African Water Governance and Legislation

The South African Constitution (1996) and its Bill of Rights state that everyone has the right to sufficient food and water and that the State must take reasonable legislative and other measures, within its available resources, to achieve this. The South African Water Services Act ensures this right by enabling access to basic water supply and sanitation. The ownership of water resources is vested in the State and the Department of Water and Sanitation (DWS) it has a legislative mandate to ensure that the country's water resources are protected, managed, used, developed, conserved, and controlled by regulating and supporting the delivery of effective water supply and sanitation. The DWS (formerly the Department of Water Affairs) manages water resources to promote equity, sustainability, and efficiency (StatsSA, 2010, p. 7). The management of water resources in South Africa is achieved at three levels, overseen by DWS:

Water Boards: The practical regulation of the water sector in South Africa is done through nine water control agencies called water boards, appointed by DWS that provide water services (bulk potable and bulk waste water) to other water services institutions, particularly municipalities, within their respective service areas.

Catchment Management Agencies: Since 1997, 19 WMAs (later reduced to nine) were designated to be managed by Catchment Management Agencies (CMAs) in collaboration with local stakeholders to ensure equity in water management (DEA, 2013). Currently, only four of the nine CMA structures have been formally established (Bourblanc and Blanchon, 2013). Swaziland and Lesotho are excluded from the formal structures of CMAs even though they are at the heart of either headwaters or are downstream users of streams and rivers originating in South Africa.

Water user associations: Water-User Associations are cooperative associations of individual water users who wish to undertake water-related activities at local level for their mutual benefit.

There are several treaties and agreements between South Africa, Lesotho, Mozambique, Swaziland, Botswana, and Namibia to manage water use in the region. The only established international research-sharing collaboration is for the basin of the Orange-Senqu River where ORASECOM was established to manage research and knowledge production for the entire basin.

In 2013 the budget demands for addressing South Africa's existing water infrastructure challenges for the entire water value chain were estimated at R670 billion over the next 10 years (DEA, 2013). In addition South Africa required an investment of R30 billion for sustainable water management programs, bringing the total 10-year water sector investment to R700 billion (DEA, 2013).

Lesotho Water Governance and Catchment Management

The entire country of Lesotho falls within the borders of the Orange-Senqu River basin and is essentially the catchment for this river basin. The international collaborative entity established to manage the basin of the Orange-Senqu river basin is ORASECOM and contributes to manage research and knowledge production for the entire basin.

Trans-boundary water resources are addressed by Lesotho in Policy Statement 4 of the Water and Sanitation Policy which states the intention to manage trans-boundary water resources on the basis of Lesotho's sovereignty in a way that ensures maximum benefits while recognizing Lesotho's obligations to downstream users under international law (ORASECOM, 2008). The Lesotho Highlands Water Project (LHWP) Treaty with South Africa addresses the protection of water quality and quantity in Lesotho in relation to the LHWP deliverables as per the Treaty of 1986 (Maseatile, 2013). Although the supply of water to Lesotho citizens is important, the supply of water from the Maloti-Drakensberg to the much bigger economy of South Africa is a strategic issue.

Ownership of all water is vested in the Basotho Nation, and water policy is implemented by the Department of Water Affairs (DWA) within the Ministry of Natural Resources following the Water Act 1978. The Lesotho Water and Sanitation Policy (2007) requires the Lesotho Government to ensure that the nation's water resource is used in a sustainable manner and to the benefit of all users, and the responsibility lies with the state to provide security of water sources and improved sanitation to its citizens (Water and Sanitation, 2007). The Commissioner of Water, within the Ministry of Natural Resources, promotes the coordination of programs and activities within the water sector and oversees two parastatals: the Lesotho Highlands Water Development Authority (LHDA) and the Water and Sewage Authority. LHDA manages catchments linked to the Lesotho Highlands Water Project.

Although Lesotho has developed a modern integrated water management framework, the implementation of its water resource management framework is constrained by several factors (ORASECOM, 2008), notably that the technical capacity within Lesotho's government agencies to implement such a framework is still somewhat limited (ORASECOM, 2008).

There are chronic catchment management challenges in Lesotho with an ongoing loss of arable land due to soil erosion, particularly in the densely populated lowlands (5.4 million tons/year in total soil loss, Compton et al., 2010) leading to reduced agricultural production, problems with sedimentation in rivers and reservoirs, and reduced quality of catchments. Compton and Maake (2007) show that the Orange River carries a relatively large suspended sediment load and ranks as the most turbid river in Africa and the fourth-most turbid river in the world and that this sediment load originates in the lowlands of Lesotho, with less sediment originating from high altitude, steep and high rainfall areas of the Drakensberg mountains in Lesotho (Compton and Maake, 2007, p. 339). Woody shrub encroachment is also one of the key factors of land and catchment degradation. Over the years, many externally funded projects have been aimed at reducing land degradation. For example, a GEF/UNDP/Lesotho/GIZ project aimed at fostering sustainable land management in Lesotho received $5.52 million in funding. Interestingly, this project identifies sustainable land management in Lesotho as a "vital ingredient of broader environmental wellbeing" because of the water the country supplies to South Africa (UNDP, 2012).

POTENTIAL CONSTRAINTS ON WATER SUPPLY

Effects of Human Activities

In South Africa, it is now expected that demand for water will exceed supply by 2025 if nothing is done to supplement current water resources (Blignaut and van Heerden, 2009; Treasury Report, 2011). The following factors contribute to this situation:

- Growing population size in South Africa, numbering 55 million in 2015 (StatsSA, 2015) and growing at a rate of 1.65% per year, reflecting rapidly increasing water needs.
- Lack of government policy to plan for lower per capita water needs (Blignaut and van Heerden, 2009).
- Soil erosion in the main high-altitude catchments of the Drakensberg/Maloti and its impact on river flows and dams, including adding to the silt load to the Katse and Mohale dams of the LHWP.
- Alien species invasions in the catchment and the threats to catchments at all levels from invasive plant species (Chamier et al., 2012) affecting soil characteristics, runoff, and transpiration.
- Climate change and the risks to basin flows, ground water recharge, and soil moisture levels if there is a change in precipitation or enhanced evaporation over the Escarpment and Highveld plateau. A potential

increase in the frequency of occurrence of droughts, including those brought about by the ENSO cycle, is a critical factor within the context of climate change.

Effects of Climate Change

A study of plausible climate change over South Africa was commissioned by the national DEA in 2013 as part of the long-term adaptation scenario (LTAS) planning. This work consisted of the downscaling of the global climate model (GCM) projections described in Assessment Reports Four (AR4) and Five (AR5) of the IPCC (Christensen et al., 2007; Niang et al., 2014). The downscalings were obtained using both statistical (Hewittson and Crane, 2006) and dynamical (Engelbrecht et al., 2009) techniques. This work (DEA, 2013) divided South Africa into six water management zones of which three include the Drakensberg.

For the statistical downscalings performed, the simulated present-day rainfall totals fall outside the range of observed variability over the northern, central, and southern Drakensberg (DEA, 2013) with the downscaled minimum and maximum temperatures showing closer correspondence to reality. The general lack of long-term homogeneous time-series data over the Drakensberg region, particularly over the high-altitude regions and Lesotho, places a severe constraint on the application of statistical downscaling techniques over this region. For dynamic downscaling, similar validation was not provided in DEA (2013) given that the simulations were bias-corrected, but Engelbrecht et al. (2009, 2015) and Dedekind et al. (2016) report general overestimations of rainfall over the eastern escarpment areas of South Africa in dynamic downscalings. However, the general west-east gradient in rainfall over South Africa, relatively high rainfall over areas to the east of the escarpment, and the dry slot of the Limpopo basin are well reflected in typical dynamic downscalings over southern Africa (Engelbrecht et al., 2009, 2015).

Under the RCP 4.5 scenario (modest to high climate change mitigation; IPCC 5th Assessment report 2013), dynamic downscaling did not give a clear indication of a significant change in rainfall until the year 2100 over the northern, central, and southern Drakensberg. With dynamic downscaling using the A2 scenario (low mitigation in the IPCC 4th Assessment Report, 2007), the forecasts are slightly different (DEA, 2013). For all three zones, covering the northern, central, and southern Drakensberg, dynamic downscaling projected a decrease of some 75 mm (around 5–10%) of annual rainfall from 2014 until 2100. However, the scatter of the individually downscaled GCMs around this median estimate is wide. In summary, the ensemble of models indicates that the annual rainfall over a larger area including the northern, central, and southern Drakensberg may plausibly decrease by some 5–10% during the next 80 years. In all of these cases, the change in precipitation is expected to follow an almost linear decreasing trend with time. However, it should be noted that in the larger set of GCM projections described in AR4 (Christensen et al., 2007) and AR5 (Niang et al., 2014), there is no correspondence across the models in terms of the projected rainfall signal over the central and southern Drakensberg regions, with some models indicative of rainfall increases and other indicative of rainfall decreases.

In order to demonstrate the nature of projected climate futures for the Drakensberg, downscalings are presented for the four main regions of the Drakensberg outlined at the beginning of this chapter. The regional model applied is the conformal-cubic atmospheric model of the Commonwealth Scientific and Industrial Research Organization (CSIRO). Six different GCMs that contributed to AR4 were downscaled for the period 1961–2100, to a resolution of about 0.5degrees in latitude and longitude over southern Africa for a low mitigation future. Full details of the experimental design of these simulations are provided in Engelbrecht et al. (2015). The regional climate model projected envelope of changes in temperature and rainfall as a function of time is presented for each of four sub-regions of the Drakensberg—the MD (Fig. 1.5A), eKangala Drakensberg (Fig. 1.5B), Maloti-Drakensberg (Fig. 1.5C), and CME (Fig. 1.5D). The model-simulated temperature and rainfall anomalies are shown for the period 1971–2100, relative to the 1971–2005 baseline period. We make use of these projections, the larger set of GCM projections described in AR4 (Niang et al., 2014), and evidence from observational studies to summarize current insights into observed trends and projected futures over these four areas in the Drakensberg.

The Mpumalanga Drakensberg

Mean temperature has increased over eastern South Africa, including the MD region, by 1.6–2 °C over the period 1961–2010 (Engelbrecht et al., 2015). Rainfall trends recorded in this region over the period 1910–2004 are not statistically significant (Kruger, 2006; also see Table 1.1). The model simulations (Fig. 1.5A) are indicative of pronounced climate variability, with annual anomalies at times exceeding 500 mm (either in the form of below- or above-normal rainfall). The region is projected to warm significantly under climate change, with temperature increases being as large as 6 °C by the end of the century under low mitigation. The regional climate regime is also projected to shift toward a significantly drier state toward the end of the century. The majority of GCM projections described in AR5 of the IPCC are consistently indicative of rainfall decreases over this part of South Africa (Niang et al., 2014).

The eKangala Drakensberg

The eKangala Drakensberg region falls in a larger part of eastern South Africa where temperature increases are estimated to have taken place at a rate of

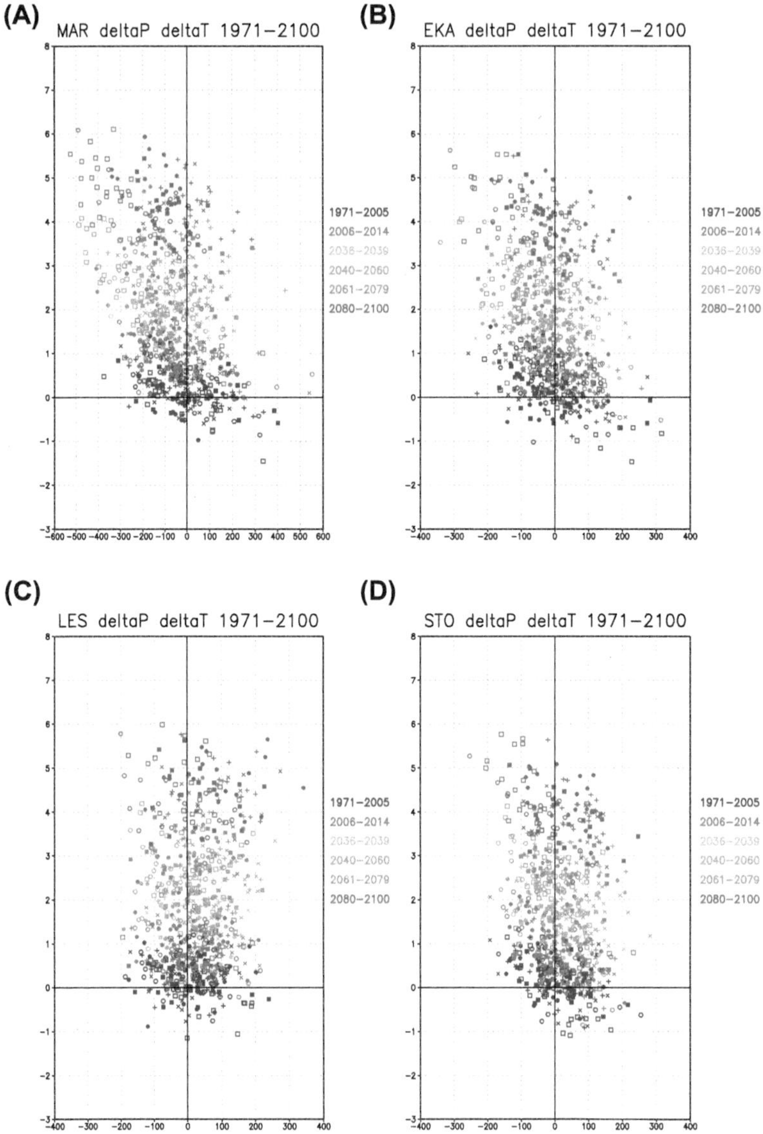

FIGURE 1.5 Projected annual temperature (°C, y-axis) and rainfall (mm, x-axis) anomalies for the period 1971–2100 over four areas of the Drakensberg under a low mitigation future. Anomalies are calculated relative to the 1971–2005 baseline climatology. (A) Mpumalanga Drakensberg, (B) Ekangala Drakensberg, (C) Maloti-Drakensberg, and (D) Eastern Cape Midlands.

1.6–2 °C over the period 1961–2010 (Engelbrecht et al., 2015). There is evidence of statistically significant decreases in rainfall totals occurring over this region during the period 1910–2004 (Kruger, 2006). The model simulations (Fig. 1.5B) are indicative of pronounced climate variability, but with smaller annual rainfall anomalies compared to the MD to the north. This variability is projected to remain a feature of the region's climate during the 21st century. Moreover, the eKangala Drakensberg is projected to warm significantly under climate change, with temperature increases being as large as 6 °C by the end of the century under low mitigation. The regional climate regime is simultaneously projected to shift toward a significantly drier state toward the end of the century.

The Maloti-Drakensberg

Temperature increases have been taking place taking place at a rate of 1.6–2 °C over an area including the northern parts of the Maloti Drakensberg over the period 1961–2010, with indications of the rate of temperature increase being somewhat less to the south of Lesotho (Engelbrecht et al., 2015). No evidence exists of pronounced trends in rainfall being present over this region during the past century (Kruger, 2006; also see Table 1.1). The climate of the region is variable, but less so than over the eKangala Drakensberg, with simulated annual precipitation anomalies reaching values as large as 200 mm. No clear signal of change is present in the regional projections performed (Fig. 1.5C). Annual rainfall is projected to remain variable, with some indications of more years with extremely high precipitation. Temperature anomalies are projected to reach values as high as 6 °C toward the end of the century, with implications for enhanced evaporation in the region's mega dams.

It should be noted that for the ensemble of GCM projections described in AR5, there is no correspondence in the projected climate change signals, in terms of whether the Maloti-Drakensberg region is plausibly to become wetter or drier. Uncertainty also surrounds the snowfall futures of this region, with no comprehensive regional climate modeling studies investigating this aspect of South African climate. However, the general southward displacement of frontal displacement over the Southern Ocean and associated decreases in frontal rainfall over South Africa (Engelbrecht et al., 2009) and the potential for cut-off low pressure systems decreasing in frequency (Engelbrect et al., 2013) are indicative of the potential for a decrease in weather events causing snowfall under climate change.

The Cape Midlands Escarpment

Over the CME, the rate of temperature increase over the period 1961–2010 has been recorded to be somewhat less than over the escarpment regions to the north, with the rate of increase being in the order of 1 °C per century

(Engelbrecht et al., 2015). There is now evidence of pronounced trends in rainfall occurring over the region during the past century (Kruger, 2006). The region exhibits natural climate variability with simulated rainfall anomalies reaching values as high as 200 mm per year for the baseline period 1971–2005. Consistent with the GCM projections of AR5, the regional downscalings are indicative of the CME becoming generally drier in a low mitigation future, while warming drastically. Temperature anomalies could plausibly reach values as high as 6 °C toward the end of the century.

What would the impact of such climate change be? Water availability might be affected, resulting from a drier climate and/or increased evapotranspiration within the context of an expected temperature increase of between 3 and 6 °C (DEA, 2013). While increased evapotranspiration contributes toward speeding up the global water cycle, there are expected to be strong regional differences in precipitation trends (Niang et al., 2014). Indeed, some climate model projections are indicative of rainfall increases over the Drakensberg region of South Africa (DEA, 2013; Niang et al., 2014; Engelbrecht et al., 2015).

Given the context of a rapidly growing South African population, this is likely to have a strong effect in per capita water availability. Secondly, increased tree cover driven by rising CO_2 and climatic changes (Bond and Midgley, 2000) could have adverse impacts on surface water flows as well as reduced grazing, arable fields, and food production in the currently grassland-dominated catchments of the Drakensberg. This situation suggests a high importance of water governance mechanisms that provide strong measures to promote within-generation equity in access to this critical natural resource.

SYNTHESIS FROM RESEARCH TO DATE

Climate Change

There is no compelling evidence that either annual rainfall in the Drakensberg mountains or annual water runoff from these mountains has significantly changed over the past 30–40 years, either increasing or decreasing (Tables 1.1 and 1.2). Although statistically there is no evidence, the between-year variability in rainfall and runoff may have increased, brought about by ENSO cycles that are more noticeable. Downscaling GCM trends yielded different regional results, some (especially those following scenario A2 in the AR4) suggesting a 5–10% decrease in rainfall in the period up to the year 2100.

Temperatures are, however, projected to increase as much as 6°C with strong potential impacts on actual evapotransporation. As shown by Schulze (1979), actual evapotransporation increases with rainfall, that is, it is not limited by energy but by precipitation. Projected temperature increases of up to 6°C will significantly increase energy delivered to the surface and with it actual evapotranspiration (AET). Without a corresponding increase in precipitation, not foreseen in any of the foregoing analyses, discharge must decline in these future scenarios. Ongoing monitoring should be established at a range of altitudes to detect this and other possible trends in the future.

Human Population Growth

South Africa has a rapidly increasing population, at 55 million in 2015, and a population growth rate of 1.65% per year (StatsSA, 2015). The projected growth in water utilization is unsustainable and will lead to a significant water shortage without even considering climate change (Table 1.5); a conclusion was also emphasized by Blignaut and van Heerden (2009).

Water Governance

The emphasis of water management has been on more efficient extraction and the increase in the water yield of river systems by increasing built infrastructure, including increasing dam capacities and building new dams, pipelines, and water transfer tunnels. Other technology options to provide more water could include the reuse of domestic waste water and desalination of seawater and acid mine water. Water governance has not emphasized demand-side management or the implementation of new technology to reduce water needs. The absence of a conservation program strongly affects the ability to respond to water shortages (Blignaut and van Heerden, 2009). In Lesotho, an IWRM system exists on paper but implementation thereof is not complete (Lesotho Government, 2011).

The effects of human population growth and of effective governance appear to be much more important than climate change in determining the short-term and medium-term demand on water from the Drakensberg.

IMPROVING THE SECURITY OF MOUNTAIN WATER CATCHMENTS IN A SEMI-ARID REGION

Improved Knowledge of the Factors Affecting Mountain Water Catchments

The measurement of climatic variables above 2000 m asl in the Drakensberg is deficient and incomplete (Table 1.1). Valuable studies like those of Schulze (1979) were restricted to a small area of the Drakensberg. Global change in southern African mountain systems includes changes to biodiversity, ecosystems, ecosystem services, natural resources, agriculture, human livelihoods, and human settlements. A complete understanding of the factors that affect these variables is lacking, even though future water security is dependent on a better understanding of these factors.

Prominent among these factors is the effect of land management on runoff. Large parts of the Lesotho mountain landscape are used for the cultivation of maize, wheat sorghum, and pulses (Makanete et al., 1998). The effects of erosion in the Maloti-Drakensberg have been described (Shower, 2005) but not quantified. ENSO events should be monitored in mountains, as well as the recovery during "la Niña" periods, so that we understand the changes that these drought events cause at high altitudes. Also needed are monitoring systems that track land use and capture track degradation and restoration.

These aims are best achieved with a system of long-term mountain observation stations (Haeberli et al., 2007). While other countries and continents are setting up mountain observatories, African countries are lagging. The South African Environmental Observation Network (SAEON) has begun to monitor mountain environments via its Grasslands-Forests-Wetlands and Ndlovu nodes (SAEON, 2010). In the Ndlovu node, research and monitoring sites have been established in the Soutpansberg and Mariepskop areas to detect change. This work should be expanded to additional sites in South Africa's other mountain regions set within other biomes, for example, the Lesotho Drakensberg-Maloti, and necessarily involves multi-country monitoring network. The need is for establishing monitoring sites in well-inventoried and studied areas that can become part of the SAEON as well as GNOMO (the Global Network of Mountain Observatories), and complement the Global Mountain Biodiversity Assessment (GMBA), GLORIA (Global Observation Research Initiative in Alpine Environments), and other international monitoring programs.

An Approach for Improved Mountain Ecosystem Governance

South Africa's mountain catchments and surrounding landscapes differ from those in other African countries in that they are not under enormous pressure from completely unrestrained human activities that include land clearing (Poulsen and Hoffman, 2015; Reusing, 2000), deforestation (Schaafsma et al., 2014; Nyssen et al., 2014; Epule et al., 2014; Getahum et al., 2013; Hall et al., 2009; Dessy and Kleman, 2007; Reusing, 2000), harvesting of bushmeat (Rovero et al., 2012; Nielsen and Treue, 2012; Fa et al., 2006), the invasion and settlement of protected areas (Petursson and Vedeld, 2015; Petursson et al., 2013; Tumusiime et al., 2011), the reduction of protected areas to "islands of forest in a degraded landscape" (Green et al., 2013), poor catchment management (Legesse et al., 2003), overgrazing and fires (Detsch et al., 2016; Jira et al., 2013; Van der Merwe and Van Rooyen, 2011), and the impacts of conflict, migration, and development (Wilkie et al., 2000). While population growth in South Africa is unlikely to abate soon, it may not show the magnitude of increase projected for other African countries, for example, future population up to 300% greater than the present by 2050 (World Population Review, 2016). South African mountain catchments could be an important locality for testing out new approaches for mountain ecosystem governance, as well as long-term ecological and climate monitoring for research purposes. This effort would require best practices in the conduct of both science (hydrology, biogeography, climatology) and governance.

The management of water catchments in South Africa is both a local and an international issue, involving South Africa and its neighbors. Lesotho and the Maloti-Drakensberg mountains supply South Africa directly with water, but South Africa has onward international water flow responsibilities, firstly to Mozambique (managed through the Tripartite Permanent Technical Commission concerning the Inkomati River)), to Swaziland (via the Joint Water Commission), to the Komati Basin Water Authority that links South Africa, Mozambique, and Swaziland, as well as via the ORASECOM commission, linking Botswana, Namibia, South Africa, and Lesotho. While these instruments enforce obligations about water use, specifically pertaining to excessive use or restriction of water supplies, there is no sharing of costs to protect the integrity of upstream catchments, whether mountains, wetlands, or grassland catchments from whence the water originates.

A number of international contractual examples exist that seek a more holistic approach toward managing and governing large complex trans-boundary and multinational ecosystems including mountain ecosystems and their natural assets (e.g., the Alpine Convention and the Carpathian Convention in the European Union). Other organizations, such as ICIMOD, the International Center for Integrated Mountain Development, conduct inter-governmental learning and knowledge sharing. ICIMOD is a regional center serving the eight regional member countries of the Hindu Kush Himalayas and is based in Kathmandu, Nepal. All member countries, as well as IDA agencies from European countries, contribute finance and expertise to the regional program. Another novel option could be to develop and use the Large Marine Ecosystems (LME) approach as a model for "Large Mountain Ecosystem" governance. The LME approaches take a strong stance on "ecosystem based management" as per the UN Convention on Biological Diversity to ensure that fisheries yields are maintained, yet is trans-boundary, multinational, and relies on science-based information for the management of productive marine ecosystems (fisheries) (Ramírez-Molsalve et al., 2016; Berg et al., 2015; Sherman and Hamukuaya, in press; Carlisle, 2014). Mountains could be similarly managed to ensure the sustainable production of a very profitable "harvest" of water across a wide, multinational landscape. This governance approach could lead to unified action toward the sustainable management of mountain water catchments in southern Africa and perhaps provide a leading example for other African mountain systems.

CONCLUSION

Food and water supply are global issues and the ongoing provision of good-quality water for multiple human needs has a strategic economic and political value. In the light of past global food price crises, nations now seek to ensure "food sovereignty," and linked to this, arid and semi-arid nations must also take a more strategic view of their own "water sovereignty." South Africa is now acknowledged to be one of the world's semi-arid, water stressed countries and must become much more effective in water management. For southern Africa, climate change has so far had a relatively small effect on mountain water supply compared to the effects of socioeconomic factors and population growth, but projections inherent in the South African Long Term Adaptation Scenario Plan (2013) show that this could change (DEA, 2013). A general pattern of increased temperatures and a risk of drier conditions to the west and south of the country and a risk of either wetter or drier conditions over the east of the country might complicate supply and demand issues, with a greater need for water transfers from catchments where it may plausibly rain more to areas of demand where rainfall may be significantly less than at present. The implementation of appropriate governance mechanisms to manage water extraction from mountains and ensure efficient water utilization is therefore critical.

Southern Africa needs a holistic mountain governance approach that governs the Great Escarpment in its totality, rather than as a fragmented collection of World Heritage Sites, trans-boundary/peace parks, national parks and Ramsar sites, and unprotected landscapes as per the high-altitude grasslands of Lesotho and overgrazed foothills of South Africa. Taking a cue from the Large Marine Ecosystem approach used to manage ocean fisheries (Carlisle, 2014), and the Ecosystem Approach proposed by the United Nations Convention on Biological Diversity, the Drakensberg escarpment should be more strategically managed as a "large mountain ecosystem" where partner and downstream nations contribute skills and finances to the research, management, compliance, and enforcement regimes necessary to manage the mountain landscapes and associated catchments sustainably and equitably into the future. Southern Africa's Benguela Current Large Marine Ecosystem expertise could be brought to bear on the management of "large mountain ecosystems" of the Drakensberg and the other components of the Great Escarpment.

ACKNOWLEDGMENTS

We thank the South African Weather Service for providing raw meteorological data for our analyses. We gratefully acknowledge the assistance of Mrs Ingrid Booysen, Cartographer, Department of Geography, Geoinformatics and Meteorology, University of Pretoria. This work was done under the auspices of AfroMont, the Mountain Research Initiative (MRI), and the Albertine Rift Conservation Society (ARCOS), and funded by the Swiss National Science Foundation.

REFERENCES

Agnew, S., 1958. The landforms of the Hogsback area in the Amatola range. Fort Hare Papers 2, 1–14.

Beniston, M., 2003. Climatic change in mountain regions: a review of possible impacts. Climate Change 59, 5–31.

Berg, T., Fürhaupter, K., Teixeira, H., Uusitalo, L., Zampoukas, N., 2015. The Marine Strategy Framework Directive and the ecosystem-based approach – pitfalls and solutions. Marine Pollution Bulletin 6 (1–2), 18–28.

Blignaut, J., van Heerden, J., 2009. The impact of water scarcity on economic development initiatives. Water SA 35, 415–420.

Bond, W.J., Midgley, G.F., 2000. A proposed CO_2-controlled mechanism of woody plant invasion in grasslands and savannas. Global Change Biology 6, 865–869.

Bourblanc, M., Blanchon, D., 2013. The challenges of rescaling South African water resources management: Catchment Management Agencies and interbasin transfers. Journal of Hydrology 519, 2381–2391.

Carlisle, K.M., 2014. The large marine ecosystem approach: application of an integrated, modular strategy in projects supported by the global environment facility. Environmental Development 11, 19–42.

Chamier, J., Schachtschneider, K., Le Maitre, D.C., Ashton, P.J., van Wilgen, B.W., 2012. Impacts of invasive alien plants on water quality, with particular emphasis on South Africa. Water SA 38 (2), 345–356.

Christensen, J.H., Hewitson, B., Busuioc, A., Chen, A., Gao, X., Held, I., Jones, R., Kolli, R.K., Kwon, W.-T., Laprise, R., Magana Rueda, V., Mearns, L., Menendez, C.G., Raisanen, J., Rinke, A., Sarr, A., Whetton, P., 2007. Regional climate projections. In: Solomon, S., Qin, D., Manning, M., Chen, Z., Marquis, M., Averyt, A.B., Tignor, M., Miller, H.L. (Eds.), Climate Change 2007: The Physical Science Basis. Contribution of Working Group I to the Fourth Assessment Report of the Inter-Governmental Panel on Climate Change. Cambridge University Press, Cambridge.

Clark, V.R., Barker, N.P., Mucina, L., 2009. The Sneeuberg: a new centre of endemism on the Great Escarpment, South Africa. South African Journal of Botany 75, 196–238.

Clark, V.R., Barker, N.P., Mucina, L., 2011. The Boschberg (Somerset East, Eastern Cape) – a floristic cross-roads of the southern Great Escarpment. South African Journal of Botany 77, 94–104.

Clark, V.R., Barker, N.P., 2014. Endemism in Mainland Regions: the role of edaphic substrate versus moisture availability in montane endemic plant distribution patterns – evidence from the Cape Midlands Escarpment, South Africa. In: Hobohm, C. (Ed.), Endemism in Vascular Plants, Plant and Vegetation, vol. 9. Springer, Dordrecht, pp. 260–281.

Clark, V.R., Dold, A.P., McMaster, C., McGregor, G., Bredenkamp, C., Barker, N.P., 2014. Rich Sister, Poor Cousin: Plant Diversity and Endemism in the Great Winterberg-Amatholes (Geat Escarpment, Eastern Cape, South Africa). South African Journal of Botany 92, 159–174.

Compton, J.S., Herbert, C.T., Hoffman, M.T., Schneider, R.R., Stuut, J.-B., 2010. A tenfold increase in the Orange River mean Holocene mud flux: implications for soil erosion in South Africa. The Holocene 20, 115–122.

Compton, J.S., Maake, L., 2007. Source of the suspended load of the upper Orange River, South Africa. South African Journal of Geology 110, 339–348.

Cullen, N.J., Mölg, T., Kaser, G., Hussein, K., Steffen, K., Hardy, D.R., 2006. Kilimanjaro Glaciers: recent areal extent from satellite data and new interpretation of observed 20th century retreat rates. Geophysical Research Letters 33, L16502.

DEA, 2013. Long-Term Adaptation Scenarios Flagship Research Programmeme (LTAS) for South Africa. Department of Environment Affairs (DEA), Pretoria, South Africa.

Dedekind, Z., Engelbrecht, F.A., van der Merwe, J., 2016. Model simulations of rainfall over southern Africa and its eastern escarpment. Water SA 42, 129–143.

Dessie, G., Kleman, J., 2007. Pattern and magnitude of deforestation in the south central rift valley region of Ethiopia. Mountain Research and Development 27 (2), 162–168.

Detsch, F., Otte, I., Appelhans, T., Hemp, A., Nauss, T., 2016. Seasonal and long-term vegetation dynamics from 1-km GIMMS-based NDVI time series at Mt. Kilimanjaro, Tanzania. Remote Sensing of the Environment 178, 70–83.

De Villiers, G.T., 1996. South Africa's water resources and the Lesotho highlands water scheme: a partial solution to the country's water problems. International Journal of Water Resources Development 12, 65–78.

Dingman, S.L., 2015. Physical Hydrology, third ed. Waveland Press, Long Grove, Illinois. 643 pp.

Du Toit, A.L., 1920. The Karoo dolerites of South Africa: a study in hypabyssal injection. Transactions of the Geological Society of South Africa 23, 1–42.

DWA, 2004. National Water Resource Strategy, first ed. Department of Water and Sanitation, Pretoria, South Africa. 271 pp.

DWA, 2012. National Water Resource Strategy. Summary Report, second ed. Department of Water Affairs, Pretoria, South Africa. 71 pp.

DWS, 2013. Draft National Water Resource Strategy 2. Department of Water and Sanitation, Pretoria, South Africa.

Engelbrecht, F., Adegoke, J., Bopape, M.-J., Naidoo, M., Garland, R., Thatcher, M., McGregor, J., Katzfey, J., Werner, M., Ichoku, C., Gatebe, C., 2015. Projections of rapidly rising surface temperatures over Africa. Environmental Research Letters 10, 085004.

Engelbrecht, F.A., McGregor, J.L., Engelbrecht, C.J., 2009. Dynamics of the conformal-cubic atmospheric model projected climate-change signal over southern Africa. International Journal of Climatology 29, 1013–1033.

Engelbrecht, C.J., Engelbrecht, F.A., Dyson, L.L., 2013. High-resolution model-projected changes in mid-tropospheric closed-lows and extreme rainfall events over southern Africa. International Journal of Climatology 33, 173–187.

Epule, E.T., Peng, C., Lepage, L., Chen, Z., 2014. Policy options towards deforestation reduction in Cameroon: an analysis based on a systematic approach. Land Use Policy 36, 405–415.

Eskom, 2011. COP 17 Fact Sheet. South Africa Electricity Supply Commission (Eskom). http://www.eskom.co.za/OurCompany/SustainableDevelopment/ClimateChangeCOP17/Documents/Renewable_energy_-_Hydroelectric_power_stations.pdf.

Fa, J.E., Seymour, S., Dupain, J., Amin, R., Albrechtsen, L., Macdonald, D., 2006. Getting to grips with the magnitude of exploitation: bushmeat in the Cross–Sanaga rivers region, Nigeria and Cameroon. Biological Conservation 129 (4), 497–510.

Getahun, K., Van Rompaey, A., Van Turnhout, P., Poesen, J., 2013. Factors controlling patterns of deforestation in moist evergreen-Afromontane forests of Southwest Ethiopia. Forest Ecology and Management 304, 171–218.

Green, J.M.H., Larrosa, C., Buirgess, N.D., Balmford, A., Johnston, A., Mbilinyi, B.P., Platts, P.J., Coad, L., 2013. Deforestation in an African biodiversity hotspot: extent, variation and the effectiveness of protected areas. Biological Conservation 164, 62–72.

Haeberli, W., Hoelzle, M., Paul, F., Zemp, M., 2007. Integrated monitoring of mountain glaciers as key indicators of global climate change: the European Alps. Annals of Glaciology 46, 150–160.

Hall, J., Burgess, N.D., Lovett, J., Mbilinyi, B., Gereau, R.I., 2009. Conservation implications of deforestation across an elevational gradient in the Eastern Arc Mountains, Tanzania. Biological Conservation 142 (11), 510–2521.

Harvey, P.M., Marker, M.E., 1992. Present-day periglacial micro-forms in the Lesotho highlands: implications for past climatic conditions. Permafrost and Periglacial Processes 3, 353–361.

Hemp, A., 2006. Continuum or zonation? Altitudinal gradients in the forest vegetation of Mt. Kilimanjaro. Plant Ecology 184, 27–42.

Hewittson, B.C., Crane, R.G., 2006. Consensus between GCM climate change projections with empirical downscaling: precipitation downscaling over South Africa. International Journal of Climatology 26, 1315–3337.

Hurni, H., Tato, K., Zeleke, G., 2005. The implications of changes in population, land use, and land management for surface runoff in the upper Nile basin area of Ethiopia. Mountain Research and Development 25, 147–154.

ILRI, 1995. FAO livestock policy analysis. In: ILRI Training Manual. International Livestock Research Institute, Nairobi, Kenya, 264 pp.

IPCC, 2007. Climate Change (2007) Synthesis Report. Summary for Policymakers. Fourth Assessment Report. Intergovernmental Panel on Climate Change (IPCC), New York.

IPCC, 2013. 5th Assessment Report: Climate Change 2013. Cambridge University Press, New York.

Jira, G., Bekele, A., Hemson, G., Mundanthra, B., 2013. Rodents in fire affected heather shrublands in Bale Mountains National Park, Ethiopia. Journal of King Saud University – Science 25 (4), 289–295.

Jury, W.A., Vaux, J.H., 2007. The emerging global water crisis: managing scarcity and conflict between water users. Advances in Agronomy 95, 1–76.

Keay-Bright, J., Boardman, J., 2007. The influence of land management on soil erosion in the Sneeuberg mountains, Central Karoo, South Africa. Land Degradation and Development 18, 423–439.

Kruger, A.C., 2006. Observed trends in daily precipitation indices in South Africa: 1910–2004. International Journal of Climatology 26, 2275–2285.

Legesse, D., Vallet-Coulomb, C., Gasse, F., 2003. Hydrological response of a catchment to climate and land use changes in Tropical Africa: case study South Central Ethiopia. Journal of Hydrology 275, 67–85.

Lesotho Government, 2011. National Range Resources Management Policy. Lesotho Ministry of Forestry and Land Reclamation, Maseru, Lesotho.

LTAS, 2013. Long-Term Adaptation Scenarios Flagship Research Programme (LTAS) for South Africa. Department of Environment Affairs (DEA), Pretoria, South Africa.

Ludwig, R., Roson, R., 2015. Climate change, water and security in the Mediterranean: introduction to the special issue. Science of the Total Environment 543, 847–850.

Maddocks, A., Young, R.S., Paul Reig, P., 2015. Ranking the World's Most Water-Stressed Countries in 2040. World Resources Institute.

Maseatile, M., 2013. Lesotho water and sanitation sector. In: Presented at the Eastern and Southern Africa Water and Sanitation Regulators Association Conference, Maseru Sun, Lesotho. 30 October 2013. http://www.lewa.org.ls/library/Publications/CoW_presentation_ESAWAS_conference.pdf.

Makenete, A., Ortmann, G., Darroch, M., 1998. Food-aid dependency in Lesotho: issues and policy implications. Development Southern Africa 15, 251–266.

Messerli, B., Viviroli, D., Weingartner, R., 2004. Mountains of the world: vulnerable water towers for the 21st Century. Ambio 29–34.

Middleton, B.J., Bailey, A.K., 2009. Water Resources of South Africa, 2005 Study. Water Research Commission Report Number TT 380/08. Water Research Commission, Pretoria, South Africa.

Morin, J., Benyamini, Y., 1977. Rainfall infiltration into bare soils. Water Resources Research 13, 813–817.

Mulder, N., Grab, S.W., 2002. Remote sensing for snow cover analysis along the Drakensberg escarpment. South African Journal of Science 98, 213–217.

Mulder, N., Grab, S.W., 2009. Contemporary spatio-temporal patterns of snow cover over the Drakensberg. South African Journal of Science 105, 228–233.

Nel, J., Colvin, C., Le Maitre, D., Snith, J., Haines, I., 2013. Defining South Africa's Water Source Areas. WWF South Africa, Cape Town, 32 pp.

Nel, W., Sumner, P.D., 2005. First rainfall data from the KZN Drakensberg escarpment edge (2002 and 2003). Water SA 31, 399–402.

Nel, W., Sumner, P.D., 2006. Trends in rainfall total and variability (1970–2000). along the KwaZulu-Natal Drakensberg foothills. South African Geographical Journal 88, 130–137.

Nel, W., Sumner, P.D., 2008. Rainfall and temperature attributes on the Lesotho Drakensberg escarpment edge, southern Africa. Geografiska Annaler A 90, 97–108.

Niang, I., Ruppel, O.C., Abdrabo, M., Essel, A., Lennard, C., Padgham, J., Urquhart, P., Adelekan, I., Archibald, S., Barkhordarian, A., Battersby, J., Balinga, M., Bilir, E., Burke, M., Chahed, M., Chatterjee, M., Chidiezie, C.T., Descheemaeker, K., Djoudi, H., Ebi, K.L., Fall, P.D., Fuentes, R., Garland, R., Gaye, F., Hilmi, K., Gbobaniyi, E., Gonzalez, P., Harvey, B., Hayden, M., Hemp, A., Jobbins, G., Johnson, J., Lobell, D., Locatelli, B., Ludi, E., Otto Naess, L., Ndebele-Murisa, M.R., Ndiaye, A., Newsham, A., Njai, S., Nkem, Olwoch, J.M., Pauw, P., Pramova, E., Rakotondrafara, M.-L., Raleigh, C., Roberts, D., Roncoli, C., Sarr, A.T., Schleyer, M.H., Schulte-Uebbing, L., Schulze, R., Seid, H., Shackleton, S., Shongwe, M., Stone, D., Thomas, D., Ugochukwu, O., Victor, D., Vincent, K., Warner, K., Yaffa, S., 2014. IPCC WGII AR5 Chapter 22, pp. 1–115.

Nielsen, M.T., Treue, T., 2012. Hunting for the benefits of joint forest management in the eastern Afromontane biodiversity hotspot: effects on bushmeat Hunters and wildlife in the Udzungwa Mountains. World Development 40 (6), 1224–1239.

Nyssen, J., Frankl, A., Maile, M., Hurni, H., Descheemaeker, K., Crummey, D., Ritler, A., Portner, B., Nievergelt, B., Moeyersons, J., Munro, N., Deckers, J., Billi, P., Poesen, J., 2014. Environmental conditions and human drivers for changes to north Ethiopian mountain landscapes over 145 years. Science of the Total Environment 485–486, 64–179.

Ohlsson, L., 2000. Water conflicts and social resource scarcity. Physics and Chemistry of the Earth B 25, 3212–3220.

ORASECOM, 2008. National Water Laws in Lesotho. Orange-Senqu River Commission. http://www.orangesenqurak.org/governance/water+governance+orange+senqu+basi/national+water+laws/lesotho.aspx.

Petursson, J.G., Vedeld, P., Sassen, M., 2013. An institutional analysis of deforestation processes in protected areas: the case of the transboundary Mt. Elgon, Uganda and Kenya. Forest Policy and Economics 26, 22–33.

Petursson, J.G., Vedeld, P., 2015. The "nine lives" of protected areas. A historical-institutional analysis from the transboundary Mt Elgon, Uganda and Kenya. Land Use Policy 42, 51–263.

Poulsen, Z.C., Hoffman, M.T., November 2015. Changes in the distribution of indigenous forest in Table Mountain National Park during the 20th Century. South African Journal of Botany 101, 49–56.

Phillipson, P.B., 1987. A checklist of the vascular plants of the Amatola mountains, Eastern Cape Province/Ciskei. Bothalia 17, 237–256.

Preston-Whyte, R.A., Diab, R.D., Washington, R., 1991. Diurnal variation of rainfall events by synoptic type in Natal. South African Geographical Journal 73, 22–28.

Preston-Whyte, R.A., Tyson, P.D., 1997. The Atmosphere and Weather of South Africa. Oxford University Press, Cape Town.

R Development Team, 2014. R: A Language and Environment for Statistical Computing. R Foundation for Statistical Computing, Vienna, Austria.

Ramírez-Monsalve, P., Raakjær, J., Nielsen, K.N., Santiago, J.L., Ballesteros, M., Laksá, U., Degnbol, P., 2016. Ecosystem approach to fisheries management (EAFM) in the EU – current science–policy–society interfaces and emerging requirements. Marine Policy 66, 83–92.

Reusing, M., 2000. Change detection of natural high forests in Ethiopia using remote sensing and GIS techniques. International Archives of Photogrammetry and Remote Sensing 33 (B7), 1253–1258.

Rijsberman, F.R., 2006. Water scarcity: fact or fiction? Agricultural Water Management 80, 5–22.

Rovero, F., Mtui, A.S., Kitegile, A., Nielsen, M.R., 2012. Hunting or habitat degradation? Decline of primate populations in Udzungwa Mountains, Tanzania: an analysis of threats. Biological Conservation 146 (1), 89–96.

SAEON, 2010. The Grasslands-Forests-Wetlands Node. South African Environmental Observation Network. http://gfw.dirisa.org/.

Schaafsma, M., Burgess, N.D., Swetnam, R.D., Ngaga, Y.M., Turner, R.K., Treue, T., 2014. Market signals of unsustainable and inequitable forest extraction: assessing the value of illegal timber trade in the Eastern Arc Mountains of Tanzania. World Development 62, 155–168.

Sene, K.J., Jones, D.A., Meigh, J.R., Farquharson, F.A.K., 1998. Rainfall and flow variations in the Lesotho Highlands. International Journal of Climatology 18, 329–345.

Sherman, K., Hamukuaya, H., 2015. Sustainable development of the world's Large Marine Ecosystem Sustainable fisheries within an LME context. Environmental Development 17 (1), 1–6.

Showers, K., 2005. Imperial Gullies: Soil Erosion and Conservation in Lesotho. Ohio University Press, Athens, Ohio.

Shultze, R., 1979. Hydrology and Water Resources of the Drakensberg. Natal Town and Regional Planning Commission, Pietermaritzburg, South Africa, 179 pp.

StatsSA, 2010. Water Accounts for South Africa: 2000. Discussion Document: D0405.1, March 2009. Published by Statistics South Africa, Private Bag X44, Pretoria 0001, p. 165.

StatsSA, 2012. The 2011 Census. Census 2011. Provinces at a Glance. Statistics South Africa. Published by Statistics South Africa, Private Bag X44, Pretoria 0001, Pretoria. 84 pp.

StatsSA, 2015. Mid-year Population Estimates (2015). Document P0302, Statistics South Africa. 20pp. Pretoria, South Africa.

StatsSA, 2016. Gross Domestic Product, Fourth Quarter (2015). Document P0441. Statistics South Africa, Pretoria, South Africa.

Tamene, L., Park, S.J., Dikau, R., Vlek, P.L.G., 2006. Reservoir siltation in the semi-arid highlands of northern Ethiopia: sediment yield–catchment area relationship and a semi-quantitative approach for predicting sediment yield. Earth Surface Processes and Landforms 31, 1364–1383.

Treasury, S.A., 2011. Water and sanitation. In: Intergovernmental Fiscal Review. Treasury, South African National Government. Department of Finance, Pretoria, South Africa.

Truswell, J.F., 1977. The Geological Evolution of South Africa. Purnell, Cape Town.

Tumusiime, D.M., Vedeld, P., Gombya-Ssembajjwe, W., 2011. Breaking the law? Illegal livelihoods from a protected area in Uganda. Forest Policy and Economics 13 (4), 273–283.

Tyson, P.D., 1986. Climate Change and Variability in Southern Africa. Oxford University Press, Oxford.

Tyson, P.D., 1991. Climatic change in southern Africa: past and present and possible Future conditions scenarios. Climatic Change 18, 241–258.

Tyson, P.D., Preston-Whyte, R.A., Schulze, R.E., 1976. The Climate of the Drakensberg. Natal Town and Regional Planning Commission, Pietermaritzburg.

UNDP, 2012. Sustainable Land Management in Lesotho. UNDP. http://www.ls.undp.org/content/lesotho/en/home/operations/projects/environment_and_energy/SLMProject.html.

Van der Merwe, H., Van Rooyen, M.W., 2011. Vegetation trends following fire in the Roggeveld, Mountain Renosterveld, South Africa. South African Journal of Botany 77 (1), 127–136.

Viviroli, D., Dürr, D.H., Messerli, B., Meybeck, M., Weingartner, R., 2007. Mountains of the world, water towers for humanity: typology, mapping, and global significance. Water Resources Research 43, W07447.

Water and Sanitation, 2007. The Water and Sanitation Policy of Lesotho, 2007. Ministry of Natural Resources. Government of Lesotho. http://washinschoolsmapping.com/wengine/wp-content/uploads/2015/10/Lesotho-Water-and-Sanitation-Policy.pdf.

Wilkie, D., Shaw, E., Rotberg, F., Morelli, G., Auzel, P., 2000. Roads, development, and conservation in the Congo basin. Conservation Biology 14 (2000), 1614–1622.

World Population Review (2016). http://worldpopulationreview.com/countries/. [accessed 29.03.2016].

Chapter 2

Mountain Area Glaciers of Russia in the 20th and the Beginning of the 21st Centuries

T. Khromova, G. Nosenko, A. Muraviev, S. Nikitin, L. Chernova and N. Zverkova
Institute of Geography Russian Academy of Sciences, Moscow, Russia

Chapter Outline

Mountain Ice and Water. http://dx.doi.org/10.1016/B978-0-444-63787-1.00002-0

INTRODUCTION

Glaciers are one of the main components of the cryosphere, and being sensitive to temperature changes, they act as natural indicators of climate fluctuations. In high-mountain areas, glaciers represent one of the most important factors of the landscape formation and may potentially trigger catastrophic phenomena hazardous to the surrounding area.

All recent data available indicate that during the past few decades, almost all glaciers in mountain areas and glaciers of Arctic islands were retreating that was manifested in a decrease in their size, volume, and ice mass. The retreat of glaciers is accompanied by increasing occurrence of natural disasters resulting in devastating effects. A research on conditions and changes of the main parameters of contemporary glaciation is required in order to mitigate the risks and minimize their effects.

Glacierization has existed on the territory of Russia for thousands of years. At present both mountain glaciers and continental ice sheets are present there. Continental ice sheets are located on islands and archipelagoes of the Russian Arctic region, and mountain glaciers are widespread on continental part of the country where such ice currently covers an area of about 3,480,000 km^2. Some 18 glacier regions occur (Table 2.1, Fig. 2.1). Twelve of them receive atmospheric moisture from the Atlantic and are included in the Atlantic Glacier Province and six are included in the Pacific Province (Krenke, 1982). Five out of twelve Atlantic regions (Khibiny, Urals, Putorana, Byrranga, and Orulgan) are situated within subarctic latitudes and seven (Caucasus, Altai, Kuznetsk Alatau, East Sayan, Kodar, Barguzin, and Baikal Ridges) within the mid-latitudes. Six Pacific glacier regions (Chukotka, Koryak and Kolyma Plateaus, Chersky Mts., Suntar-Hayata Mts., Kamchatka) are located relatively close to the Pacific Ocean within 68–62°N. No uniformity occurs in studies of the glacier systems of Russia, whereas current data availability varies for the different glacier regions.

The compilation of the Glacier Inventory of the USSR (1965–1982) was the one and only attempt to collect consistent data on glacier conditions nationwide within a definite time frame. Even though uniform rules and guidelines were set for the purpose of the compilation of the inventory, its quality varied from one region to another, depending upon the availability and quality of data used as well as the level of authors' accuracy while following the guidelines. The digital version of the Glacier Inventory of the USSR, as a part of the World Glacier Inventory, is available on the Website of the National Snow and Ice Data Center (https://nsidc.org/).

TABLE 2.1 Mountain Glacier Systems on the Territory of Russia

		Glacier System	North Latitude, °	Longitude, °	Number	Area, km^2	Average Size, km^2	Sources
Atlantic Glacier Province	Subarctic latitudes	Khibiny	67.8–68.0	*E*-33.4–33.9	4	0.1	0.02	Glacier inventory of the USSR 1965–1982, V.1. Ch.1
		Ural	64.8–68.2	*E*-59.7–67.0	143	28.66	0.20	Glacier inventory of the USSR 1965–1982, V.3. Ch. 3
		Putorana Mts.	69.0–70.0	*E*-90.0–92.3	22	2.54	0.12	Glacier inventory of the USSR 1965–1982, V.16. Is.1. Ch.6
		Byrranga Mts.	75.7–76.0	*E*-107.4–108.0	96	30.5	0.32	Glacier inventory of the USSR 1965–1982, V.16. Is.1. Ch.2
		Orulgan ridge	67.4–69.0	*E*-127.8–128.9	74	18.38	0.25	Glacier inventory of the USSR 1965–1982, V.17. Is.5. Ch.2
		Caucasus	39.07–43.95	*E*-39.90 - 48.00	2080	1427.12 (988.61 – on the territory of Russia)	0.69	Glacier inventory of the USSR 1965–1982, V.8. Ch.1–12; 1975 V.9. Ch.1–4
	Temp-latitudes	Altai	47.0–50.4	*E*-83.7 0–90.15	1517	923.1 (800 – on the territory of Russia)	0.61	Glacier inventory of the USSR 1965–1982, V.15. Is.1. Ch.1–8; V.16. Is.1. Ch.4
		Kuznetsk Alatau	53.7–54.7	*E*-88.3–89.32	91	6.79	0.07	Glacier inventory of the USSR 1965–1982, V.15. Is.2. Ch.1
		East Sayan	51.72–56.00	*E*-96.68– 100.62	105	30.3	0.29	Glacier inventory of the USSR 1965–1982, V.16.

TABLE 2.1 Mountain Glacier Systems on the Territory of Russia—cont'd

		Glacier System	North Latitude, °	Longitude, °	Number	Area, km^2	Average Size, km^2	Sources
								Is.1. Ch.3; 5. V.16. Is.2. Ch.1
		Kodar Ridge	56.8–57.0	*E*-117.2–117.7	30	18.8	0.63	Glacier inventory of the USSR 1965–1982, V.17. Is.2. Ch.1
		Barguzin Ridge	51–51.5	*E*-110.5–111.00	7	0.37	0.05	Kitov et al. (2014)
		Baikal Ridge	51–51.5	*E*-108.00–108.3	2	0.43	0.21	Aleshin (1982)
Pacific Glacier Province		Chukotka Upland	64.6 -67.8	*E*-175.47–W-171.42	47	13.53	0.29	Sedov (1988, 1992, 1995, 1996, 1997a)
		Koryak Upland	60.17–62.3	E-166.7–173.35	1451	303.46	0.59	Glacier inventory of the USSR 1965–1982, V.20. Ch.1
		Northeastern Koryak Upland	62.50–63.00	*E*-176.00–176.60	116	43.96	0.38	Sedov (2001)
		Kolyma Upland	60.38–61.99	E-152.00–160.10	19	3.61	0.19	Sedov (1995, 1997b)
		Cherskii Mts.	64.9–67.7	E-138.6–149.2	372	155.4	0.42	Glacier inventory of the USSR 1965–1982, V.17. Is.7. Ch.2,4; V.19. Ch.4
		Suntar_Khayata Mts.	61.8–62.9	*E*-140.6–143.2	208	201.6	0.97	Glacier inventory of the USSR 1965–1982, V.17. Is.3. Ch.1; V.17. Is.7. Ch.3; V.19. Ch.3
		Kamchatka	51.3–62.1	*E*-156.7–173.4	405	874.0	2.16	Glacier inventory of the USSR 1965–1982, V.20. Ch.2–4

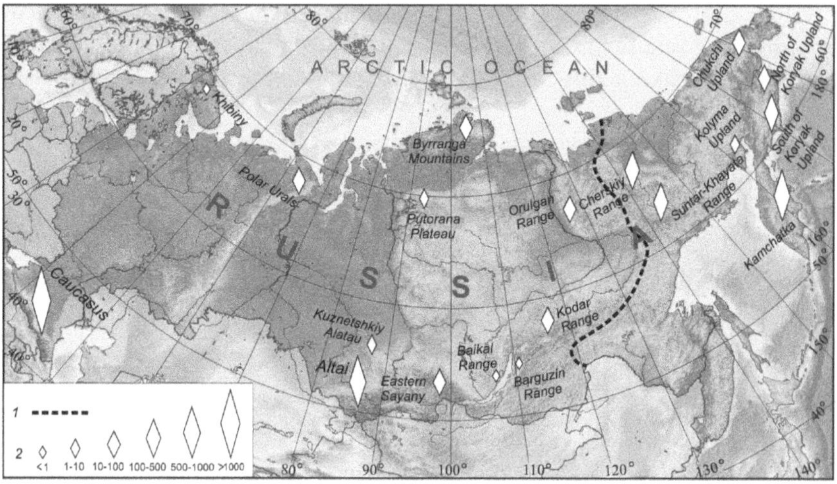

FIGURE 2.1 Mountain glacier systems of Russia. (1) Boundaries between the Atlantic and Pacific Glacier provinces. (2) Glacier system area (km^2) in accordance with the Glacier inventory of the USSR (1965–1982).

The Glacier Inventory of the USSR was compiled during 1960–80s based on special researches and data of aerial survey conducted in the mid-20th century and it was limited by the state of knowledge at the time. After all, 69 volumes of the inventory had been published, several glaciers were discovered on Chukotka and Kolyma Upland, and also the number of glaciers on the Koryak Plateau and their sizes had to be amended. Glaciers of the Baikal and Barguzin Ridges located to the southwest of the Kodar glacier system were not included in the Glacier Inventory of the USSR (Table 2.1).

The last third of the 20th century was marked by tremendous development of space research that has opened new possibilities in studying global environment. Satellite data on glacier areas became available that have provided an opportunity not only to research existing conditions of glaciers, but also to study their dynamics as well. Repeated satellite data have indicated glacier changes during the past few decades, whereas the location of terminal moraines has made it possible to study past changes in glaciers for the whole duration of the Little Ice Age.

By the beginning of the second decade of the 21st century, the implementation of satellite data for the territory of continental Russia has resulted in detailed research of the largest glacier areas: the Caucasus, Altai, and Kamchatka. At present the glacier systems of the Caucasus and Altai are divided by national boundaries; however, for the purpose of this study, we have examined them as integral natural formations using a physiographic approach, which allowed the maintenance of a consistency with previous research papers.

Among the small glacier systems, glacial regions of Polar Ural, Suntar-Hayata Mountains, Kodar, and East Sayan Ridges have the best data coverage and level of research. The issues in research of the small glacier systems using satellite images are primarily related to the small sizes of certain glaciers that make it difficult to use available satellite images with low and medium resolution and requires incorporation of additional data with high resolution. Our findings give an opportunity of evaluating common factors and regional features of changes in the glaciation of continental Russia during the past 100 years.

GLACIER SYSTEMS RECEIVING MOISTURE FROM THE ATLANTIC. SUBARCTIC LATITUDES: KHIBINY, URALS, PUTORANA, BYRRANGA, AND ORULGAN

Khybiny

Khibiny is a small, medium-altitude mountain group, the westernmost mountain region of Russia where glaciers still could be detected in the 1970s. Due to meteorological conditions—average annual temperature of −3.4°C, average annual precipitations of 954 mm, and intensive snow drifting—a small-scale glaciation has remained until the 21st century. Four glaciers occur with a total area of 0.1 km^2: three hanging glaciers and one near-slope glacier elevated at 900−1000 m (Perov, 1968). Currently, the existence of this glacier system is open to question. No special researches have been conducted in recent decades.

Urals

The northern part of the Ural Mountains is full of glaciers; their distribution is uneven and they form isolated clusters located in the most elevated and rugged parts of the range. Two most significant areas where glaciation occurs are the Subpolar Urals (64.25−65.75°N) and the Polar Urals (66.50−68.25°N).

Back in the middle of the past century, the occurrence of glaciers in the Ural Mountains was considered to be "climatically unjustified," and only by the end of 1950s, a hypothesis was advanced that Ural glaciers were not relics of the past, but naturally developing, contemporary glaciers and their occurrence could be explained by particular orographic-climatic conditions (Dolgushin, 1960). No glaciers occurred on the westerly slope facing the main moisture-bearing air flows (Glaciation of the Urals, 1966). All glaciers are located on the opposite slopes, in east-, southeast-, and northeast-facing cirques as increased snow concentration on the leeward and shady slopes played a significant role in their occurrence. Nevertheless, the evolution of glaciers in such conditions is essential in understanding their response to climate changes.

Contemporary conditions and dynamics of glacier boundaries have been assessed in the Institute of Geography of the RAS, based on field surveys and data of satellite imagery. Field work included terrestrial survey of sample glaciers, measurements of meteorological parameters with automatic sensors, snow survey, and geodetic measurement of glaciers using Differential global positioning system (DGPS).

The Subpolar Urals

According to the Glacier Inventory of the USSR based on aerial photographs dated 1953 and 1960, 48 glaciers occurred in the area. In August 2002, the Institute of Geography of the RAS organized a field survey to study Gofman Glacier on Sablinsky Ridge and Manaraga Glacier in the vicinity of Mt. Narodnaya (1894 m).

Within the Subpolar Urals in the central part of the Ural Mountains, the volume of solid precipitation decreases toward the east whereas the elevation of mountain base increases. Thus, on Sablinsky Ridge, the first meridional ridge within this mountain system standing in the way of westerly air masses, glaciers are located up to 600−800 m, and in the area of Mt. Narodnaya in the middle of Subpolar Urals, they are located at 1000−1200 m.

The cirque-valley Gofman Glacier is the largest in the Subpolar Urals (Fig. 2.2). Its accumulation area is located in the cirque on the eastern slope of Sablya Ridge. The elevation of mountain border reaches 1497 m, whereas the cirque bottom is located at 750 m. When A.N. Aleshkov had visited the glacier in 1932, its area was 0.37 km^2 and length 1000 m. The glacier tongue was protruding from the cirque and terminating at 600 m. Even then, while comparing the data from 1929 to 1932, Aleshkov had established the retreat of the glacier tongue with a rate of about 2 m/y (Aleshkov, 1935). According to the survey conducted in 2002, in comparison with 1932, the glacier has retreated by 150 m whereas the glacier tongue surface has lowered by 15−20 m. No significant changes occurred in the accumulation area. The upper borders of the avalanche cones in the accumulation area have stayed almost at the same places as on the photographs made in 1932.

Similar changes took place within other glaciers of the Sablinsky Ridge—glacier tongues were retreating toward the cirque walls with lakes forming within cleared surfaces (Fig. 2.3).

A survey of Manaraga Glacier located near Mt. Narodnaya has been conducted to study the evolution of glaciers in the middle part of the Subpolar Urals. The first time that the cirque glacier was visited in 1932 by S.G. Botch, its area was 0.28 km^2 and the length 650 m. On the aerial photograph dated 1957, these parameters have stayed almost the same. However, in 2002 that glacier could not be located (Glazovsky et al., 2005). A lake was established in its place whereas residues of ice were left only on the northeast wall of the cirque (Fig. 2.4).

A similar instance has been recorded at a distance of 7 km to the west from Manaraga Glacier, at the location of Balban Glacier. In 1957, that glacier had still existed, its dimensions were 400 m width and 450 m length, but by 2002 it has almost disappeared. These examples suggest that "climatically unjustified"

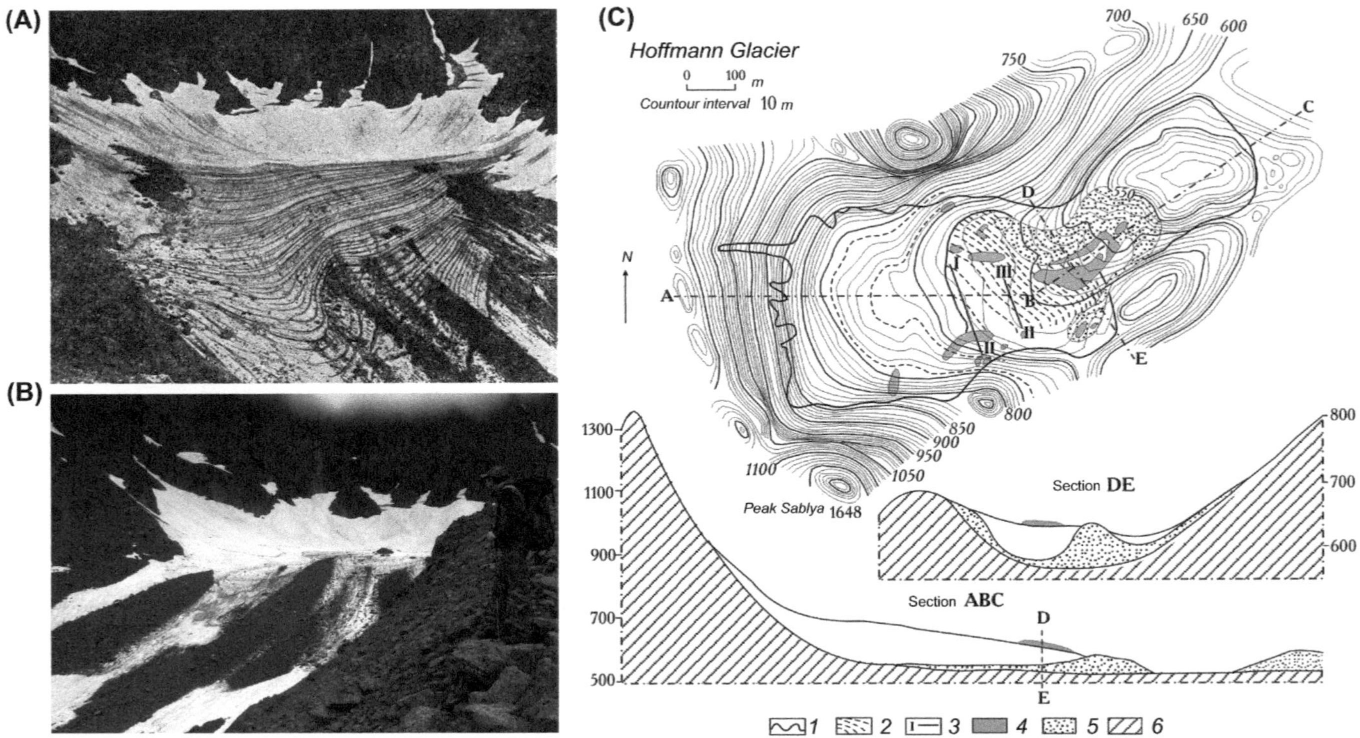

FIGURE 2.2 Hoffmann Glacier: (A) in 1935 *(Photo by AN Aleshkova.)*; (B) in 2006 *(Photo by A.G. Nosenko.)*; (C) glacier scheme based on the results of geodetic survey expedition by A.N. Aleshkov (1935).

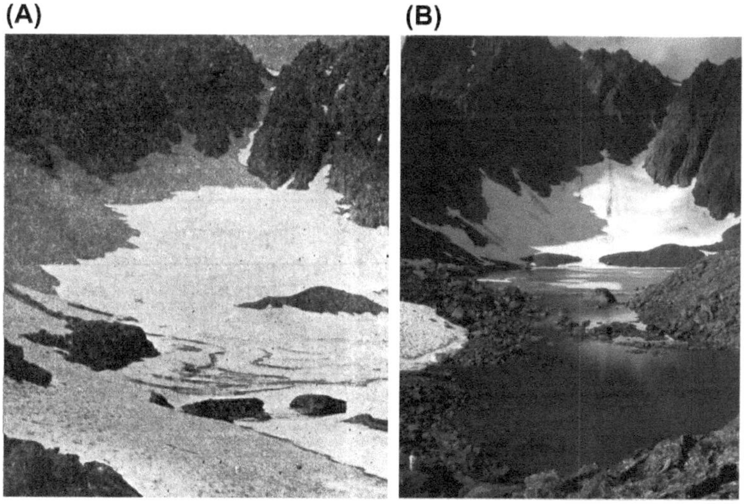

FIGURE 2.3 The cirque glacier number 4 on the eastern slope of the Sablinskay ridge: (A) in 1932 *(Photo by A.N. Aleshkova.)* and (B) in 2006 *(Photo by G. A. Nosenko.)*.

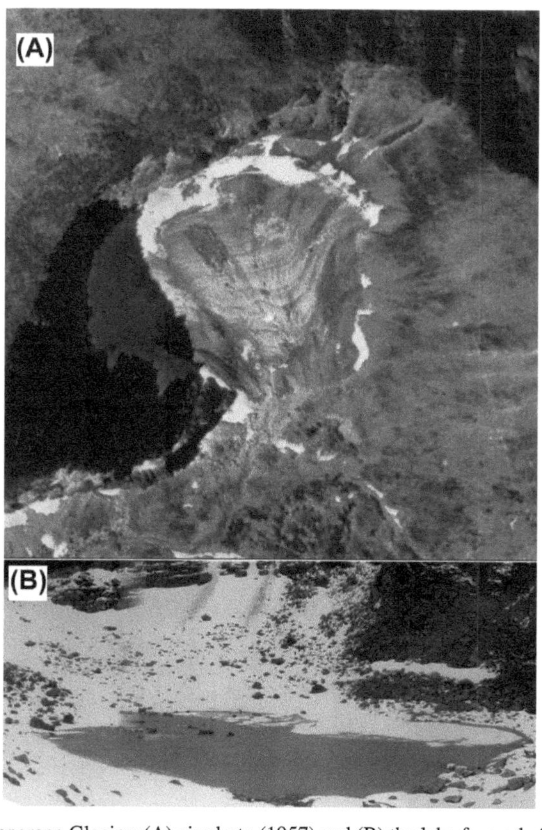

FIGURE 2.4 Manaraga Glacier: (A) air photo (1957) and (B) the lake formed after glacier retreat *(Photo by G. A. Nosenko, August 12, 2002.)*.

Urals glaciers obviously respond to contemporary climate changes. Certain limits occur to favorable oroclimatic conditions, but beyond them a stable balance of a glacier has become disrupted, so that the glacier may disappear.

According to the records of the Pechora meteorological station (Fig. 2.5), the closest location to the glaciers of the Subpolar Urals, during the 50-year period, the average summer temperatures increased by 2.5°C, while the total precipitation has increased as well. At the end of the winter, the depth of snow cover was about 50 cm, and it increased closer to the mountain ranges. According to snow-survey data, in 2006 and 2007, the snow depth on Loptopay Plateau in Sablinsky Ridge area was over 2 m.

Glaciers are still retreating despite the considerable amount of solid precipitation at present. It is apparent that an increase in summer air temperatures is the main cause of glacier disappearance.

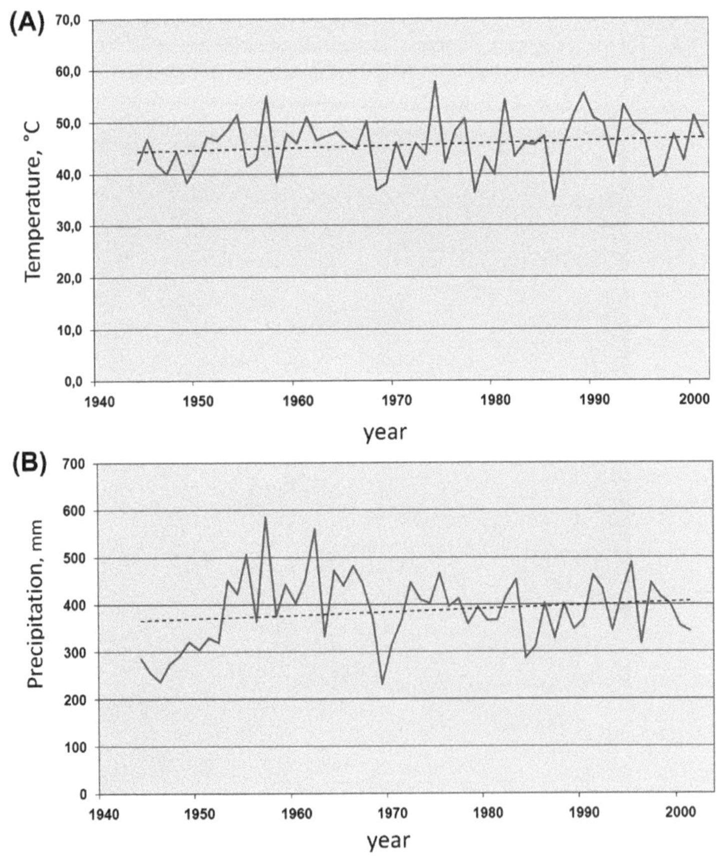

FIGURE 2.5 Monthly averaged summer temperatures (A) and winter precipitation (B) from the weather station Pechora.

Small sizes and hard-to-reach locations of glaciers in the Subpolar Urals make their study a rather challenging task. Therefore, only a limited number of glaciers were studied during the field survey in 2002. Bad weather conditions and seasonal snow cover pose the main difficulties for assessing the satellite imagery of high-latitude glaciers; for that reason it was not yet possible to receive good high-definition, satellite images for that area. Based on the data of a limited number of ASTER images available, glaciers of the Subpolar Urals have almost disappeared, giving way to newly formed lakes.

The Polar Urals

A significant cluster of glaciation is located in the northern part of the Polar Urals in the vicinity of Hadatinsky and Shuchyi lakes. Cirque and niche glaciers are the most common there, among them are the largest in the Urals, which are thoroughly studied cirque-valley glaciers IGAN and MGU. According to the Glacier Inventory of the USSR, 63 glaciers occurred in the 1950s with total area of 15.09 km^2. IGAN and MGU were only two glaciers with an area over 1 km^2 (1.25 and 1.17 km^2 respectively), whereas most of glaciers had an area from 0.3 to 0.1 km^2.

The location at high latitudes determines low average-annual temperatures in the region, even though its elevation is relatively low. The nearest meteorological station in Salehard (66.538°N, 66.678°E; 16 m a.s.l.) recorded positive average monthly air temperatures from June to September. However, they do not exceed 14°C. Air temperatures in Salehard correlate well with temperatures on glaciers of the Polar Urals. The correlation coefficient between daily air temperature in Salehard and temperatures measured on IGAN Glacier by an automatic meteorological station in 2008–2009 was 0.89, which is quite high.

Glaciers of predominately cirque and niche types are located at elevations between 400 and 1400 m. Cirque glaciers are generally located in well-developed east-facing cirques and occupy the whole cirque bottom or its part. Their ends are surrounded by ice-cored moraine lines, commonly impounding shallow glacier lakes. Niche glaciers are characteristically located on downwind cliffs of plateaus and high plain terraces. Those are mainly glaciers with thicknesses of up to 15–20 m; their length hardly reaches 200–400 m down-hill, with widths of 0.5–1.00 km, and in some places up to 2 km along the slope.

Glacier termini vary from 400 to 800 m a.s.l. The terminus of Obruchev Glacier descends down to 388 m a.s.l. (the lowest position in the region), whereas the terminus of the largest IGAN glacier is located at 835 m. The elevation of the firn line, characteristically located near the equilibrium line, varies within a wide range of altitudes: 500–550 m at Obruchev Glacier, 800–900 m at MGU Glacier, and 900–1000 m at IGAN Glacier (Troitsky and Kemmerikh, 1966).

With the beginning of the international project GLIMS in the 2000s, satellite imagery ASTER and Landsat ETM+ have become available. High-quality satellite images of the Polar Urals glaciers have been received for the first time due to favorable meteorological conditions in summer of 2000. Those images were used to assess the dynamics of glacier boundaries in the region during the 50 years period (Kotlyakov and Nosenko, 2001; Nosenko and Tsvetkov, 2003) (Fig. 2.6).

Thirty glaciers have been selected. They included 28 glaciers with areas of over 0.1 km^2 and two glaciers with areas of 0.08 and 0.09 km^2 (Table 2.2). Glaciers of smaller sizes were not considered due to the uncertainty of identifying their borders on satellite image. Aerial photographs with resolution 1–3 m made in July 1953 and August 1960 under conditions of clear weather and the lowest seasonal snow cover were used as historic data.

ASTER satellite imagery did not cover the south-west sector of the Polar Urals with the glaciers of IGAN, Anuchin, Oleniy, and Bolshoi Usinsky; therefore, panchromatic image Landsat ETM+ with the same geometric resolution (15 m) has been used. The main conclusion is that glacier cover is decreasing steadily (see Table 2.2). From the 1960s to 2008, the 30 glaciers studied in the Polar Urals on average have lost 23% of their area. In comparison with the period from 1981 until 2000, the rate of the area shrinkage in 2000–2008 has increased.

The results of terrestrial land surveys of IGAN and Obruchev glaciers performed in 1963 were preserved as 1:5000 topographic maps (Tsvetkov, 1970); they were used to evaluate the changes in the surface elevation and

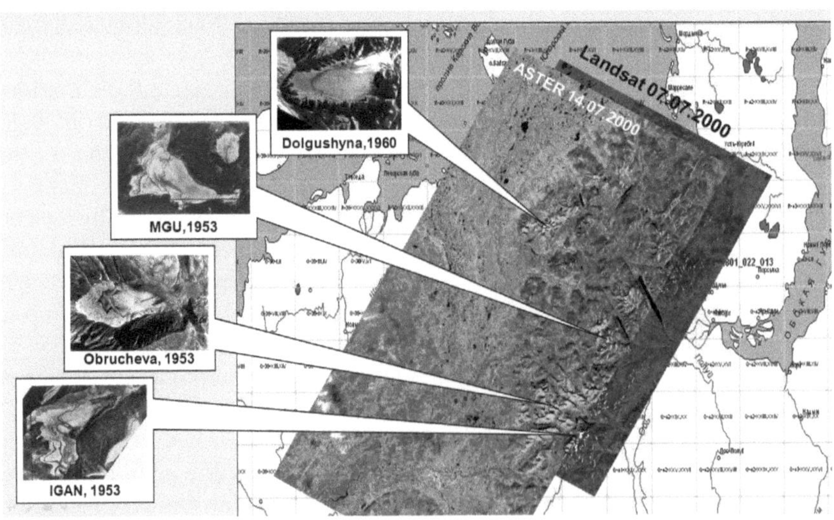

FIGURE 2.6 Aerial photographs for 1953, 1960, ASTER and Landsat imagery for 2000 were used to estimate changes in surface area of 30 glaciers.

TABLE 2.2 Changes in Surface Areas of Polar Urals Glaciers in 1953–2008, km^2

								Area Decrease During 1953/ 1960–2000 years	
Glacier	**1953**	**1960**	**1968**	**1973**	**1981**	**2000**	**2008**	**km^2**	**%**
IGAN	1.17	1.11			1.06	0.89	0.74	0.28[a]	23.9 ± 2.7
Bolshoy Usinskiy	0.70					0.67		0.03[a]	4.3 ± 2.6
Obruchev	0.34	0.33	0.30		0.30	0.25	0.24	0.09[a]	26.5 ± 2.7
Chernov		0.26			0.25	0.16	0.15	0.10	38.5 ± 3.2
Berg		0.26				0.24		0.02	7.7 ± 3.2
Kovalskiy	0.20					0.13		0.07[a]	35.0 ± 3.4
Lepehin	0.10					0.08		0.02[a]	20.0 ± 4.0
Shumskiy	0.17					0.11		0.06[a]	35.3 ± 4.1
Avsyuk	0.11					0.10		0.01[a]	9.1 ± 5.3
Kalesnik	0.18					0.16		0.02[a]	11.1 ± 3.4
Khabakov	0.12					0.09		0.03[a]	25.0 ± 5.5
Skrytniy	0.08					0.05		0.03[a]	37.5 ± 6.7
Avgevych	0.09					0.07		0.02[a]	22.2 ± 5.4
Oleniy	0.12				0.10	0.10		0.02[a]	16.7 ± 5.2
Anuchin	0.11				0.10	0.10		0.01[a]	9.1 ± 5.2
MGU	0.98	0.93	0.86		0.78	0.53	0.45	0.45[a]	45.9 ± 2.7

Continued

TABLE 2.2 Changes in Surface Areas of Polar Urals Glaciers in 1953–2008, km^2—cont'd

Glacier	1953	1960	1968	1973	1981	2000	2008	Area Decrease During 1953/1960–2000 years	
								km^2	%
Synok	0.19		0.19			0.12		0.07[a]	36.8 ± 3.1
Karskiy	0.60		0.61			0.51		0.09[a]	15.0 ± 2.6
Palgov		0.15				0.10		0.05	33.3 ± 6.1
Markov		0.19				0.16		0.03	15.8 ± 3.4
Malysh		0.13				0.12		0.01	7.7 ± 4.1
Fedorov		0.25				0.23		0.02	8.0 ± 2.7
Shuchiy		0.47				0.37		0.10	21.3 ± 2.1
Tronov		0.22				0.20		0.02	9.1 ± 2.8
Aleshkov		0.31				0.26		0.05	16.1 ± 2.4
Terentyev		0.15				0.14		0.01	6.7 ± 5.6
MGG		0.47		0.46		0.38		0.09	19.1 ± 3.3
MIIGAIK		0.33		0.30		0.25		0.08	24.2 ± 3.3
Dolgushin		0.67		0.63		0.49	0.46	0.18	26.9 ± 2.8
Bocha		0.10		0.10		0.09		0.01	10.0 ± 6.2
Total		9.17				7.13		2.04	22.3 ± 3.9

[a] *1953 is used as reference year.*

volume of the aforementioned glaciers. During a joint field trip of the Institute of Geography of RAS and the University of Reading, a DGPS survey has been performed with the help of Topcon Hiper Pro (Fig. 2.7).

The geodetic mass balances of the IGAN and Obruchev glaciers were calculated for the period of 1963–2008. They were -13.54 ± 2.57 and -20.66 ± 2.91 m w.e., respectively (Shahgedanova et al., 2012).

Due to significant impact of topographic and local meteorological conditions on small glaciers, it is rather difficult to explain changes in their sizes and thickness in the context of climate change. For instance, the highest rate of decrease in area in the Polar Urals has been observed for cirque-valley glaciers such as IGAN, Obruchev, and MGU glaciers (32.48, 26.47, and 45.92%, respectively), and the lowest for slope and cirque glaciers such as Bolshoi Usinsky, Berg, and Anuchin (4.3, 7.7, and 9.1% respectively).

A comparison of IGAN Glacier with Obruchev Glacier provides a good example of the influence of morphological factors on the rate of area decrease. The area of IGAN Glacier within the studied time frame was shrinking faster than the one of Obruchev's, because IGAN Glacier was not completely transformed into a cirque glacier, whereas its residual thin valley part was shrinking faster despite more favorable temperature conditions in comparison with Obruchev Glacier. Obruchev has a lower elevation (390–650 m: Obruchev Glacier; 790–1180 m: IGAN) and, therefore, is located in the zone

FIGURE 2.7 Topoplan of year 1963 and a network of historical reference sites were used for the DGPS survey in 2008 to assess changes in glacier mass balance using geodetic method.

of higher temperatures, contributing to its increased melting. However, at the beginning of our survey, Obruchev Glacier had almost lost its characteristics of a valley glacier and met the definition of "cirque" glacier. Even though its melting was more intensive, the loss of its mass was reflected more in the change of the surface elevation rather than in the size of its area. A similar feature was observed in MGU Glacier where a newly formed lake was contributing to the destruction of the glacier's valley part (Fig. 2.8).

The trend of summer temperatures and winter precipitation (Fig. 2.9) in the Polar Urals suggests that the increased melting of glaciers in those mountains occurred during the past decade and is linked to the increase in summer temperatures by 1–2°C, whereas winter precipitation has increased by 30 mm.

It is quite possible that during the next few decades, a distinctive snow cover of the Polar Urals and small glaciers within negative landforms will remain (Fig. 2.10). With the current climatic trends, cirque-valley glaciers may transform into cirque glaciers. However, cirque and niche glaciers will remain for considerably long periods of time, perhaps even until contemporary warming gives way to another climatic period.

Putorana Mountains

The Putorana Mountains are located within the Polar Circle in the northwest part of the Central Siberian Plateau. The mean elevation of these mountains is

FIGURE 2.8 The destruction of the valley part of the MGU Glacier.

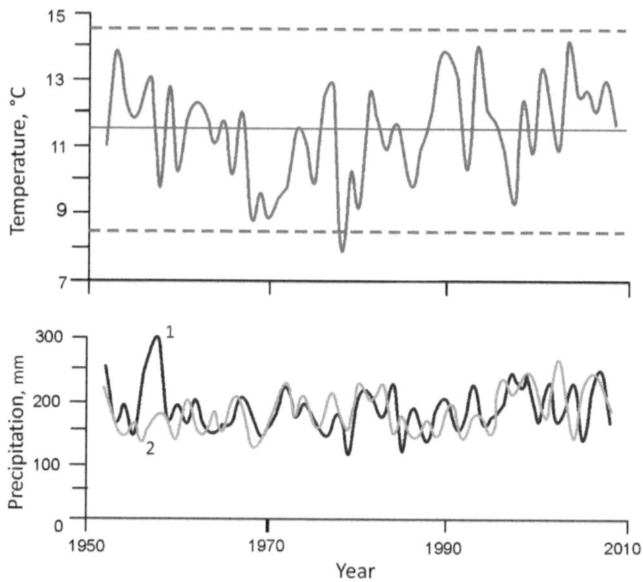

FIGURE 2.9 Summer temperatures and winter precipitation in the Polar Urals, according to the meteorological station Salekhard (1) and reanalysis NCEP/NCAP (2) (NCEP, National Centers for Environmental Prediction; NCAP, National Center of Atmospheric Research).

over 700 m with certain summits reaching 1400–1700 m a.s.l. Contemporary glaciers are located in cirques below mountain crests.

Glaciation of the region is represented by small glaciers, snow fields, and transitional forms between those two (Sarana, 2004). According to the data of the Glacier Inventory of the USSR (1965–1982), the glaciation is represented

FIGURE 2.10 The Polar Urals ridges. View from the top of Mount Howe-Naurdy-Key. *Photo by G.A. Nosenko, April 2009.*

by 22 glaciers with total area of 2.54 km^2. Some 67, mostly cirque-hanging and niche glaciers with a total area of 7.18 km^2, have been identified based on the field survey data (1997–1999) and results of interpretations of aerial photography dated 1972 and 1981, as well as satellite images Landsat 7 received in 1999 (Sarana, 2004, 2005). After studying morphological features of glaciers, the author was able to suggest a relative stability of glacier sizes during the past 40–50 years.

Byrranga Mountains

The Byrranga Mountains are located in the northern part of the Taymyr Peninsula and stretch over 700 km from southwest to northeast as a complex system of ranges and plateaus with a peak elevation of 1146 m a.s.l. in its north-eastern part (Mt. Lednikovaya). The glacial system of Byrranga Mountains consists of 96 glaciers with a total area of 30 km^2, among them, 30 with an area less than 0.1 km^2. Glaciers are densely located in the highest northeastern part of that low-mountain area. Eight of the largest glaciers of twinned-valley, valley, and cirque-valley types form three quarters of the glacial surface (Glacier Inventory of the USSR, 1965–1982).

According to the data of aerial photography for 17 years (1950–1967), the total area of glaciation and the size and number of glaciers have decreased (Govorykha, 1971; Glacier Inventory of the USSR, 1965–1982). In 2007, previously unknown glaciers were discovered during a field survey: two additional centers of mountain glaciation represented by small glaciers have been identified in the northeastern part of the Byrranga Mountains. During 40 years (1967–2007), glaciers have been retreating (Sarana, 2007).

Based on the results of satellite imagery (Landsat-4) and the data from the Glacier Inventory of the USRR, an attempt was made to assess the changes in glaciers (Ananicheva and Kapustin, 2010). According to the data, during 36 years (from 1967 to 2003), the total area of 56 glaciers studied has decreased by nearly 17%. In order to evaluate the conditions and dynamics of glaciers in that area, more detailed analysis using satellite imagery is required.

The Orulgan Ridge

The Orulgan Ridge is the highest part (2000–2300 m) of the northern offshoot of the Verkhoyansky Range. The highest point of the Ridge (2389 m) is situated in the headstream of the Sobopol River. According to the data of the Glacier Inventory of the USSR, the glacier systems of the Orulgan Ridge consist of 74 glaciers with an area of 18.38 km^2; they are located along the main dividing range at two sites of 112 and 25 km in the north and in the south, in the way of Western Arctic cyclones. Most of the glaciers are of cirque and twinned-valley types. The biggest glacier is twinned-valley Kolosov Glacier, with an area of 4.22 km^2 (Fig. 2.11).

FIGURE 2.11 Kolosov Glacier in the Orulgan ridge. Google Earth (October 4, 2013).

Additionally, eight glaciers with a total area of 0.5 km^2 have been discovered in 1991 (Sedov, 1997b, p. 172). According to the data of D.K. Bashlavin (1970), few changes occurred in the glaciers during 1951–1967. According to estimations (Ananicheva and Karpachevsky, 2015) based on the data of the Glacier Inventory of the USSR and satellite imagery ASTER (2008), from 1972 to 2010, the area of glaciation in the region has been reduced approximately by 62%; however, this figure requires clarifications based upon additional studies.

TEMPERATE LATITUDES: THE CAUCASUS, ALTAI, KUZNETSKY ALATAU, VOSTOCHNY SAYAN, BARGUZIN AND BAIKAL RANGES, KODAR RANGE

The Caucasus

The biggest center of mountain glaciation in Russia is situated in the Caucasus. The Greater Caucasus is a complex mountain system stretching to about 1100 km. The area of glaciation is conventionally divided into three parts—Western, Central, and Eastern Caucasus; the Central Caucasus consists of the part of the mountain system from Elbrus to Kazbek including both volcanic massifs (Fig. 2.12).

Due to significant elevation, the Greater Caucasus stands on the way of flows of moist air brought by Mediterranean and Atlantic cyclones from the

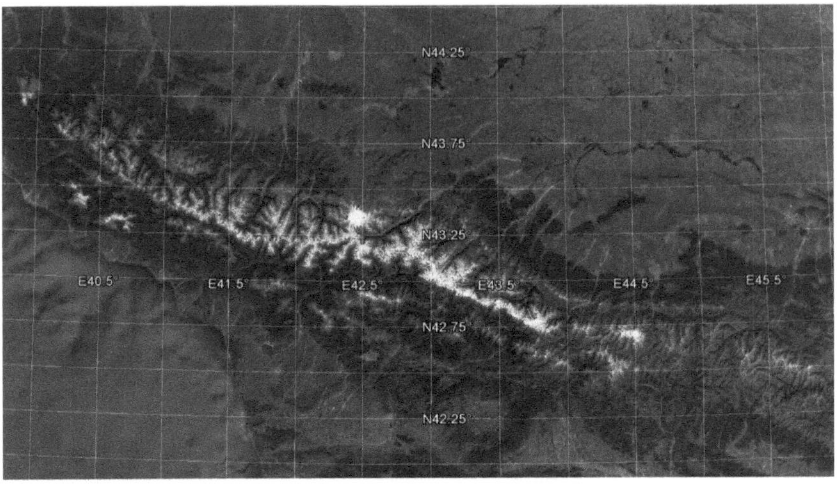

FIGURE 2.12 The Caucasus Mountain system from space (Google Earth).

west and southwest. The maximum precipitation occurs in the Western Caucasus and on the South macroslope of the Greater Caucasus in general. The continentality of climate increases toward the east, with the elevation of the equilibrium line of the glaciers increasing accordingly: from 2500 to 2700 m in the westernmost catchment areas of Belaya, Laba, and Mzymta rivers to 3700–3950 m in the easternmost catchment areas of Samur and Kusarchai rivers. The maximum elevation range of glaciation is almost 4 km—from the lowest glacier in the catchment area of the Belaya River (1710 m) to the eastern peak of Elbrus (5642 m), the highest point in Russia and Europe.

Half of the glaciers and about 70% of the glacial area of the Greater Caucasus are located in the Central Caucasus. The most significant "nodes" of the modern glaciation and the largest glaciers—Bezengi, Dyh-Su, and Karaugom in the northern slope and glaciers Lekzyr, Tzanner and Tvyber on the southern slope—are located here. The development of glaciation in this area mostly occurs due to the high elevation of the mountains. The height of many peaks there exceeds 4000–4500 m and 5000 m in some instances.

The contribution of glacier melting into the runoff formation in the Caucasus is presented in Fig. 2.13. The estimate is based on a series of observations from 1930 to 1975 and referenced to the hydrological stations for which the values have been determined. The area of circular diagrams is proportionate to the value of the annual river runoff, and the contributions of snow and glacier melt runoff are presented as sectors, the area of which is proportionate to their volume (World Atlas of Snow and Ice Resources, 1997).

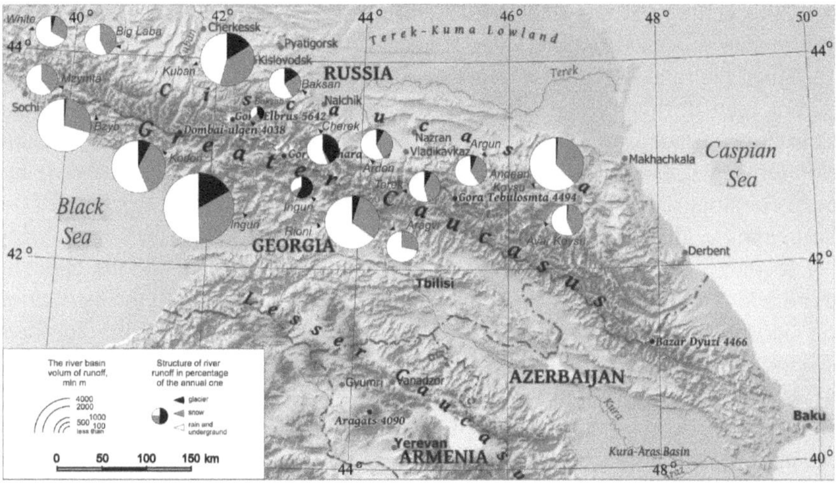

FIGURE 2.13 Melt snow and glacier runoff. Caucasus.

Decrease in Glacier Areas of the Greater Caucasus Mountain Range

In recent years, the data on repeated satellite imagery, DTM, Ground-penetrating radar (GPR) survey, terrestrial DGPS surveys, and ongoing land surveys of the mass balance of two tested glaciers Garabashi and Dzhankuat are widely used for assessing the dynamics of glaciers and climate changes in the Caucasus. The results of comparison of Glacier Inventories of 1911, 1957, and contemporary surveys indicate that during the 20th century and at the beginning of the 21st century, the glaciation in the Caucasus was decreasing.

At the beginning of 2000s, the overall picture of glacier dynamics of the Caucasus has been obtained in the Institute of Geography of the RAS based on the data of satellite imagery. According to satellite imagery data, at the beginning of the 21st century, about 1706 glaciers occurred in the Caucasus with a total area of 1225.8 km^2. The Glacier Inventory of the Caucasus (Volumes 8 and 9 of the Glacier Inventory of the USSR) has recorded 1557 glaciers (with sizes from 0.1 km^2 and above) with a total area of 1397.25 km^2. During the satellite image interpretation, 129 glaciers have not been found and 127 glaciers were fragmented into 308 parts.

To expand the time frames of the analysis, the first existing inventory of Caucasus glaciers, the Inventory by Podozersky published in 1911, has been digitized. The assessment of the dynamics of glacier area in the Caucasus between the beginning, middle, and end of the 20th century has been made as a result. From the beginning until the middle of the 20th century, the glacier area of the Caucasus has decreased by 24.7%, and from the middle until the end of the 20th century, it has decreased additionally by 17.7%. For Elbrus, these parameters were 14% and 6.28%, respectively.

Thus, the rate of glacier retreat during the first half of the 20th century was higher than during the second half. These results correspond well with the estimations made by other researchers of the Caucasus. However, a noticeable difference exists between the dynamics of glaciers on the southern and northern slopes. With the general trend of retreating, during the first half of the past century, glaciers on the northern slope have decreased by 30% and during the second by 17.9%, whereas glaciers on the southern slope had the opposite trend and retreated by 12% and 28%, respectively. According to the estimation of Georgian glaciologists, from 1960 until 2014, glaciers of the southern slopes of the Nenskra and Nakra rivers catchment basins have decreased by 47% and 45%, respectively (Tielidze et al., 2015).

A comparison of the results of modern surveys with data of the aforementioned inventories allows assessment of the dynamics of glaciers within a significant time frame of several decades. However, the dynamics of the contemporary situation suggests a more detailed research of nival–glacial processes in time. The past decade is of special interest as this comparatively short period of time was marked by a rapid melt of glaciers and an increase in the number of natural disasters in the high-mountain area of the Caucasus—an increase in the number and capacity of snow avalanches and occurrence of glacial mudflows, among them the Karmadon disaster of 2002. These studies have been performed by the Institute of Geography of the RAS and the University of Reading (UK) with the help of satellite imagery ASTER and Landsat (Nosenko et al., 2013; Shahgedanova et al., 2014). Repeated satellite imagery has provided a uniformity of imaging conditions and compatibility of different time data on glaciers, and the availability of the Terskol meteorological station operating in a high-mountain area has allowed the performance of comparative analysis of variability of the main meteorological parameters within the period of assessment.

The area of research included the Greater Caucasus Mountain Range from Klukhorsky passage in the Eastern Caucasus up to the Mestyisky passage in the Central Caucasus, including Mt. Elbrus. The main requirement in selection of research area was the availability of repeated satellite imagery at the end of the ablation period. ASTER satellite imagery (2001–2012) and Landsat ETM+ (1999–2001) were used as initial data sources.

In total, 478 glaciers were digitized and used for comparison. The total area of those glaciers in 2001 was 407.3 km^2 (Table 2.3). By 2010, it had reduced by 19.2 km^2 or 4.7 $\pm$ 2.1%. Moreover, glaciers of the Central Caucasus have retreated by 8.5 km^2 (5.0 $\pm$ 2.4%) and glaciers of the Western Caucasus by 4.9 km^2 (4.1 $\pm$ 2.7%). Differences in the decreased ratio between glaciers of the central and western parts of the Greater Caucasus Mountain Range, as well

TABLE 2.3 Caucasus Glaciers Area Changes for the Period 1999–2012

Region		Area (km^2)		Area Reduction	
		1999/2000/ 2001	2010/ 2012	km^2	%
Central Caucasus	Southern	97.22	91.79	5.43	5.7 ± 2.3
	Northern	73.41	70.28	3.31	4.3 ± 2.3
	Total	171.19	162.32	8.87	5.0 ± 2.4
Western Caucasus	Southern	55.15	53.05	2.1	3.8 ± 2.7
	Northern	63.15	60.38	2.77	4.4 ± 2.7
	Total	118.3	113.4	4.9	4.1 ± 2.7
Elbrus		118.42	112.34	5.91	4.9 ± 1.2
Total		407.3	388.1	19.2	4.7 ± 2.1

as the northern and southern slopes, are insignificant and fall within the measurement accuracy. Nevertheless, the highest rate of retreat of 5.6 ± 2.3% was registered for glaciers in the southern slope of the Central Caucasus and the lowest on the southern slope of the Western Caucasus. It duly corresponds with the deteriorating conditions of glacier alimentation from the west to the east along the Greater Caucasus Mountain Range.

The decrease in area in 20 glaciers has exceeded 20%; 41 glaciers have lost from 10% to 20%, and 59 glaciers did not change their sizes because their changes either were within the limiting resolution of the images or were difficult to determine due to morphological features or ground moraine; in such cases they were set to zero.

Changes in Mt. Elbrus Glaciers

Mt. Elbrus is located to the north of the line of the Greater Caucasus Mountain Range on one of its offshoots (Fig. 2.14). Glaciation of this volcanic massif differs from glaciers of the Greater Caucasus Mountain Range by its morphology, sizes, and conditions of alimentation. The areas of accumulation of Mt. Elbrus glaciers are located at higher elevations under conditions of practically free atmosphere. The average height of Elbrus glaciers is 3980 m, whereas glaciers of the northern and southern slopes of the range are 3130 m and 2810 m, respectively.

The dynamics of Elbrus glaciers was estimated using imagery from Landsat ETM+ 1999 and ASTER 2012 (Fig. 2.14). To define the location of

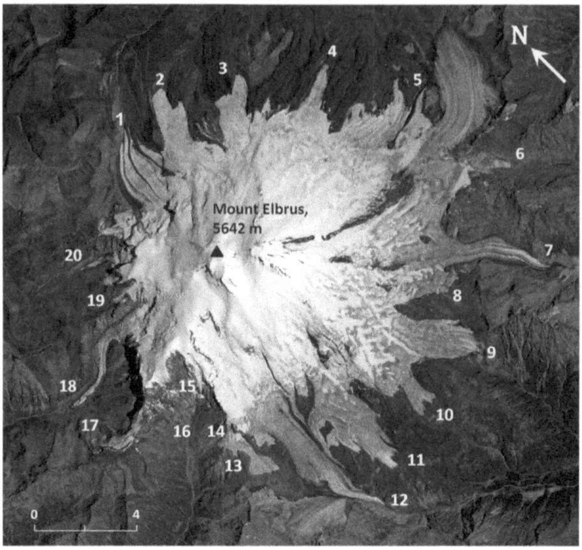

FIGURE 2.14 Elbrus from International Space Station. August 20, 2010. The numbers refer to the glaciers in Table 2.10.

glacier boundaries in alimentation areas, a digital elevation model from ASTER GDEM 2 was used. The estimation has shown that in 1999, the total area of glaciers was 118.4 km^2. By 2012, Elbrus had lost 5.8 km^2 or 4.9%. The annual rate of glacier area retreat was 0.4% per year despite the aforementioned differences in morphology and features of the regime of glaciers (Table 2.4).

Retreat of glaciers in Elbrus was accompanied by a shift of ice-formation zones, a melt-out of new parts of lava ranges, and an increase in the area of ice-free surfaces below the equilibrium line located generally at an elevation of 4000 m (Figs. 2.15 and 2.16).

The surface-elevation decrease near the equilibrium line of the test glacier Garabashi during the past 10 years has reached 9 m. Only during the summer of 2010, the glacier has lost about 2.5 m w.e. of ice. This figure is almost twice as high as the average within the past decade. Annual mass balance of Garabashi Glacier during the past decade has decreased dramatically, which was determined primarily by a decrease in accumulation. According to the data of the Terskol meteorological station located in the valley near the glacier, a consistent drop in winter precipitation has occurred (Fig. 2.17).

This drop was particularly noticeable after a maximum in 2006 with winter snow precipitation of 950 cm w.e. During 2012–2014, the average winter precipitation was only 470 cm w.e. This is close to the lowest value in a 30-year series of observations corresponding to the latest climatic period in

TABLE 2.4 Glacier Area Changes on Mt. Elbrus: 1999–2012

	Glacier		Area		Area Change	
Number (Fig. 2.19)	Name	WGI ID	1999	2012	km^2	%
1	Ulluchiran	SU4G08005001	11.63	11.13	0.5	4.3
2	Karachaul	SU4G08005002	7.44	7.22	0.22	3.0
3	Ullukol and Ullumalienderku	SU4G08005003	4.85	4.54	0.31	6.4
4	Mikel'chiran	SU4G08005005	5.05	4.86	0.19	3.8
5	Dzhikiugankez	SU4G08005006	26.12	24.22	1.90	7.3
6	Irikchat	SU4G08005018	1.41	1.31	0.10	7.1
7	Irik	SU4G08005020	9.18	8.90	0.28	3.1
8	No 25	SU4G08005025	0.60	0.56	0.04	6.7
9	Terskol	SU4G08005026	7.99	7.71	0.28	3.5
10	Garabashi	SU4G08005027	3.26	3.02	0.24	7.4
11	Malyi Azau	SU4G08005028	8.84	8.27	0.57	6.4
12	Bol'shoy Azau	SU4G08005029	19.70	18.53	1.17	5.9
13	311	SU4H08004311	0.46	0.39	0.07	15.2
14	312	SU4H08004312	0.26	0.22	0.04	15.4
15	313[a]	SU4H08004313	1.07	1.05	0.02	1.9
16	Ullukam	SU4H08004313	0.65	0.64	0.1	1.5
17	317	SU4H08004317	0.55	0.55	0	0
18	Kyukyurtlyu	SU4H08004318	6.92	6.83	0.09	1.3
19	319	SU4H08004319	0.46	0.46	0	0
20	Bityuktyube	SU4H08004320	1.98	1.90	0.8	4.0
	8 separate ice bodies (Fig. 2.20)		–	0.3	–	–
Total			118.4	112.3	5.8	4.9

[a]*Glacier N 313 is now a disconnected part of Ullukam glacier with the same WGI identification number.*

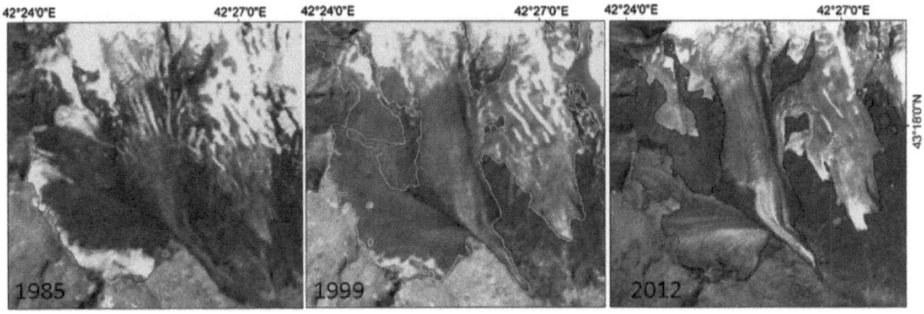

FIGURE 2.15 The increase in ice-free surface on the southern slope of Mount Elbrus (Bolshoi and Malyi Azau) on Landsat images in 1985, 1999, and 2012.

FIGURE 2.16 The exposed outputs of lava ridges near the "Shelter of 11" on the Garabashy glacier (altitude 4000 m). *Photo by Nosenko G., September 21, 2011.*

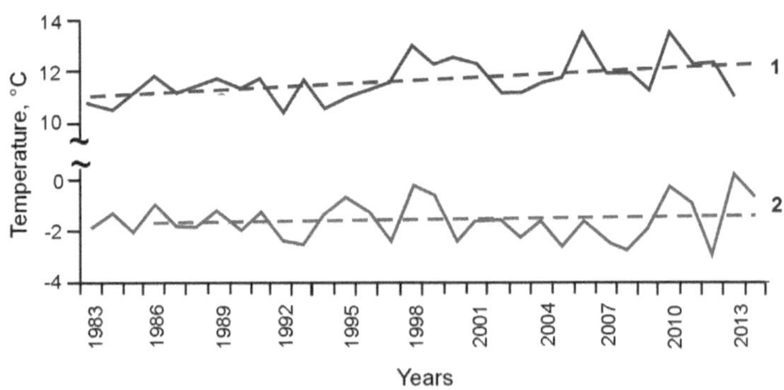

FIGURE 2.17 Summer (2) and winter (1) precipitation time series from meteorological station Terskol. The linear (*dashes*) and the polynomial (*dashes and dots*) trends.

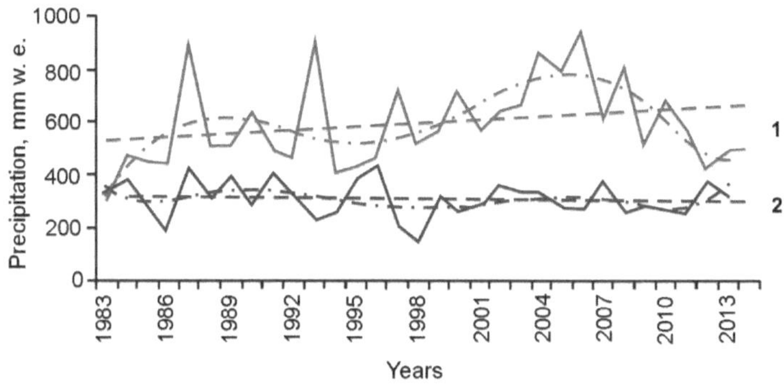

FIGURE 2.18 Summer (1) and winter (2) air temperature time series from meteorological station Terskol. The *dashed line* indicates the linear trend.

the whole. At the same time, melting processes are slightly delayed due to increase in summer snowfalls.

In general, glacier mass balance in summer has decreased during the whole period of observation. According to the data of the Terskol meteorological station, during the past 30 years, summer temperatures have increased by 1.5°C; however, in recent years, after the abnormally hot summer 2010, temperatures have eminently decreased (Fig. 2.18). This proves once again that during the past 3–4 years, abruptly negative values of annual mass balance were not much determined by summer melt, but by small quantity of winter snow on the glacier. Due to the small volume of snow, the snow cover on the glacier melts early thus opening an ice surface that has a higher coefficient of melting in comparison with the snow surface. This results in additional melt in the second part of summer.

Retreat of Glacier Fronts

Changes in glacier borders within two periods of time—1987–2001 and 2001–2010—have been examined in order to determine the dynamics of glacial retreat (Nosenko et al., 2013). The location of a glacier front is a rather responsive indicator of external conditions within such a relatively short period of time as a decade. The period from 2001 to 2010 has been covered with repeated ASTER satellite imagery. Aerial photographs taken on September 25 and 26, 1987, have been used in order to estimate the rate of retreat of frontal parts of glaciers during the previous period. The estimation was made based on 28 valley glacier samples located on the northern and southern slopes of the Greater Caucasus Mountain Range and on the south-eastern slope of Mt. Elbrus. Samples predominantly consisted of valley and compound-valley glaciers with lengths from 2 to 12 km.

The estimations have shown that all 28 glaciers included in the sampling were retreating in those periods of time (Table 2.5). Accumulative values of retreat from 1987 up to 2010 fall within the range from 50 to 500 m depending on sizes of glaciers, their morphology, and the elevation of their tongues. Compound-valley and large valley glaciers with tongues descending down valley to the low elevations had the most significant retreat; for example, the Bolshoi Azau Glacier has retreated by 500 m.

A tributary glaciers cutoff in a complex glacier deprives the main ice stream of a part of the alimentation that results in its degradation, for example, Glacier Schelda (retreated by 300 m), Chalaat (retreated by 240 m), and Lekzyr (retreated by 490 m). At the termini of such tongues, large masses of dead ice are being formed, and their melt results in abrupt loss of considerable area by a glacier. The results of field survey of Georgian glaciologists confirm the data received (Tielidze et al., 2015).

The front of glaciers with high elevated tongues retreats more steadily. For example, Garabashi Glacier ends at 3250 m (Fig. 2.19). Cirque and hanging glaciers are relatively stable: the former due to regularity in snow drift and avalanche alimentation and the latter due to predominantly high elevation. However, they also deteriorate in years of intensive melt, although changes in their borders are less evident due to their morphological characteristics.

While comparing average rates of glacier-tongue retreat within the given time frame it was found that from 1987 till 2001 the glaciers of the Baksan River catchment retreated with an average velocity of 6.4 m/year and from 2001 up to 2010 the velocity of retreat has increased almost two times up to 13.0 m/year. The retreat of glaciers on the southern slope of the Greater Caucasus Mountain Range in the Ingury River catchments was slower, but the velocity ratio of retreat during that period was higher at 2.9 and 9.9 m/year.

To identify the reason of those trends in glaciers, an analysis has been performed to identify changes of the main climatic parameters: air temperature and precipitation as well as mass balance of the Garabashi test glacier, which had a continuous series of surveys from 1983 and reconstructed data for the previous period starting from 1905 (Rototaeva et al., 2003).

According to the diagram of cumulative balance dynamics of the mass of Garabashi Glacier (Fig. 2.20), a similar instance had occurred in the 1950s. Those years were marked with the lowest values of negative balance of glacier mass starting from the beginning of the century. According to the data from meteorological station Terskol and other stations in the Caucasus, the record-breaking high temperatures of the summer months had been observed also during that decade.

The 1960s were marked by more favorable conditions for glaciers in the Caucasus—considerable increase in snowiness and decrease in summer air temperatures. According to surveys in the Elbrus (Garabashi Glacier) and the northern slope of Mt. Kazbek (Kolka Glacier), in the 1960–70s, winter

TABLE 2.5 Change in Glacier Length for the Periods 1987–2001 and 2001–2010: Baksan and Ingury River Basins (Nosenko et al., 2013)

No	Glacier Name/WGMS Number	S_{2001}, km^2	L_{2001}, km	$\Delta L_{1987-2001}$, m	$\Delta L_{2001-2010}$, m	$\Delta L_{2001-2010}$, %
Baksan River Basin						
1	Irikchat	1.89	2.60	85	100	3.85
2	Irik	9.98	9.80	75	120	1.22
3	Terskol	7.39	7.20	80	90	1.25
4	Garabashi	3.70	3.90	85	100	2.56
5	Malyi Azau	9.49	8.20	70	150	1.83
6	Bolshoi Azau	16.58	10.20	300	200	1.96
7	SU4G08005037	2.62	1.80	20	30	1.67
8	SU4G08005044	2.71	4.60	40	80	1.74
9	SU4G08005049	2.22	3.30	50	100	3.03
10	Shkhelda	6.83	9.70	150	150	1.55
11	SU4G08005056	1.92	2.80	140	120	4.29
12	Kashkatash	2.85	4.60	60	90	1.96
13	Bashkara	4.06	4.30	60	180	4.19
14	Dzhankuat	2.41	3.10	40	130	4.19

Continued

TABLE 2.5 Change in Glacier Length for the Periods 1987–2001 and 2001–2010: Baksan and Ingury River Basins (Nosenko et al., 2013)—cont'd

No	Glacier Name/WGMS Number	S_{2001}, km^2	L_{2001}, km	$\Delta L_{1987-2001}$, m	$\Delta L_{2001-2010}$, m	$\Delta L_{2001-2010}$, %
Inguri River Basin						
15	SU5T09105177	1.21	2.40	50	70	2.92
16	SU5T09105178	1.61	2.50	20	70	2.80
17	SU5T09105179	0.43	0.80	20	50	6.25
18	SU5T09105210	1.47	3.00	20	80	2.67
19	Северный Лядешт	3.61	2.60	30	60	2.31
20	SU5T09105214	0.84	2.00	10	40	2.00
21	Lyadval	1.88	3.60	20	30	0.83
22	Tsalgmil	2.40	2.50	60	90	3.60
23	Lakra	2.63	3.10	65	160	5.16
24	Kvish	7.42	6.10	150	200	3.28
25	Dolra	6.59	5.50	15	60	1.09
26	Ushba	8.78	5.80	30	80	1.38
27	Chalaat	9.42	7.30	40	200	2.74
28	Lekzyr	31.38	8.30	40	200	2.42

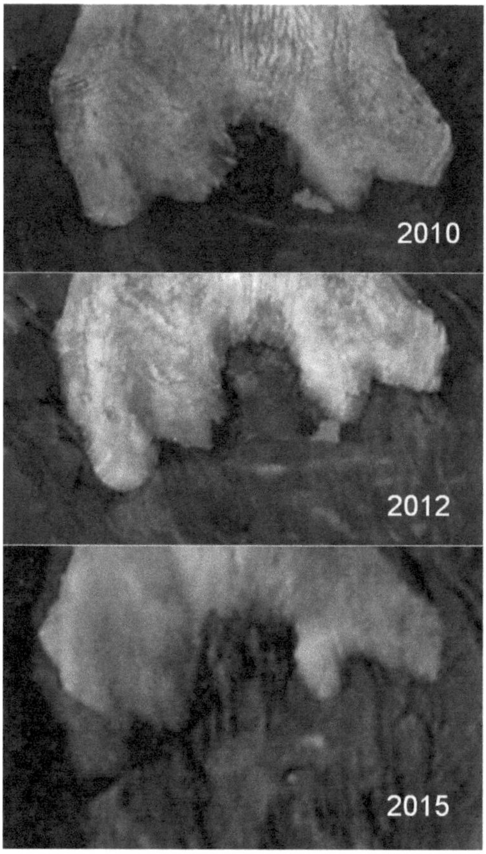

FIGURE 2.19 The Garabashy glacier front position in 2010, 2012 (ASTER), and 2015 (ISS).

precipitation was the highest during the century with a low rate of melting, and annual glacier mass balance retained positive values. During the next two decades, the mass balance of Garabashi Glacier was on average close to zero. Glaciers of the southern slope of Mt. Elbrus received a considerable amount of precipitation, and Garabashi and Terskol glaciers were hardly retreating during 1960–90s (Rototaeva et al., 2003).

During 1965–1990, with a general trend of continuing retreat of glaciers in the Caucasus, it was indicated that many of them were stable or their fronts have advanced insignificantly. Although in the 1930–60s, an average velocity of glacier retreat in the central part of the Northern Caucasus exceeded 14 m/year, in the 1970–80s, it was less than 6 m/year (Panov, 1993). Due to the delayed response on improving conditions of alimentation in the big compound-valley glaciers, their velocity has decreased later, in the 1980–90s.

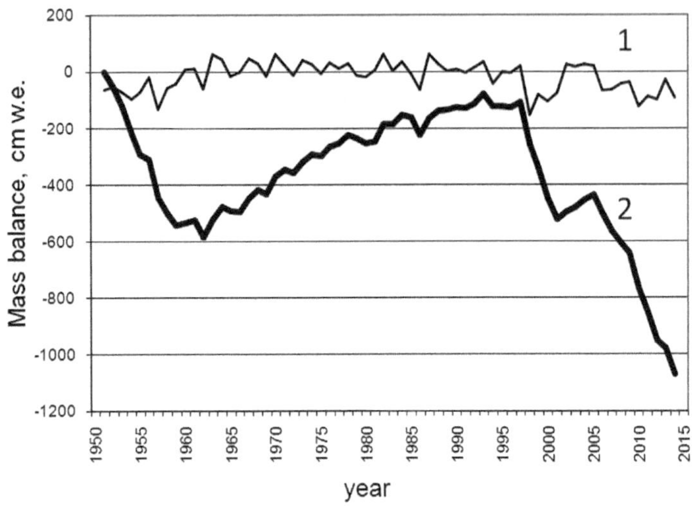

FIGURE 2.20 Changes in Garabashy mass balance. The annual balance (1) and the cumulative balance (2).

However, by the end of 1987–2002s, estimations based on satellite imagery from ASTER and the International Space Station had indicated a considerable retreat of tongues of almost all the glaciers of Elbrus as well as of 65 large glaciers of the Central Caucasus. The most significant retreat occurred in the glacier tongues Karaugom (by almost 600 m); Shaurtu, Rtzyvashky, Midagrabin (about 300 m); Schhelda and Bashyl (more than 200 m); Tzeya and Skazka (160 m). This period has included the years of anomaly of 1998–2001 with catastrophic glacier melt all over the Caucasus and tremendous ice loss in the Elbrus glaciers (Rototaeva et al., 2005).

During the past years of the 20th century, vast stable anticyclones spread all over the European part of Russia have caused extreme high summer-air temperatures in the Caucasus Mountains during four subsequent years (1998–2001). The average summer temperature in Terskol has reached 12.5°C, which could be only compared with the 1950s. During those summer seasons, a catastrophic melt took place. Annual mass balance of Garabashi Glacier during four abnormal years was on average 104 cm w.e.; on the whole, the glacier has lost a layer of 4 m w. e. and its tongue more than 8 m. Within the accumulation area, the boundaries of the ice-formation zones have shifted, and new parts of the lava ranges have melted out.

The period of abnormal melting in 1998–2001 has been briefly interrupted by more snowy years with cold summer periods (2002–2005) that has resulted in positive values of glacier mass balance, even though they were close to zero (Rototaeva et al., 2009). However, in subsequent years, glacier mass balance continued to remain negative. At present, the value of cumulative balance has dropped below the minimum recorded in the past century in the 1950s.

Surging Glacier Kolka

Glacier Kolka, one of the most well-known surging glaciers in the world, is situated in the Caucasus. Its surges took place in the years 1834, 1902, and 1969. At the beginning of the 21st century, the Glacier Kolka has drawn attention to itself yet again: this time it was an unparalleled glacial catastrophe of tremendous scale. In September 2002, a glacier mass with a huge amount of water suddenly broke out of the cirque and turned into a giant ice-water-rock mudflow that rushed 16 km across the valley and filled in the Karmadon basin. Hitherto a lot of theories explained the cause of that disaster (Huggel et al., 2005, Tutubalina et al., 2005, Panov et al., 2002); however, it is quite obvious that a volcanogenic factor had played its role in this event (Haeberli et al., 2004, Kotlyakov et al., 2014).

The next year after the mudflow, the Institute of Geography of the RAS started regular surveys of the changes in the area of ice-rock debris. The main purpose was to study the process of regeneration of Kolka Glacier. The formation of the new glacier had begun in 2003 as a result of ice mass flow from its former tributaries, the hanging glaciers on the wall Djimaray-Khoh. By summer 2014, all of the back part of the cirque of the former Kolka Glacier had been filled by ice.

The location of the boundary of the new glacier in 2006–2014 had been mapped with the help of the data of terrain GPS surveys and satellite imagery (Fig. 2.21). Those data provide a quantitative characteristic of the dynamics of filling a glacier bed with ice, which demonstrated that by 2010 the rate of change in area had slowed down and stayed almost constant—0.015 km^2/year—during the past 4 years.

The results of ground radar survey performed during the field work by the Institute of Geography of the RAS in July 2014 gave an opportunity to estimate the thickness of a newly forming glacier. While conducting a radar survey, a significant amount of large debris made it difficult to move on the surface and interfered with the quality of radio signal by increasing its scattering. Nevertheless, the information on the occurrence and thickness of the ice layer was obtained for key sites including the so-called "riegel," a part of the bed of the former glacier not yet filled with ice, as well as the regenerating glacier (Fig. 2.22).

Radar measurements performed on the tongue of the regenerating glacier have shown that on the margins near the walls of the valley, its thickness was 10–20 m, and in its central part it had reached 50 m. Measurements across the whole area were restricted by Health and Safety issues due to regular rockfalls and icefalls from hanging glaciers. If, based on the data received, we could

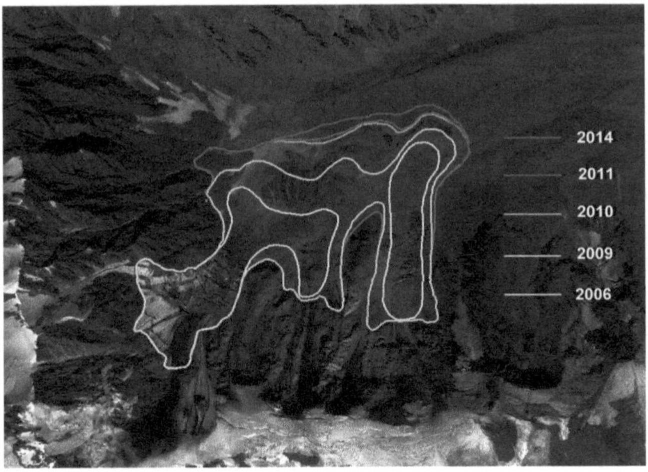

FIGURE 2.21 Changes in the reviving Kolka glacier boundaries from 2006 to 2014. Satellite image (2010 years) from Google Earth was used as background.

suggest that the average thickness of the newly formed glacier was about 20 m, then by 2014, the volume of ice accumulated was not less than 0.021 km^3 or nearly 16% of the volume of Kolka Glacier measured in 1988, 19 years after the surge of 1969 (Nikitin et al., 2005).

The glacier tongue responds to pressure increase from the ice incoming from above by forming radial crevasses and by moving along the valley axis by several meters per annum, although insignificant but detectable by GPS

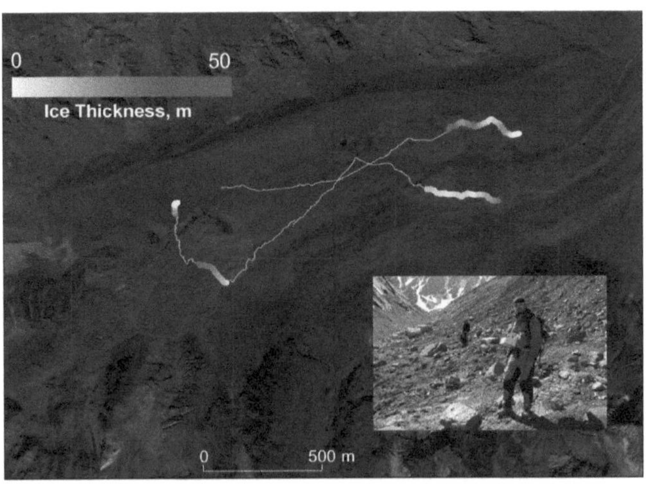

FIGURE 2.22 The radar survey of the Kolka glacier in 2014.

FIGURE 2.23 The Kolka glacier front in July 2014. *Photo by S.A. Nikitin.*

survey (Fig. 2.23). Its front margin has a form of an ice cliff with a steepness of up to 70°. The elevation of the front reached 40 m and its width about 350 m.

The further development of processes of glacier regeneration depends upon a number of factors including the temperature regime of the surrounding area. The release of the ice from the cirque on September 20, 2002 has resulted in an abrupt change in temperature conditions of melting. The cooling effect of glaciers located in the bed and on the slopes of the cirque has disappeared and an enormous empty basin covered with black debris has emerged. A general drop in the surface elevation to the level of the former bed of Kolka Glacier, heating of the open space of moraine and rocks, and considerable decrease in the total albedo and hence a general change in the heat balance in the cirque of the Kolka Glacier led to an increase in snow and ice melt on the former bed. In comparison with the previous surge in 1969, the rates of glacier regeneration are considerably lower. Most probably it was determined by dry winters of the past years and high summer temperatures in the Northern Caucasus. The findings of this research are detailed in a separate publication (Kotlyakov et al., 2014).

The Altai

The highest part of Altai-Sayan Mountain System, the Mountain Altai, is located in the heart of Asia, stretching from 81° to 106° E within 48–52°N at the junction of borders of four countries: Mongolia, China, Kazakhstan, and Russia. The main topographic features are vast erosion surfaces with high mountains of the alpine type. Mountain ranges and massifs in its central part are elevated up to 3000–4000 m a.s.l. and have numerous and diversified forms of contemporary glaciation. The Central Altai has mostly valley and cirque-valley glaciers with vast high-angled firn basins and less-steep glacier tongues located on the beds of ancient cirques and troughs. Glaciers have uneven distribution across the territory and form clusters around the highest mountains and massifs (World Atlas of Snow and Ice Resources, 1997; Tronov, 1925).

Three main factors influence the climate in the Altai: its position within the mid-latitudes of Northern Hemisphere, the dominance of Westerlies from the Atlantic, and the influence of the strong winter Asian Anticyclone with predominantly clear freezing weather. The amount of precipitation changes from the west to the east, whereas the elevation of the equilibrium line increases in that direction from 2200 to 3200 m. The accumulation at the elevation of the equilibrium line subsequently decreases from 400 to 100 g/ cm^2 (Krenke, 1982). The data of three meteorological stations, Akkem (2050 m), Kara-Tyurek (2600 m), and Aktru (2025 m), located in a close proximity of the glacier area, allow the assessment of changes of climatic parameters during the past few decades. According to that data, summer temperatures and annual precipitation were steadily increasing in the second half of the 20th century.

Contemporary glaciation in the Altai spreads over 1500 km^2 (Galakhov and Mukhametov, 1999; Kotlyakov et al., 2012). Detailed glacier inventories have been compiled for the territories of Russia, Kazakhstan (Glacier Inventory of the USSR, 1965–1982), and China (Concise Glacier Inventory, 2008). About 800 km^2 of glaciation is located in Russia in the northeastern part of the mountain area and 100 km^2 in the northwest lies within Kazakhstan. A detailed characteristic of Altai glaciers located on the territory of modern Russia and Kazakhstan are given in the Glacier Inventory of the USSR. According to those data, 97% of glaciers are situated in the catchment area of Ob River and 3% in the catchment area of the Yenisei River on the Shapshalsky Range and Tzagan-Shibetu Range. Their biggest cluster (78% of the surface of glaciation) is located in the Central Altai on the Mt. Belukha (4506 m), the highest in the Altai, and in its close proximity (World Atlas of Snow and Ice Resources, 1997, map № 196b), on Katunsky Range, Severo-Chuisky, Yuzhno-Chuisky, and Kara-Alakhinsky Ranges (Nikitin, 2009). The contribution of glaciers melting into runoff formation in the Altai is shown in Fig. 2.24.

Dynamics of Glaciers of Katunsky, Severo-Chuisky, and Yuzhno-Chuisky Ranges

Satellite images from ASTER (2000–2012), ALOS PRISM (2008), and Landsat TM+ (2000–2002) were used for interpretation of contemporary location of glacier boundaries and analysis of contemporary glaciation in the Khatunsky, Severo-Chuisky, and Yuzhno-Chuisky Ranges (Fig. 2.25). In certain instances, images received in the International Space Station in 2006 were used for interpretation of glacier boundaries. Historic data were obtained from the Glacier Inventory of the USSR, 1957 topographic maps at a scale of

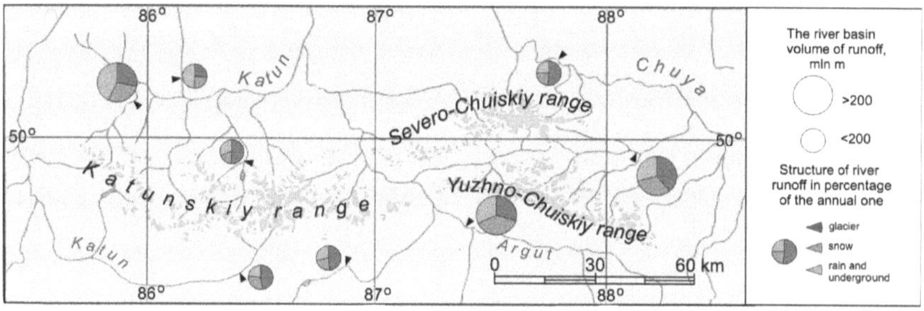

FIGURE 2.24 Melt snow and glacier runoff, Altai.

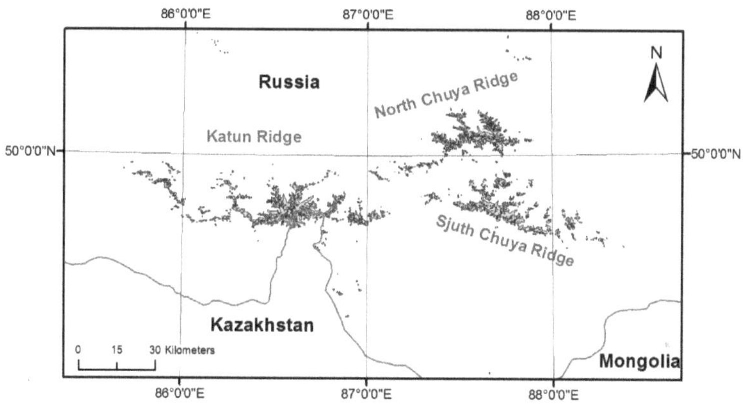

FIGURE 2.25 Altai glaciers. Glacier outlines derived from space images.

1:50,000, aerial photographs from August 24 to September 3, 1952, and CORONA satellite imagery from 1968.

On the territories of the Severo-Chuisky and Yuzhno-Shuisky Ranges, covered by the quality satellite imagery of ASTER, the location of 256 glaciers has been identified and their digitized margins were created. Among them 223 have stayed unchanged and had the same position in the Inventory, whereas 16 glaciers have been fragmented into 34 separate parts. In comparison with the data of the Glacier Inventory, the general area recession was 82 km^2 or 24.4 ± 5.8% (Nosenko et al., 2010).

Glaciers with areas less than 0.5 km^2 (50.8%) have retreated the most, for them the errors in the Glacier Inventory overlapped with interpretation problems at the limit of resolution of the ASTER satellite imagery. If we discard this group, then for the rest of the glaciers, we will have 19.7 ± 5.8% (instead of 24.4 ± 5.8%). It seems that this value more adequately characterizes the level of glacier retreat in this region.

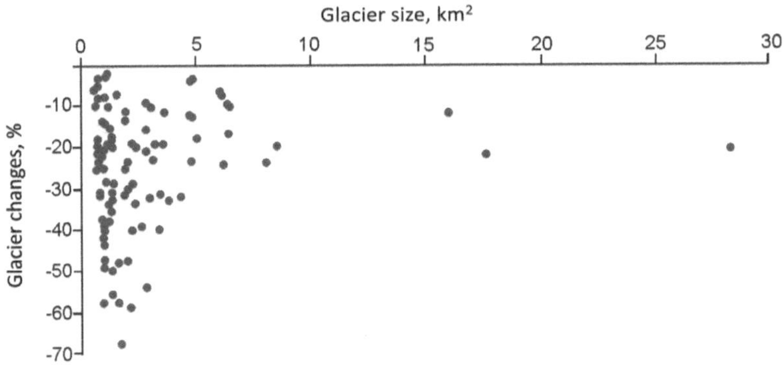

FIGURE 2.26 Glacier area changes in accordance with glacier size. For glaciers with area more than 0.5 km^2.

A distribution of the rate of retreat of glaciers with the area of over 0.5 km^2 from 1952 to 2004 depending upon their sizes is given in Fig. 2.26 and examples of changes in the location of large glaciers boundaries during that period are shown in Figs. 2.27 and 2.28.

According to the Glacier Inventory of the USSR, which used aerial photography from 1952, 336 glaciers occurred on the Katunsky Range with a total area of 281.5 km^2 (excluding glaciers with areas less than 0.1 km^2) (Fig. 2.29). The results of interpretation of the CORONA images have shown that by 1968 the number of glaciers have decreased to 322. Also, 14 glaciers with a total area of 2.1 km^2 have disappeared and the area of the remained glaciers has decreased to 264.6 km^2 (Table 2.6). Based on the results of ALOS PRISM image interpretation, by 2008, 313 glaciers with a total area of 208 km^2 on the Katunsky Range (nine more glaciers with a total area of 1.36 km^2 could not be discovered on the images) (Nosenko et al., 2014).

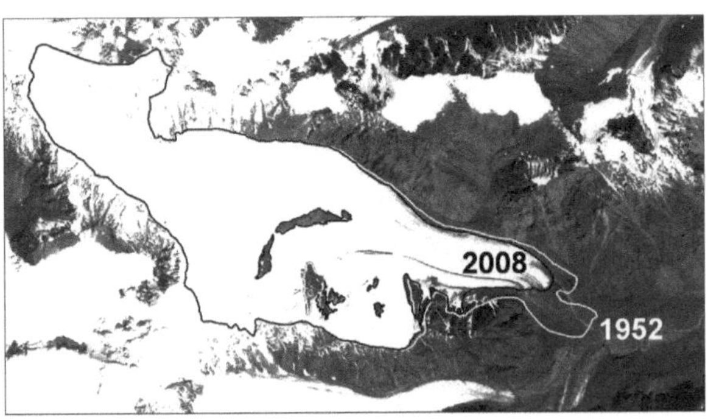

FIGURE 2.27 Changes in the front position of Dzhelo Glacier.

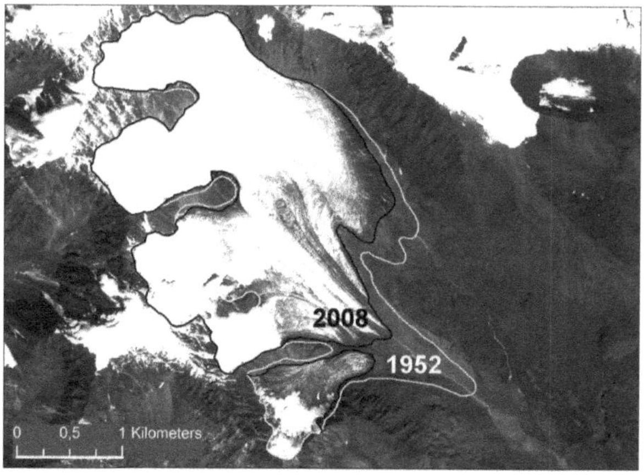

FIGURE 2.28 Changes in the front position of Levyi Aktru Glacier.

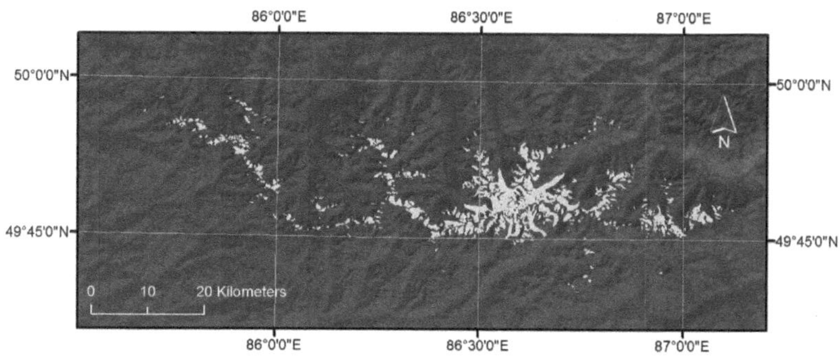

FIGURE 2.29 Glacier outlines in Katun Ridge derived from space imagery.

TABLE 2.6 Glacier Area Changes in Katun Ridge: 1952–2008 years

Glaciers With Area >0.1 km^2	1952 Glacier Inventory of the USSR	1968 CORONA	2008 ALOSPRISM	1952–1968	1968–2008
Number	336	322	313	14 glaciers with total area of 2.1 km^2 disappeared	Nine glaciers with total area of 1.36 km^2 disappeared
Area	281.5	264.6	208.0	−5.9%	−21.4%

TABLE 2.7 Glacier Area Loss According to Glacier Size in the Katun Ridge

Glacier Area, km^2	S_{1952}, km^2	S_{1968}, km^2	S_{2008}, km^2	$S_{1952}-S_{2008}$		$S_{1968}-S_{2008}$		Number of Glaciers
				km^2	%	km^2	%	
0.5–1	47.9	39.7	29.4	18.5	38.6	10.3	25.8	57
1–2	59.5	46.7	35.8	23.7	39.8	10.8	23.2	34
2–5	68.5	64.3	51.9	16.6	24.2	12.3	19.2	18
>5	62.8	73.3	67.79	−5.0	−7.9	5.5	7.5	6
Total >0.5	238.7	223.9	185	53.7	22.5	38.9	17.4	115
<0.5	42.8	40.6	22.9	19.9	46.5	17.7	43.5	206
Total	281.5	264.5	207.95	73.6	26.1	56.6	21.4	322

Thus, based on the Inventory data, from 1952 to 2008, the area of the whole glacial system of the Katunsky Range has decreased by 73.3 km^2 (26.1%). The contribution of every particular glacier to that value depends upon its size. The distribution of the rate of recession in five groups of glaciers has been calculated (<0.5 km^2, 0.5–1 km^2, 1–2 km^2, 2–5 km^2, >5 km^2) to estimate this contribution (Table 2.7).

Considerable variability in glacier retreat (up to complete disappearance) depending upon morphological types, sizes, and local conditions of alimentation of glaciers has been identified for the assessed territory. The distribution of the rate of retreat of glaciers with areas of over 0.1 km^2 from 1968 to 2008, depending upon their size, is shown in Fig. 2.30.

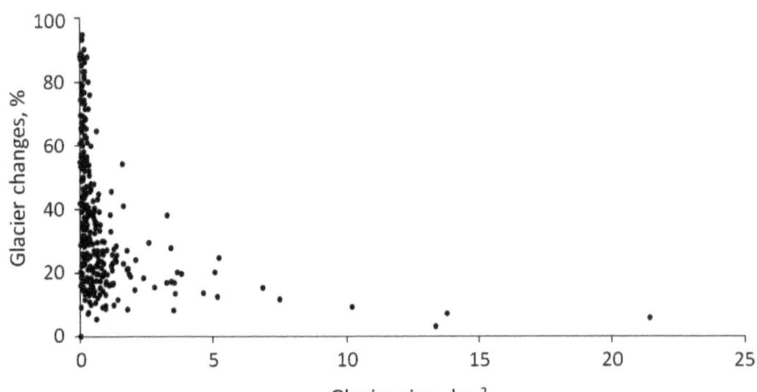

FIGURE 2.30 Glacier area changes according to glacier size for glaciers with area more than 0.1 km^2 in the Katun Ridge.

The highest rate of retreat was indicated for the group of glaciers with areas less than 0.5 km^2 (43.5%), the lowest rate was observed for the group of glaciers with areas over 5 km (7.5%), whereas glaciers with areas of 1–2 km have retreated by 23.2%, which is the closest to the changes in the glacial system on the whole (21.4%) (Fig. 2.31).

The glaciation of the Katunsky Range is represented by six morphological types of glaciers that had different responses to climate changes occurring over 40 years (Table 2.8). Hanging and cirque glaciers have the highest rate of retreat (more than 40%) and valley glaciers have the lowest rate (13.2%). Not many of them occurred, but they represent more than 50% of the total area of glaciation; the largest glaciers of the Katunsky range are among them: Sapozhnikov (21.45 km^2), Rodzevich (13.81 km^2), and Bratya Tronov (13.39 km^2). Cirque-valley glaciers are the closest to the average rate of retreat of the glacier system, in general (25.2%) (Nosenko et al., 2014).

If it was necessary to monitor the condition of glacier systems by land survey methods, these figures could be used as a criteria of the rate of representativeness of a glacier for a particular glacier system. In the case of the Katunsky Range, the glacier could be found among valley or cirque-valley glaciers with areas from 1 to 5 km^2.

Presumably the decrease in area of glaciers with different aspects (Fig. 2.32) reflects the general conditions of alimentation of glaciers in the Katunsky Range that is determined by complex interconnection between air masses from the west and southwest and orography of the Range. Glaciers facing moisture-bearing air flows have retreated the least, and to the contrary, the glaciers with north and northeast aspects sheltered from the winds have retreated the most.

The data obtained on the decrease in glacier areas characterize the changes of the whole glacier system of the Katunsky Range. However, it stretches

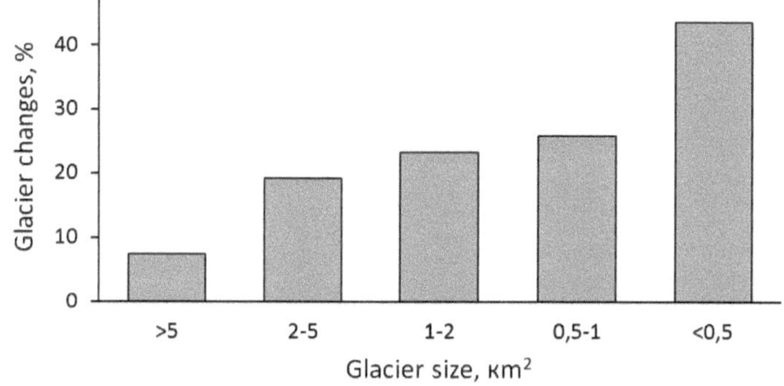

FIGURE 2.31 Glacier area changes in the Katun Ridge for different glacier size groups.

TABLE 2.8 Glacier Area Loss According to the Morphological Type in the Katun Ridge

Morphological Type	Number of Glaciers		Glacier Area, km^2		ΔS, %
	1968	2008	1968	2008	
Valley	37	37	143.31	124.37	13.2
Cirque-valley	80	79	67.96	50.81	25.2
Cirque	147	142	36.07	21.63	40.0
Cirque-hanging	18	17	4.63	3.03	34.6
Hanging	38	35	5.50	3.08	44.0
Hollow	2	2	7.03	5.02	28.5
Total	322	312	264.5	207.9	21.4

considerably (over 130 km) in the east-west direction and the conditions of the glaciers existence are diverse due to complex orography and high elevations. The highest eastern part including Mt. Belukha appears to be a great obstacle for moisture-bearing air masses that result in the formation of a significant node of glaciation in this area.

With the 2012 ASTER satellite imagery available for this part of the Range, we have gotten an additional possibility to evaluate the dynamics of glacier boundaries in that region. The estimation of glacial area in 1968, 2008,

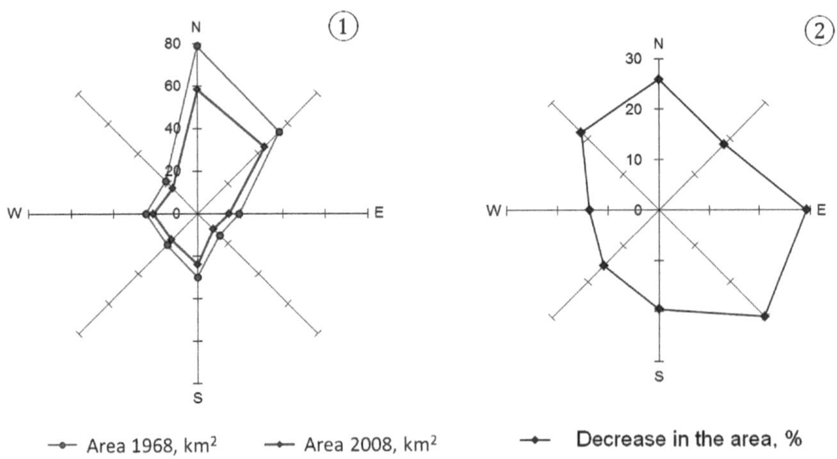

FIGURE 2.32 Glacier area (1) and glacier area change (2) in the Katun Ridge according to aspects.

and 2012 (Table 2.9) has been performed for 60 glaciers within the area of overlap of three diachronous satellite images. It was equal to 145.43, 122.29, and 118.33 km^2, respectively.

Since large valley glaciers of Mt. Belukha has been included in the number of those glaciers, the index of retreat for this area in 1968–2008 (15.9%) is rather different from the figure for Katunsky Range in general (21.4%). An additional period of time from 2008 to 2012, when the area of those glaciers has decreased by another 2.7%, allows the comparison of average annual rate of its change. From 1968 to 2008, the rate was 0.58 km^2 (0.40%) per annum and for the past 4 years, it has increased more than one and a half times up to 0.99 km^2 (0.68%) per annum. During the past decade, a similar increase in deterioration of the glaciers has also been observed in different mountain regions in the Caucasus, Polar Urals, and Suntar-Khayata Mountains (Galanin et al., 2013; Nosenko et al., 2013; Shahgedanova et al., 2012).

Differences in the dynamics of boundaries of the six largest glaciers on the Mt. Belukha massif demonstrate how their morphology, main direction of moisture flow, and facing can influence this process (Table 2.10, Fig. 2.33). All those glaciers have their heads on the slopes of Mt. Belukha and belong to the same valley type, but at the same time they have a rather wide range of their area reduction during the whole observation period—from 5.1% to 17.9%. Area recession mostly occurs due to the retreat of the glacier tongues. The areas of the Bolshoi and Maliy Berelski glaciers have shrunk at the highest rate. Their accumulation areas have a form of steep-walled cirques situated in the "wind shadow" of Mt. Belukha, since the main moisture-bearing air flows arrive from the west. Their southern facing adds a negative impact increasing the intensity of melting processes.

From 1952 to 2008, the glaciers of the Katunsky Range have retreated by 28.3 ± 5.8% and the glaciers of Severo-Chuisky and Yuzno-Chuisky Ranges by 27.1 ± 5.8% (Table 2.11). For glaciers with a size over 0.5 km^2, those parameters are equal to 22.5 ± 5.8% and 21.0 ± 5.8%, respectively. The difference of 1.5% could be the result of errors and assumptions, as well as it may indicate the actual ratio of the processes of glacier retreat in those areas determined by different orographic-climatic conditions.

The data on glacier areas received have been used to estimate their volumes in corresponding years. Correlation dependences obtained from the ground

TABLE 2.9 Area Decrease of 60 Glaciers in the Eastern Part of Katun Ridge

Glacier Area, km^2			Glacier Area Decrease, %		
1968	**2008**	**2012**	**1968–2008**	**2008–2012**	**1968–2012**
145.43	122.29	118.33	15.9	2.7	18.6

TABLE 2.10 Area Decrease of the Largest Valley Glaciers Located on the Mount Belukha 1968–2012

WGMS ID	Glacier Name	1968, km^2	2008, km^2	2012, km^2	1968–2008, km^2	2008–2012, %
SU5A15105391	Sapozhnikov	21.45	20.20	20.12	5.8	0.4
SU5A15104205	Rodzevich	13.81	12.83	12.67	7.1	1.2
SU5A15104013	Gebler Геблера	10.22	9.29	9.08	9.1	2.0
SU5A15103122	Bolshoi Berelski	7.52	6.64	6.18	11.7	6.2
SU5A15103116	Malyi Berelski	6.89	5.85	5.65	15.1	2.8
SU5A15104189	Tronov Brothers	13.39	12.98	12.48	3.1	2.0

FIGURE 2.33 Position of large glaciers outlines on Mount Belukha massif in 1968, 2008, and 2012.

TABLE 2.11 Glacier Area Decrease in Katun and North and South Chuya Ridges: 1952–2008

	Total Area, km^2				Area Decrease, %		
Ridges	1952	1968	2004	2008	1968–2008	1952–2004	1952–2008
Katun	289.9	264.5		208.0	21.4 ± 5.4		28.3 ± 5.8
North and South Chuya	336.0		254.5	245.1		24.4 ± 5.8	27.1 ± 5.8

GPS survey of 131 Altai glaciers, including 46 valley, 36 cirque-valley, 29 cirque and cirque-hanging, and 6 hollow basin and flat-summit glaciers, were used (Nikitin, 2009). Dependences between volumes and areas of different types of glaciers were approximated by ratios of exponential functions (Fig. 2.34).

The estimations have shown that by 2008 glaciers of the Katunsky Range lost 4.22 km^3 (28.1 ± 5.8%) of their ice (Table 2.12). The volume of glaciers on Severo- and Yuzhno-Chuisky Ranges during the same period reduced by 4.67 km^3 (26.9 ± 5.8%). Exclusive of glaciers with areas less than 0.5 km^2, those parameters will be 3.23 km^3 and 3.47 km^3, respectively.

The results of studies of satellite images have shown that glacier retreat in the Altai Mountains was taking place during the second part of the 20th century and at the beginning of this century. By 2008, the area of glaciers in the Katunsky and Severo- and Yuzhno-Chuisky Ranges has reduced by 172.4 km^2 and their volume by 8.9 km^3 or 27.4%. A comparison of satellite images from 2004 to 2012 with historic data of the past century brings a conclusion that the rate of glacier retreat during the past decade has increased 1.5−2 times.

The study of glaciers in the southeast Altai, based on satellite images interpretation and field observations, also has shown a retreat of glaciers in the 20th century and at the beginning of the 21st century (Ganyushkin et al., 2016). From 1989 to 2011, the area of glaciers on the massifs of the Talduayar, Mongun-Taiga-Malaya, and the southern part of the Chikhachev Range has decreased by 29%, 45%, and 31%, respectively.

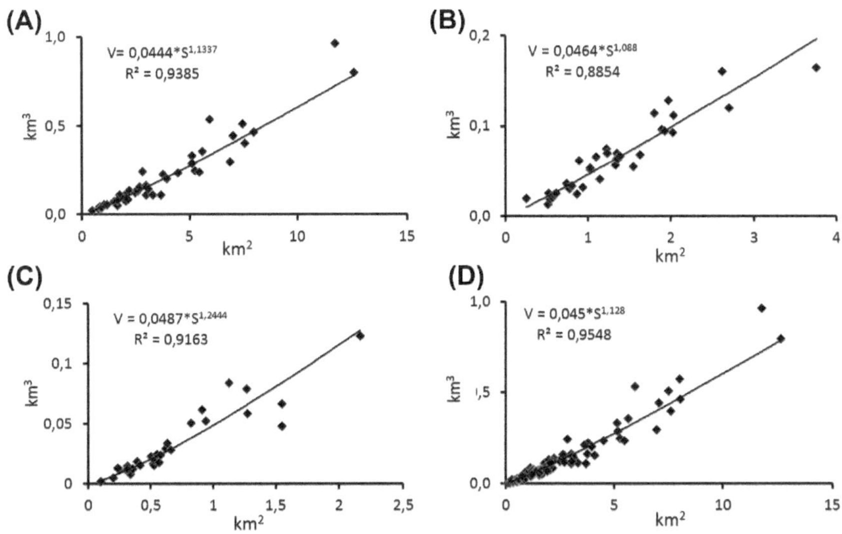

FIGURE 2.34 Correlation dependences between glacier area and volume for different morphological types: (A) valley, (B) cirque-valley, (C) cirque, and (D) all glaciers.

TABLE 2.12 Glacier Volume Decrease in Katun and North and South Chuya Ridges: 1952–2008

	Total Volume, km^3				Glacier Volume Decrease, %		
Ridge	**1952**	**1968**	**2004**	**2008**	**1968–2008**	**1952–2004**	**1952–2008**
Katun	15.02	13.67		10.80	20.5 ± 5.4		28.1 ± 5.8
North and South Chuya Ridges	17.37		13.17	12.70		24.1 ± 5.8	26.9 ± 5.8

The retreat of a particular glacier depends on a combination of factors of topography and climate that defines its morphology, size, elevation range, and position in relation to moisture-bearing air flows. During 1968–2012, certain large valley glaciers of the Mt. Belukha massif have retreated by 3.1–15.1 km^2. Valley or cirque-valley glaciers with area from 1 to 5 km^2 are the most representative for this glacier system.

According to the data from three meteorological stations, Akkem (2050 m), Kara-Turek (2600 m), and Actru (2025 m), located close to the glacier area, the main climatic parameters—average summer temperature and precipitation—have increased constantly during the past few decades (Fig. 2.35) (Surazakov et al., 2007). Also, a similar situation is currently

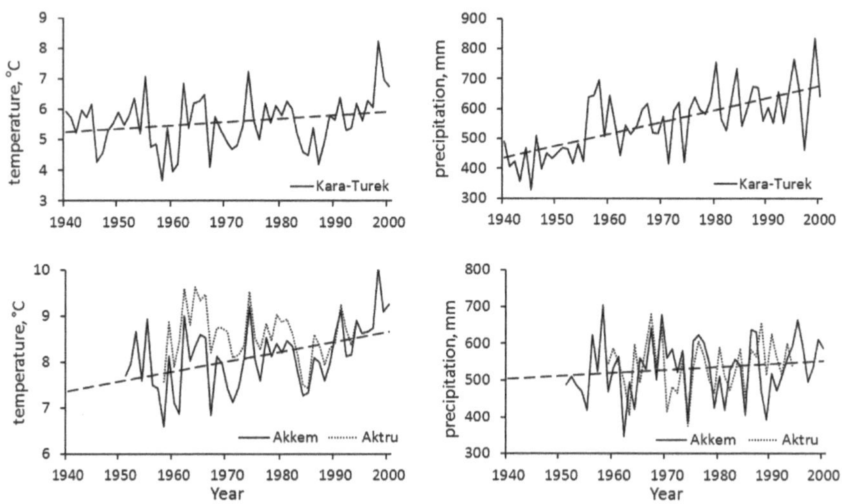

FIGURE 2.35 Time series of averaged summer air temperature and averaged annual precipitation for Kara-Tyurek, Akkem, and Aktry weather stations.

observed in the other mountain areas of Russia (the Caucasus, Polar Urals, non-volcanic regions of Kamchatka), where an increase in summer temperatures is not balanced by increase in precipitation. It appears that in this case the increase in summer temperatures is critical for glacier retreat in the Mountain Altai.

Kuznetsk Alatau

Kuznetsk Alatau is a part of Altay-Sayan mountain region and consists of numerous ridges with different orientation, massifs, and ranges. The climate of Kuznets Alatau is significantly determined by its location in the center of Eurasia. Continental air of the mid-latitudes is predominant there throughout the year. Average annual temperatures on the western macro-slope of the Range and in the high-mountain area are negative. Average temperature in January is −15°C and in July +16.5°C. Average summer temperature is +9 to +12°C. According to the data of the field survey, on the western slope at the level above 1200–1400 m, precipitation is 3000–3500 mm per annum, while half of it is solid precipitation. On the eastern macro-slope, precipitation decreases dramatically with distance from the main divide (Shpin, 1980).

The glacial system of Kuznetsk Alatau consists of 91 glaciers with a total area of 6.79 km^2. It is a northern offshoot of the larger Altai glacier system. Three glacier regions occur: (1) northern—glaciers near Mt. Bolshoi Taskhil (1448 m) and Mt. Bobrov (1673 m), where only three glaciers occur with a total area of 0.04 km^2; (2) central—37 glaciers with a total area of 2.65 km^2; (3) southern—glaciers at Tegir-Tyz Range, with 51 glaciers having a total area of 4.1 km^2. According to the data available (Kovalenko, 2008), glaciers have retreated during 1975–2003. In general, recession of the glacier margin varied from 30 to 200 m. Table 2.13 summarizes the data on the condition of six glaciers in 1975–2006. During that period they have lost from 25% to 57% of their area.

East Sayan

East Sayan is a highland stretching mainly to the southeast; it is situated in the southern margin of Siberia between the Yenisei and Angara river valleys approximately 1000 km to the east from the Altai. The climatic conditions of the territory are defined by its continental position. In winter, the region falls under the influence of a persistent Siberian High (mainly clear, low-wind weather, with frequent ground inversions). In spring and autumn, a zonal transportation of air masses increases and the number of cyclones arriving from the west and northwest goes up. In summer, latitudinal transportation weakens, and a flat pressure gradient with slight winds prevails. Average annual surface air temperature is about −8°C. Maximum precipitation occurs within three summer months, with up to 70% of the annual sum. On average,

TABLE 2.13 Area Decrease of Six Glaciers Located in Kuznetski Alatau

	Geographical Coordinates		Glacier Area, km^2			Glacier Area Decrease	
Glacier Name	North Latitude, °	East Longitude, °	1975 (Shpin, 1980)	2003 (Kovalenko and Kitov, 2011)	2006 (Kovalenko and Kitov, 2011)	km^2	%
№ 16	54.23	88.37	0.09	0.06		0.03	33
Geogrphers (№ 22)	54.22	88.38	0.06	0.04		0.02	33
NGPI (№ 23)	54.21	88.39	0.04	0.03		0.01	25
Cherno-Iyussky (№ 83)	54.36	88.42	0.19	0.10	0.08	0.11	57
Tolmachev (№ 86)	54.37	88.38	0.21		0.09	0.12	57
№ 87	54.38	88.37	0.2		0.10	0.10	50

in the mid-mountains of East Sayan, the rate of precipitation is over 400 mm per annum and on the highland it is more than 1000 mm (according to the data of Olenya Rechka meteorological station it is 1200 mm per annum). At an altitude of 2000 m, solid precipitation occurs during 10 months, from September to June.

Glaciation of the Vostochny Sayan is represented by 107 glaciers of mainly cirque, cirque-valley, and niche types, with a total area of 30.8 km^2, located predominantly on the leeward of mountain ridges facing east, northeast, and north. Groswald (1965) has defined three regions of contemporary glaciation: the Northwestern, Southeastern, and Central.

The detailed studies of glacier dynamics based on satellite imagery have been performed for two centers of mountain glaciation on the southeast of Vostochny Sayan: Munku-Sardyk Range and Peak Topography (Osipov et al., 2013). High-resolution satellite images were obtained from Quick Bird (August 2006, resolution 0.6 m), WorldView-1 (July–August 2008, resolution 0.5 m), image Landsat-7 (sensor ETM+, August 2001, spatial resolution 15–30 m), and topographic maps 1:100,000 (published in 1981) and 1:50,000 (published in 1992, situation in 1978). To analyze climatic conditions of the past decades, the data of four nearby meteorological stations (Mondy, Orlik, Olenya Rechka, and Tulun) as well as long-term records of the meteorological station of Irkutsk were used.

Overall, a study of 13 glaciers with a total area of 5.1 km^2 has been conducted, including three glaciers (with a total area of 0.63 km^2) in the Munku-Sardyk area and 10 glaciers (with a total area of 4.47 km^2) in the Peak Topography area. The area of only one glacier exceeds 1 km^2. The length of glaciers varies from 0.28 to 2.01 km (average length 0.90 km).

Based on the results of interpretation of the satellite images of terminal and flank moraines, glacial borders at the end of the Little Ice Age have been reconstructed. From the middle of the 19th century, the glacier area has recessed by 26–68% (on average by 49%). Small glaciers have decreased in area more than large glaciers.

The comparison of obtained results with the data of the Inventory based on aerial photographs in 1953 and 1958 (Glacier Inventory of the USSR, 1965–1982) has shown that during the past 55 years the area of surveyed glaciers has receded by 36% (0.68 per annum), which represents about a half of deglaciation of the surveyed area during 160 years.

Long-term data on average summer temperature from meteorological stations located close to glaciers have shown quasi-cyclic changes. The reduction in summer temperature in 1850–1910 by 1.5°C has changed to an increase in temperature by 0.6°C in 1910–1955 and this was followed by a new reduction in temperature by 0.5°C that lasted approximately until the mid-1980s. Starting from that time, the temperature was gradually increasing (by 1.5°C) until 2005, but after that the trend has changed to the opposite. Unlike summer temperatures, the sum of solid precipitations at every station did not show any distinctive trends during the past 45 years. Thus, the most unfavorable climatic conditions for glaciation in the mountains of the southern part of East Sayan were in the1980–1990s. These data have allowed the suggestion that climatic cooling observed during the past decade with an insignificant increase in the sum of solid precipitation may contribute to the restoration of glacier mass balance and slow down the rate of deglaciation.

Baikal Ridge

Baikal Ridge stretches 300 km along the western shore of Lake Baikal within 54–56°N. Modern glaciers are located on its top—Mt. Chersky (2572 m a.s.l) at 55°N in the upstream of the Kurkula River. The climate in the area of the Baikal Range is severe and distinctly continental with a prolonged winter season (8–9 months) and a short cool summer. The glaciers of Baikal Ridge on the slope of Mt. Chersky were studied in 1979–1981 (Aleshin, 1982).

In the study by Kitov et al. (2014), data were obtained on Chersky glacier conditions in 1960 (topographic map) and in 2010 (satellite image Landsat-7). According to those data, the area of the glacier has retreated only by 9%—from 0.45 km^2 in 1960 to 0.41 km^2 in 2010. It is necessary to consider a small size of the glacier and an accuracy of baseline data that usually may lead to significant errors in estimation. A detailed study of glaciers of this region with the use of satellite imaging data is yet to be performed.

Barguzin Ridge

The total length of Barguzin Ridge from the downstream of Barguzin River to Verhnaya Angara River reaches 300 km. The average height of peaks of the Ridge is 2400–2500 m. The highest point of the ridge and the Baikal region in general is Baikal Peak, 2841 m. Average air temperatures in January vary from −24 to −28°C, in July +8°C, in April and October −8°C, and annual precipitation reaches 1200 mm.

Initially, glaciers of Barguzin Ridge became known at the beginning of 1980s (Plusnin, 2003). Field surveys have been conducted in 2011–2013 (Kovalenko and Kitov, 2011; Kovalenko et al., 2012); then after studying map data and interpretation of satellite imagery with implementation of GIS technologies, the first-ever database of glaciers and snowfields of Barguszin Ridge has been compiled and their inventory has been completed (Kitov et al., 2014). For the purpose of analysis, topographic maps at a scale 1:50,000 were published in 1960, as well as satellite images Landsat (2003, 2007–2013) and WorldView-1 (2008) were used. Out of the 187 objects identified as nival–glacial, seven were, in fact, glaciers. The total area of nival–glacial formations in the middle of the 20th century (1960) was 9.2 km^2. The contemporary area of those formations is 2.3 km^2. Cirque glaciers have receded by 48% and niche glaciers by 82%. Based on preliminary estimations, during the past 50 years the total area of nival–glacial objects in Barguzin Ridge has decreased four times; however, the area of cirque glaciers has decreased two times. The major glacier of Barguzin Ridge retreated with a velocity of 0.002 km^2/year, which is compatible with the rate of recession of glaciers in East Sayan (Kitov et al., 2014).

Kodar Ridge

Kodar Ridge lies in the southeast of Siberia (56°45′–57°15′N, 117°–118°E) between two tributaries of the Lena River—Vitim and Olekma—500 km to the northeast from Lake Baikal. Local mountains have distinctive narrow deep valleys, whereas the altitudes are the highest in Transbaikal that supports the occurrence of glaciers, unlike other areas to the east of Baikal located at the same latitude. The highest mountain in the area, Baikal-Amur Magistral Peak (BAM), reaches 3073 m. The climate of the area is distinctly continental. The Siberian High determines the situation in November and March, resulting in low temperatures and low precipitation (Panagiotopoulos et al., 2005).

According to the data of the Glacier Inventory of the USSR, in 1955–1963, 30 glaciers had a total area of 18.8 km^2 (whereas glaciers with a size less than 0.1 km^2 were not considered). Glaciers with areas less than 1 km^2 represent 70% of the glacier area; four glaciers have an area over 1 km^2. In the 1970–2000s, 10 more glaciers were discovered in the Kodar system, located mainly to the northwest from the main massif, with a total area of 2.35 km^2 (Plastinin, 1998; Plusnin et al., 2008). According to E.Y. Osipov (2010), in 2001, 42 glaciers were there with a total area of 11.9 km^2; the exposed ice area of the glaciers was 60%.

The contemporary changes of glaciers in the Kodar Ridge based on satellite images were studied in the work of Stokes et al. (2013). Satellite images from Landsat (1995, 2001, 2010), CORONA (1964), and the digital elevation model (the ASTER Global DEM) were used for glacier identification.

To evaluate climatically variable records of temperatures (from 1960) and precipitation (from 1951), data from the Chara meteorological station were used. The total area of open ice has reduced from 11.72 ± 0.72 in 1995 to 9.53 ± 0.29 in 2001 and 7.01 ± 0.23 km^2 in 2010. The study has demonstrated that a decrease in the area of open ice has occurred in glaciers of all sizes. The biggest losses have occurred in large glaciers, although in terms of percentage, small glaciers have receded the most, which is typical for other regions as well.

The study by Shahgedanova et al. (2011) shows that size reduction of the Azarova glacier from 1979 to 2007 took place under conditions of a steady rise of average summer temperatures (according to the data of Chara meteorological station). From 1980, temperatures were increasing during every season. The rise in temperature during summer months is particularly important as it prolongs the rather short ablation period in this area. The warmest decade was observed during 2000–2010 when average temperature reached 15.2°C, exceeding the average value for the whole period of survey by 1°C. The warmest summer in this area occurred in 2008 when the average temperature of June reached 17.8°C.

Thus, according to the data by Stokes et al. (2013), the area of open ice on Kodar glaciers stayed almost unchanged from the 1960s to 1995 and decreased by almost 40% during the next 15 years (1995–2010) with a considerable rise in the average summer temperature starting from the 1980s. In the opinion of the authors of this research, some glaciers in that area may soon be transformed into rock glaciers.

GLACIER SYSTEMS RECEIVING MOISTURE FROM PACIFIC. CHUKOTKA, KORYAK, KOLYMA, CHERSKY, SUNTAR-HAYATA, KAMCHATKA

Chukotka Upland

The Chukotka Upland is a system of medium-mountain ranges and low-mountain massifs stretching from Chaun Bay in the Chukchi Sea to the east. The territory is affected by the Arctic Ocean, the Aleutian Low that is responsible for cyclonic activity, and the cold Bering Sea. The duration of winter there is 8 months. Average temperatures of January in the west vary from −30 to −32°C, and in the east from −20 to −23°C, and the lowest temperatures reach −55°C. The summer is short and cool (average temperature of July is 3–8°C). Average precipitation in the west is about 250–300 mm and in the east – up to 500–650 mm.

The glacier system of Chukotka Upland was studied during 1980–2000 after publishing of the Glacier Inventory of the USSR (Sedov, 1988, 1990, 1992, 1995, 1996, 1997c; Chernova et al., 2011). Forty seven small glaciers of cirque and niche types, with a total area of 13.53 km^2, are dispersed over the vast territory and gathered into five groups (Kotlyakov et al., 2011). Glaciers located in the mountains near the Pacific coast occur in the highest massifs and occur along the major watersheds stretching across the main water-barring air flows (Sedov, 1997a). Cirque glaciers are predominant there. All glaciers have the signs of retreat. The glacier termini and lateral edges are covered with surface moraine. The surface of the glaciers gradually decreases.

The estimation of changes in glacier sizes with the use of satellite images was not done there. The implementation of high-resolution satellite images will allow the verification of the number, sizes, and dynamics of glaciers in Chukotka at some time in the future.

Koryak Upland

The Koryak Upland occupies a vast territory at the very far northeast of the Asian territory of Russia. In the southeast, it is bounded by the Bering Sea, in the southwest by the waters of the shallow Penzhinsk Bay of the Sea of Okhotsk, and in the northeast also by the shallow Anadyr Bay. The Koryak Upland consists of parallel ranges oriented mainly from the northeast to the southwest. Their altitude varies from 600 to 1800 m (the highest point of the central part is Mt. Ledyanaya, 2562 m). Climate in the Upland area is cold oceanic. Annual precipitation on the southeastern slope exceeds 700 mm and on the northwestern side 400 mm.

The glacier system of Koryak Upland consists of three isolated areas. The first two of them are a group of several dozen glaciers in the South on Malinovsky Range (west coast of Olyutorsky Bay of Bering Sea) and a group of several hundred glaciers situated near the Bering Sea coast from 60° to 62°N (Glacier Inventory of the USSR, 1965–1982). Inconsistencies in data exist in those areas: according to the data of A.P. Vaskovsky (Vas'kovsky, 1955), 461 glaciers and snow fields occurred with a total area of 185 km^2; according to the data of the Inventory, 1335 glaciers occurred with a total area of 259.5 km^2, among which 715 glaciers were over 0.1 km^2 with a total area of 233.1 km^2.

The third area is located on Meinipilgin Range near lake Pekulneisky at 63°N. According to the data of Sedov (2001), 116 glaciers occurred, with a total area of 43.96 km^2. The rate of glacier area dynamics for this region has been evaluated at 30% (Ananicheva et al., 2010). The data of the Catalogue of Glaciers prepared by R.V. Sedov (2001), based on the aerial photographs of 1967 and field surveys of 1984–1985, as well as satellite images from Landsat TM ETM+ dated 2008–2010, were used for comparison.

Kolyma Upland

The Kolyma mountain area occupies the far eastern areas of northeastern Siberia. The eastern boundary passes near mountain massifs that form a watershed of rivers of the Arctic and Pacific Oceans. The southern areas form the Verkhne-Kolyma Upland. The average altitudes of the Upland are 1000–1200 m, but certain massifs rise above 2000 m. Climate is continental and subarctic.

The system of the Kolyma Upland stretches in the south of the area to the northeast and then to the north. The south boundary—mountains of the Okhotsk-Kolyma watershed, Yukagir upland approaches in the north. Average altitudes of the Upland are 1300 m. The data on the Kolyma Upland glacier system have been received after compilation of the Glacier Inventory of the USSR (Sedov, 1995; Kotlyakov et al., 2011; Chernova et al., 2011). Some 19 glaciers are located there, predominately cirque and niche types, with a total area of 3.61 km^2. The first group of 14 cirque glaciers facing north and northwest is located on Taigonos Peninsula at the far northern part of Okhotsk Sea. The second group consisting of five glaciers is located on the eastern slope of Kolyma Plateau near the western coast of Okhotsk Sea. Glaciers there are bigger (0.2–0.5 km^2), and they are located slightly higher and face the north. The data on changes in glaciers are yet to be received.

Chersky Mountains

The ranges of the Chersky highland extend parallel to each other from the northwest to southeast from the middle reach of Yana River to the headwaters of Kolyma River. The highest point is located on Buordakh massif (Mt. Pobyeda, 3147 m) with significantly rugged, alpine relief. The climate of the region is distinctively continental, with very low temperatures during winter when average monthly temperatures may go below −40°C, with an absolute minimum of −71°C. In July the average monthly temperatures change from 12°C in the north to 16°C in the south. During the warm season, average precipitation reaches 150–200 mm and during the cold season 30–40 mm. The amount of precipitation increases with altitude.

According to the data of the Glacier Inventory of the USSR, the Chersky mountain system consists of 372 glaciers with a total area of 155.4 km^2 spreading over the territory of several thousand square kilometers (Glacier Inventory of the USSR, 1965–1982). About 55% of the area of glaciation (85.2 km^2) is located in the highest Buordakh massif on the right bank of the Indigirka River.

Glaciers in the Buordakh massif were studied based on the satellite images from Landsat 7 ETM+ dated 2001 and field survey data (Gurney et al., 2008). Altogether 80 glaciers with sizes varying from 0.1 to 10.4 km^2 have been discovered with a total area of 70 km^2. Also, the sizes of glaciers during the Little Ice Age have been estimated. It was shown that glacier area since the time of the last maximum before 2001 has reduced by 14.8 km^2 (17%).

Suntar-Khayata Mountains

The Suntar-Khayata Mountains, the large south branch of Verkhoyansky Range, are located in a close proximity to the "Pole of Cold" Oymyakon to the south and represent the second largest area of mountain glaciation in the northeast of Asia (Galanin et al., 2013). The Severny massif, with the highest peak of 2959 m (Mt. Mus-Khaya), has the best data coverage. Even before the discovery of glaciation over there, some researchers have absolutely denied the possibilities of glaciation development at such altitudes (Koreysha, 1963).

During the Holocene Period, a cirque-valley glaciation of the Suntar-Khayat Mountains was developing in the area of active interaction of moist cyclonic air masses of the Okhotomorsk basin with the Yakutsk anticyclone (Galanin et al., 2013). Short summers with low temperatures and heavy solid precipitation both in winter and in summer provide favorable conditions for the occurrence of glaciers.

According to the Glacier Inventory of the USSR, some 208 glaciers occur, with a total area of 201 km^2. The predominant types of glaciers are cirque, cirque-valley, valley, twinned valley (piedmont glacier type), and hanging glaciers (Glacier Inventory of the USSR, 1965–1982). They are grouped into three main massifs: northern (90 km^2), central (50 km^2), and southern (61 km^2) (Ananicheva and Krenke, 2007).

According to the data of Ananicheva and Koreysha (2005), the Little Ice Age glaciers of the northern massif have retreated by 50–100 m and glaciers of the southern massif by 100–200 m. The area of recession of the most studied glacier № 31 was 5%. Based on the results of the reconstruction (Solomina and Filatov, 1998), the length of glaciers during that time has decreased by 250 m and termini of glaciers have retreated upward in altitude by 50 m. The study by Ananicheva et al. (2006) shows the results of the analysis of the Landsat satellite image dated 2003; according to it, at that time 195 glaciers occurred in the Suntar-Khayata Mountains with a total area of 163 km^2. In comparison to the data of M.M. Koreisha for the period from 1945 to 2003, the area of glaciation has receded by approximately 20%.

Various research scholars have provided different estimations for the total area of glaciation in the Suntar-Khayata Mountains: L.L. Berman (1947), 360 km^2; Y.N. Popov (1947), 380 km^2; Y.N. Popov (1948), 160 km^2; A.P. Vaskovsky (1955), 246 km^2; M.M. Koreisha (1963), 206 km^2; and the Glacier Inventory of the USSR (1965–1982), 199.4 km^2. Slightly modified data from the Glacier Inventory of the USSR were included in the World Digital Glacier Inventory, and the given figure for the Suntar-Khayata Mountains was 201.6 km^2. In her work dated 2009, M.D. Ananicheva has estimated the area of glaciation in Suntar-Khayata Mountains for the period of survey (1943–1947) as 210 km^2. The data published in general indicate the recession of glaciers during the past 60 years; however, those data are based mainly on the results of low-resolution satellite images and are rather contradictory.

Changes in the contemporary glaciation of the northern massif of the Suntar-Khayata Mountains during the past 60 years were studied with the data of the glaciers of the USSR and satellite imagery Landsat 2010 (Galanin et al., 2013). Some 87 glaciers were identified with the help of 2010 Landsat satellite images. Due to the low resolution of the images, it was decided to interpret only the open ice surfaces. The area of glacier surfaces without moraine in the northern-most massif in 1954 was about 84.2 km^2 and in 2010 it was 61,64 km^2, that is, it has reduced by 22.56 km^2 or 26.8%. Cirque glaciers with an area less than 1.5 km^2 have retreated the most. These estimations correspond with the results of other studies (Ananicheva et al., 2006), according to which by 2003 the area of glaciers in the Suntar-Khayata Mountains has decreased by approximately 20%.

The chart of glacier retreat in the northern part of Mt. Mus-Khaya has been compiled based on GIS analysis. The total area of the group of glaciers studied from 1945 to 2011 has receded by 12.74 km^2 (from 34.6 to 22.16 km^2) or 36%. During that period, one cirque glacier has disappeared and several glaciers have been fragmented. The rate of retreat of individual glaciers varies from 17% to 96%. In general, small cirque glaciers with areas less than 1.5 km^2 have retreated the most.

Kamchatka

Data and Research Methodology

Contemporary research on Kamchatka glaciers is based on a combination of field surveys with the use of remote-sensing data. In general, glaciological field surveys take place in volcanic areas of the peninsula (e.g., Klyuchevskaya and Avachinskaya groups of volcanoes, Mutnovsky and Shiveluch volcanoes). Generally this type of work includes field survey of glacier tongues, taking their photographic images and identifying the location of glacier fronts with GPS. This allows estimation of contemporary conditions of glaciers and changes in the location of their fronts (compared with maps and data of aerial photography, satellite imaging, and geodetic survey). Glaciers of the Sredinny Range and Kronotzky peninsula with difficult access are studied mostly by remote-sensing methods.

Contemporary Condition of Glaciers in Kamchatka and Its Variations

Sredinny Range

According to the data of the Glacier Inventory of the USSR (Glacier Inventory of the USSR Kamchatka, 1968), 52.8% (461.7 km^2) of glaciation area in Kamchatka and 59% of total number of its glaciers are located within the Sredinny Range. Most of the glaciation of the Sredinny Range occurs in three areas: the northern part of the Sredinny Range (the Ostraya-Khuvkhoitun massif and further to the north), the Alney-Chashakondja massif, and the Ichinsk volcano (Fig. 2.36).

The change in glacier areas of the northern part of Sredinny Range (from 57°15′ to 59°10′N) from 1950 to 2002 has been estimated in the study (Muraviev and Nosenko, 2013). According to the data of the Glacier Inventory of the USSR, some 197 glaciers occurred (total area 371 km^2). By 2002 the area of the largest glaciers in the region has receded insignificantly (2.1% for the group of glaciers with an area over 5 km^2). Glaciers with areas less than 0.5 km^2 have retreated the most (49.3%) (Table 2.14). The areas of glaciers facing south, southeast, and northwest have decreased the most—23.1, 40.4, and 23.0%, respectively (Muraviev and Nosenko, 2013) (Table 2.15).

Glaciers facing north and northeast have retreated the least—8.1% and 8.8%, respectively. In general, from 1950 to 2002, glaciation of the northern part of the Sredinny Range has retreated by 16.6%. For glaciers with an area over 0.5 km^2, this parameter is 12.5%.

According to the data of the Glacier Inventory of the USSR, some 26 glaciers occurred within the Alney-Chashakonja massif with a total area of

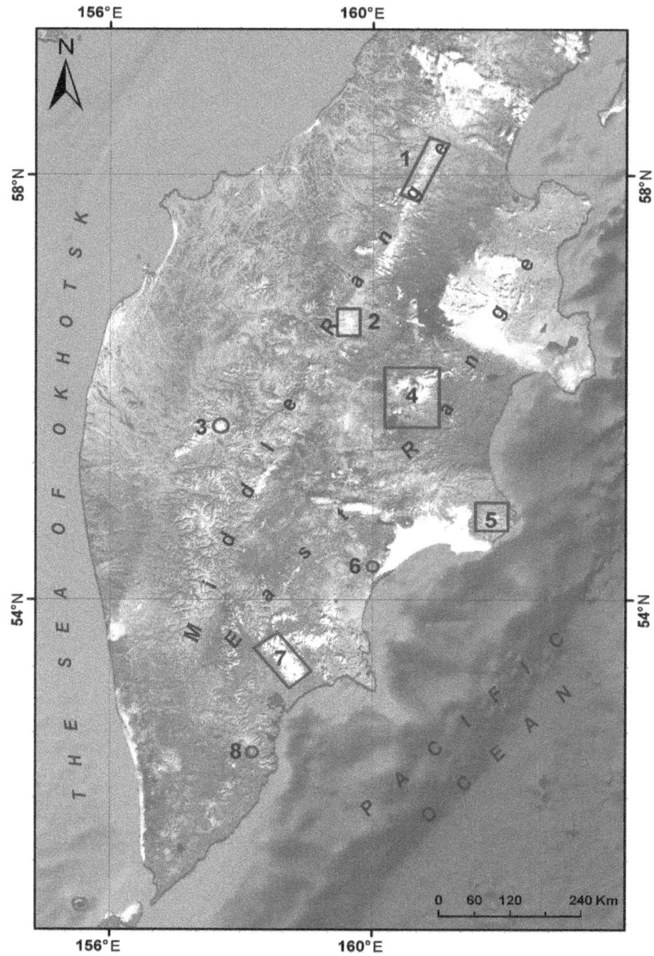

FIGURE 2.36 Nodes of modern glaciation in Kamchatka: (1) Ostraya-Huvhoytun massif; (2) Alney-Chashakondzha massif; (3) Ichinsky volcano; (4) Kluchevskaya group of volcanoes; (5) Kronotsky Peninsula; (6) volcano Big Semyachik; (7) Avachinskaya group of volcanoes; and (8) Mutnovskiy volcano.

61.4 km^2 (Muraviev, 2014). Interpretation of the satellite image of WorldView-2 (July 20, 2010) has revealed 45 glaciers within the Alney-Chashakonja massif. The lowest rate of retreat was observed for glaciers with areas of 0.5 −1 km^2 (4.5%) and 1−2 km^2 (8.9%). The highest recession (51.0%) was observed for glaciers with areas less than 0.5 km^2 (Table 2.16). The area of glaciers facing northeast and northwest was found to be reduced the most (by 35.8% and 31.9%, respectively). On the other hand, the area of glaciers facing southwest and east was found to be increased (by 9.9% and 2.8%, respectively) (Table 2.17). In general, from 1950 to 2010, the glaciation of the Alney-Chashakonja massif decreased by 19.2%.

TABLE 2.14 Glacier Area Loss According to the Glacier Size in the Northern Part of Sredinny Range

Glacier Size, km^2	Glacier Area in 1950, km^2	Glacier Area in 2002, km^2	Glacier Area Loss 1950–2002, km^2/%	Number of Glaciers
>5	143.7	140.74	−3.0/−2.1	15
2–5	89.4	80.76	−8.6/−9.7	44
1–2	46.8	27.01	−19.8/−42.3	27
0.5–1	29.7	22.52	−7.2/−24.2	38
Total >0.5	309.6	271.03	−38.57/−12.5	124
<0.5	38.7	19.62	−19.1/−49.3	74
Total	348.3	290.65	−57.7/−16.6	198

TABLE 2.15 Glacier Area Loss According to Aspects in the Northern Part of Sredinny Range

Glacier Aspect	Glacier Area in km^2: 1950/2002	Glacier Area Changes in km^2/%: 1950–2002	Number of Glacier, 1950	Average Glacier Size in km^2: 2002
North	40.3/36.76	−3.54/−8.8	24	1.53
North-East	52.1/47.86	−4.24/−8.1	14	3.42
East	69.8/55.64	−14.16/−20.3	27	2.06
South-East	31.8/18.96	−12.84/−40.4	22	0.86
South	11.6/8.92	−2.68/−23.1	11	0.81
South-West	44.8/38.74	−6.06/−13.5	7	5.53
West	61.3/55.57	−5.73/−9.3	25	2.22
North-West	36.6/28.20	−8.40/−23.0	31	0.91
Total	348.3/290.65	−57.65/−16.6	161	1.81

TABLE 2.16 Glacier Area Loss According to the Glacier Size in the Alnei–Chashakondzha Massif

Galcier Size, km^2	Glacier Area During 1950, km^2	Glacier Area During 2010, km^2	Glacier Area Changes During 1950–2010, km^2/%
>5	13.10	9.62	−3.48/−26.6
2–5	31.60	26.17	−5.43/−17.2
1–2	9.40	8.56	−0.84/−8.9
0.5–1	3.00	2.86	−0.14/−4.5
Total >0.5	57.10	47.21	−9.89/−17.3
0.1–0.5	3.40	1.67	−1.73/−51.0
Total	60.50	48.88	−11.62/−19.2

TABLE 2.17 Glacier Area Loss According to Aspects in the Alnei–Chashakondzha Massif

Glacier Aspect	Glacier Area in km^2: 1950/2010	Glacier Area Changes in km^2/%: 1950–2013	Number of Glacier: 1950	Average Size in km^2: 2002
North	6.0/4.85	−1.15/−19.2	5	0.97
North-East	12.2/7.83	−4.37/−35.8	3	2.61
East	8.8/9.05	0.25/2.8	3	3.02
South-East	3.6/3.41	−0.19/−5.2	2	1.71
South	2.7/1.98	−0.72/−26.6	4	0.50
South-West	5.4/5.93	0.53/9.9	2	2.97
West	18.2/13.38	−4.82/−26.5	2	6.69
North-West	3.6/2.45	−1.15/−31.9	3	0.82
Total	60.5/48.88	−11.62/−19.2	24	2.04

The data for the period 1950–2006 from the meteorological stations of Klyuchi and Ust-Voyampolka, closest to the areas of study, have been analyzed in order to evaluate climatic changes (Muraviev and Nosenko, 2013). In comparison with 1951–1980, there was a general increase in summer temperatures by 1°C at the Klyuchi meteorological station and by 0.6°C in the Ust-Voyampolka meteorological station (Fig. 2.37).

Compared to 1966–1980, the sum of frozen precipitation at those meteorological stations during 1989–2006 has decreased in general by 3.5% and 20.8%, respectively (Fig. 2.38). Therefore, we can conclude that the retreat of glacier areas in the northern part of the Sredinny Range and the Alney-Chashakonja Massif corresponds with changes in the main climatic factors.

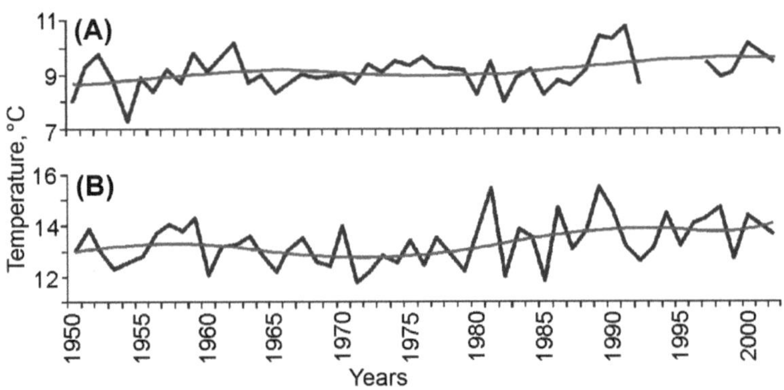

FIGURE 2.37 Average summer air temperature time series from meteorological station Ust-Voyampolka (A) and Kluchi (B).

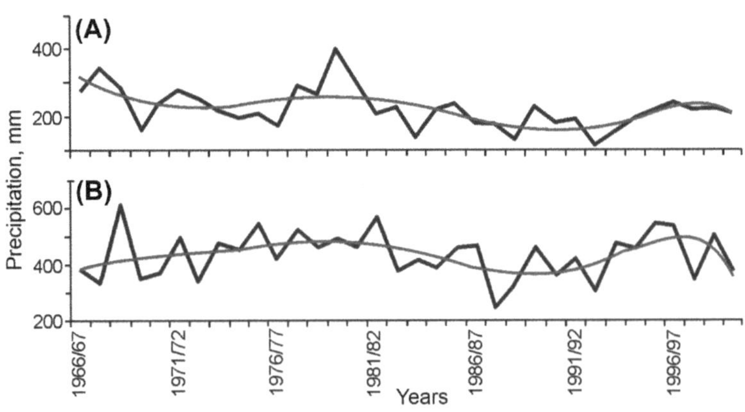

FIGURE 2.38 Winter precipitation sum time series from meteorological station Ust-Voyampolka (A) and Kluchi (B).

The glaciers of the Sredinny Range in Kamchatka are retreating. From 1950 to 2002, the area of glaciers of its northern part has receded by 57.7 km^2 or by 16.6%. The glacier area of the Alney-Chashakonja Massif has reduced by 11.6 km^2 or by 19.2%. This is consistent with retreat of such glacier systems of Northern Eurasia such as the Altai, Tien Shan, or Caucasus and correlates with changes in the main climatic indices—the rise in summer temperatures and decrease in winter precipitation.

Kronotsky Peninsula

Glaciers of this region are rather responsive to climatic changes. According to the data of the Glacier Inventory of the USSR, some 32 glaciers with a total area of 91.9 km^2 occurred on the Kronotsky Peninsula. By 2013 six of them were fragmented (Muraviev, 2014). Fifty glaciers have been discovered in the process of interpretation of a Landsat (02.09.2013) satellite image for the territory of the Kronotsky Peninsula. The lowest recession has been indicated for large glaciers (19% for a group of glaciers over 5 km^2). The highest recession (76.7%) has been indicated for glaciers with an area of less than 0.5 km^2 (Muraviev, 2014) (Table 2.18).

Glaciers facing the northeast, southeast, and southwest have decreased the most—79.3%, 38.2%, and 31.8%, respectively (Table 2.19). The average area of glaciers facing the northeast (0.31 km^2) is small in comparison with other groups and is represented by glaciers with an area less than 0.5 km^2, which have the highest rate of measurement errors.

An analysis of data records for 1950–2001 in the Kronoky meteorological station (Muraviev, 2014) has shown a rise in summer temperatures (by 0.3°C during 1989–2001 compared to 1951–1980) and a decrease in the sums of

TABLE 2.18 Glacier Area Loss According to the Glacier Size in the Kronotskii Peninsula

Glacier Size, km^2	Glacier Area in 1950, km^2	Glacier Area in 2013, km^2	Glacier Area Loss in 1950–2013, km^2/%
>5	34.7	28.10	−6.6/−19.0
2–5	31.0	23.91	−7.09/−22.9
1–2	11.8	8.13	−3.67/−31.1
0.5–1	4.6	3.12	−1.48/−32.2
Total >0.5	82.1	63.26	−18.84/−22.9
<0.5	7.0	1.63	−5.37/−76.7
Total	89.10	64.89	−24.21/−27.2

TABLE 2.19 Glacier Area Loss According to Aspects in the Kronotskii Peninsula

Glacier Aspect	Glacier Area, km^2: 1950/2013	Glacier Area Changes, km^2/%: 1950–2013	Number of Glacier: 1950	Average Glacier Size, km^2: 2013
North	10.5/7.76	−2.74/−26.1	2	3.88
North-East	4.5/0.93	−3.57/−79.3	3	0.31
East	1.1/1.18	0.08/7.3	1	1.18
South-East	14.2/8.77	−5.43/−38.2	3	2.92
South	8.9/7.02	−1.88/−21.1	3	2.34
South-West	16.6/11.32	−5.28/−31.8	5	2.26
West	10.2/9.86	−0.34/−3.3	4	2.47
North-West	23.1/18.05	−5.05/−21.9	8	2.26
Total	89.1/64.89	−24.21/−27.2	29	2.24

solid precipitations (by 32% during 1989–1997 in comparison with 1966–1980). In general from 1950 until 2013, glaciation of the Kronotsky Peninsula has retreated by 27.2%. For glaciers with an area less than 0.5 km^2, this parameter was 22.9%.

The twinned-valley glacier Koryto on the Kronotsky Peninsula has been studied the most. According to the data of the study by Yamaguchi et al. (2003), from 1960 to 2000, the terminus of the Koryto Glacier has retreated by 450 m, its surface lowered in general by 30 m, and its volume decreased by 2.4×10^8 m^3. The estimated cumulative mass balance from 1940 to 2000 was −32 m w.e. (−0.5 m w.e. per year) (Glaciation of the North and Central Eurasia in Modern Epoch, 2006).

Volcano Group Klyuchevskaya (Fig. 2.39)

Glaciers of this region are considerably affected by significant volcanic activity; they are sensitive indicators of seismic and volcanic processes reflecting their behavior. Many of them have distinctive features of dynamic instability that are generally related to the considerable volcanic activity (Muraviev et al., 2010). As a result of eruptions, pyroclastic material accumulates on the surface of glaciers, considerable amounts of water accumulate at the border between a glacier and its bed, and seismic activity increases, which in the context of

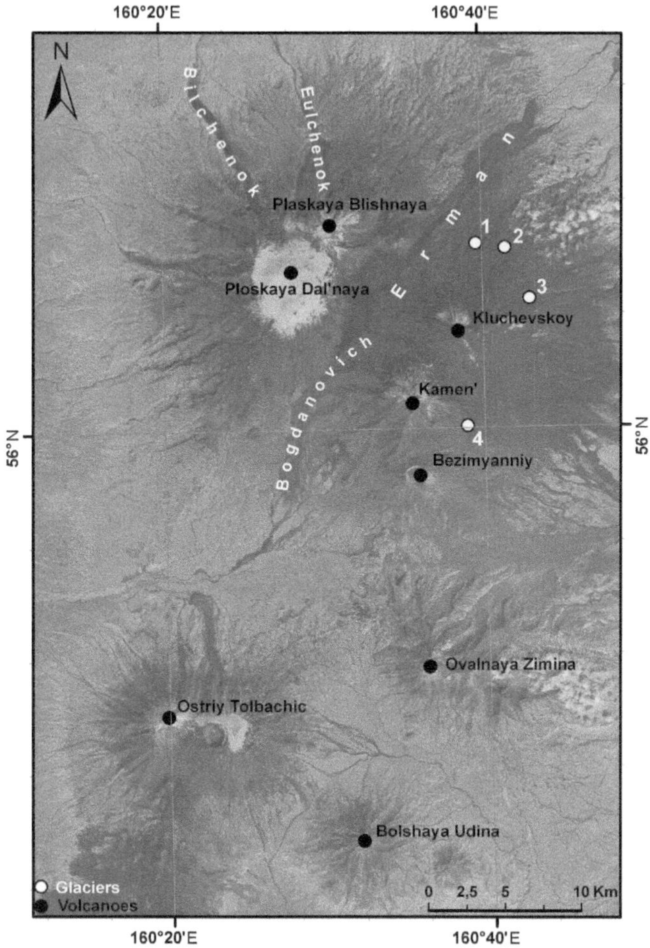

FIGURE 2.39 Kluchevskaya group of volcanoes. Glaciers—(1) Vlodavtsa, (2) Sopochnii, (3) Kellya, and (4) Shmidta.

climatic conditions unfavorable for glaciers of Kamchatka results in surges of the glaciers.

A considerable part of glaciers is being destroyed quite often during volcanic eruptions. The adventive eruption of the Klyuchevskoy volcano in 1974 was accompanied by destruction of a part of the accumulation area of Bogdanovich Glacier by a lava flow and due to formation of scoria cone (Vinogradov et al., 1977). During the great Tolbachinsky fissure eruption in 1975–1976, as a result of subsidence and collapse of the inner caldera of Plosky Tolbachik, two-thirds of Tolbachinsky glacier was destroyed (Vinogradov and Muraviev, 1982). As a result its area has decreased from 1.54 to 0.5 km^2.

FIGURE 2.40 Glacier Erman front in river Sukhaya valley. August 2015. *Photo by A. Ya. Muraviev.*

The advance of the Erman Glacier was initiated in 1945 during the eruption of the Klyuchevskoy volcano after a volcanic-glacial landslide along the Krestovsky defile of Klyuchevskoy volcano with a volume of 0.3 km^3 (Muraviev and Şalamatin, 1993). Further collapses of smaller volumes have been repeatedly observed during volcanic eruptions in the 1980s and 2000s. During all that time, the glacier was constantly advancing (Fig. 2.40).

The analysis of the data of field surveys and comparison of the location of the front of the glacier on the 2009 Landsat satellite image with a plane-table measurement in 1935 have shown that the glacier terminus has descended down to 1000 m (from 1650 m in 1935); during 74 years, the length of the glacier has increased by approximately 3.2 km and its area by 6 km^2 (Muraviev et al., 2010). After adventive eruptions of the Klyuchevskoy volcano in 1953 and 1967–1968, the glaciers of Sopochny and Vlodavetz started to advance and kept advancing until the present (Muraviev et al., 2010). In 2009 the tongue of the Shmidt Glacier (it descends to the southeast from the saddle between Klyuchevskoy and Kamen volcanoes) advanced on the previous surface of the same glacier remaining since the surge in 1978–1987. In the second half of the 20th century, the glacier of Eulchenok that descends to the north from the volcano Krestovsky (Mt. Ploskaya Blizhnaya) has been advancing (Fig. 2.41).

From 1949 to 2000, its front has moved 2.4 km forward. According to the data from the analysis of Landsat satellite images, in 2000–2009, the location of the glacier tongue was quasi-stationary. According to the estimation of the authors (Vinogradov and Muraviev, 1989), from 1938 to 1987, the glacier area of Klyuchevskoy volcano has increased by 4.8 km^2 (3.6%). All known cases of surges or retreats of glaciers on its slopes more or less related either to summit or to adventive eruptions.

FIGURE 2.41 Eulchenok Glacier. August 2015. *Photo by A. Ya. Muraviev.*

One of the most-studied glaciers of the Kluchevskaya group of volcanoes is the caldera-valley glacier Bilchenok (Fig. 2.42) (№ 162, area 21.8 km^2), descending along the northern slope of the Ushkovsky volcano (Mt. Ploskaya Dalnaya). It is one of the largest surging glaciers in Russia (Muraviev et al., 2012). During the second part of the 20th century, two surges of that glacier have been recorded. During a surge in 1959–1960, its terminus has advanced by 2 km (up to 615–630 m). After its surge, the glacier started a rapid retreat. By July 1975, it had retreated by almost 2 km (up to 930–950 m), the surface of its valley part has reduced by 60–80 m, and its area has decreased by 2.6 km^2. The surge in 1982–1984 had a smaller scale—the advance of the glacier front was 700–800 m. After the second surge, the glacier was visited

FIGURE 2.42 Bilchenok Glacier. August 2014. *Photo by A.Ya. Muraviev.*

almost every year. The survey of the glacier in July 2011 has shown a general increase in its surface, a development in areas of crevassing on its valley part, and an advance of its front during the previous year by 20–30 m.

Southeast Kamchatka

Some of the main centers of glaciation in this region are volcanoes of Avacninsky group (Avachinsky, 2741 m; Koryaksky, 3456 m; and Kozelsky, 2189 m), Mutnovsky volcano (2322 m), and Zhupanovsky volcano (2923 m). Volcanic craters, calderas, and explosion cirques form a suitable reservoir for accumulation of snow and ice. Together with a significant amount of solid precipitation, they create favorable conditions for existence of contemporary glaciation. Contemporary glaciation of the volcano group Avacninckaya is located in an area of nearly 21.2 km^2 (Fig. 2.43).

According to the reconstruction described in a research paper (Vinogradov and Muraviev, 1992), in the middle of the 19th century, its area was 35.2 km^2. Since then it has decreased almost 1.7 times. The reference glacier, Kozelsky,

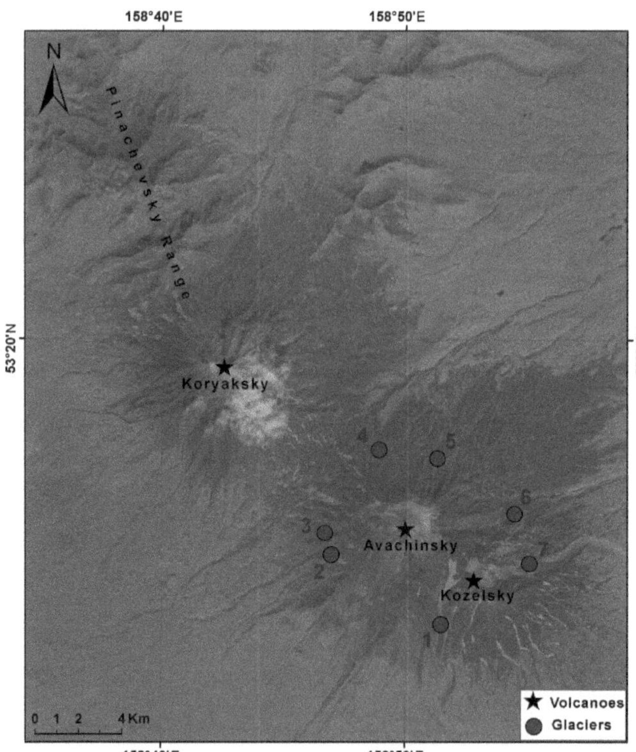

FIGURE 2.43 Avachinskaya group of volcanoes. Glaciers—(1) Kozelsky, (2) Kambalny, (3) Yelizovsky, (4) Novograblennova, (5) Arsenyeva, (6) Zavaritsky, and (7) Dietmar.

one of the most extensively studied glaciers in Kamchatka, is located in this region. The glacier retreated in 1909–1933. The rates of recession were small. The average estimated mass balance during that period was 380 mm w.e. per annum. In 1934–1940, Kozelsky Glacier has degraded significantly despite relatively high accumulation (estimated mass balance −1100 mm w.e. per annum). The period 1941–1952 was the most unfavorable for the glacier, with average estimated mass balance of 1470 mm w.e. per annum. The lowest level of humidity that time in the 20th century was recorded. In February 1945 during the Avachninsky volcano eruption, the glacier was covered with pyroclastic debris material of 1 m thickness; thus as a result, its mass balance was equal to zero for the next 15 years. Further, a "new" glacier has begun its formation above the layer of volcanic deposits. By 1967, its size had reached 0.7 km^2, while during 1968–1976 when the climatic conditions were favorable, the accumulation area of the glacier was completely restored. Snowy winters coincided with cold summer seasons resulted in glacier advance by 330 m in 1971–1976, while its area increased by 0.09 km^2 (5.3%). The period 1978–1981 was stable and only a slight retreat of the glacier occurred (Vinogradov and Muraviev, 1992). Then until the recent years, the glacier tongue covered with a thick (up to 1.5–2 m) layer of volcanogenic material (Fig. 2.44) and was advancing. From 1977 to 2004, the Kozelsky Glacier was advanced by 250 m (Glaciation of the North and Central Eurasia in Modern Epoch, 2006).

FIGURE 2.44 Kozelsky Glacier. August 2015. *Photo by A.Ya. Muraviev.*

A field survey of a number of glaciers descending from the Avacninsky and Kozelsky volcanoes was conducted in August 2011. Compared to the 2007 ASTER satellite images, the analysis of GPS survey of glaciers termini has shown that during the past 4 years, the front of Novograblenny Glacier has advanced approximately by 40 m. The Kambalny Glacier has advanced as well (Fig. 2.45). The glaciers Arcenyev, Zavaritsky, and Ditmar were in quasi-stable condition or slowly retreated. The Yelizarovsky Glacier ends with a mass of dead ice dissected by water steams and has all the indications of degradation. All glacier tongues on Avachinsky and Kozelsky volcanoes are densely covered with volcanic debris. Therefore, despite unfavorable climatic conditions for glaciers, their melting is restricted.

Crater glaciers on the Mutnovsky volcano have retreated considerably during the eruption in 1945. After the eruption they started regeneration. By 1980, the area of Severo-Vostochny Glacier had increased by 0.24 km^2 and Yugo-Zapadny by 0.16 km^2 (Muraviev, 1985). In total, the area of glaciation in this region for the given period has increased by 17%. A considerable amount of solid precipitation occurred in the area of the Mutnovsky volcano. Accumulation in crater glaciers during 1979–1984 has been 2900–3000 mm w.e. Temperatures in craters were on average 2°C lower than that within surrounding areas. This might be due to the appearance of gas–vapor cloud in the atmosphere above the crater that has considerably reduced incoming solar radiation on glacier surfaces. A positive ice balance has occurred as a result, despite intensive melting of glacier tongues advancing at thermal areas in the craters.

Kropotkin Glacier, located in the northern cirque of Bolshoy Semyachik volcano (Fig. 2.36), together with its moraine complex was surveyed in 2001–2002 (Golub and Muraviev, 2005). The glacier reached its largest sizes

FIGURE 2.45 Kambalny Glacier. August 2012. *Photo by A.Ya. Muraviev.*

during the Little Ice Age and smallest size at the end of 1940s and 1990s. The retreat of Kropotkin Glacier in the first half of the 20th century has changed for glacier advance in the second part of 1960s—beginning of the 1970s. Summer air temperature during that period was 10.3°CC (average annual 10.5°C) and precipitation of the cold season 945 mm on average (20% above normal). During the next 25 years, the glacier front retreated by 100 m. A formation of contemporary moraine started at the end of the 1990s. In the 20th century, 23–28 years periods of changes in the glacier regime have been recorded that were particularly connected with changes in accumulation. A similar precipitation cycle during the cold season of the year has been recorded for a number of meteorological stations in Kamchatka (Glazyrin, 1985). A reconstructed range of components of the mass balance of the Kropotkin Glacier has been compared with similar ranges for other glaciers in the eastern coast of Kamchatka (Golub and Muraviev, 2005). The trend of mass balance of Kozelsky Glacier was similar to the one of Kropotkin Glacier (Fig. 2.46), whereas the trend of the mass balance of the Koryto Glacier was rather different. The Kozelsky Glacier is considerably influenced by eruptions of the Avachinsky volcano that makes it different from the other two.

Fluctuations of glaciers in the southeast of Kamchatka are rather heterogeneous. Contemporary climate conditions do not contribute to the glaciation in the area. Some glaciers retreat or maintain a quasi-stationary condition (e.g., the glaciers of Elizovsky, Arsenyev, Zavaritsky, and Ditmar). Other glaciers (e.g., Kozelsky, Kambalny, and Novograblenny) commonly advance due to contemporary volcanic activity.

In general, changes in glaciation in Kamchatka vary. Glaciation of the nonvolcanic areas retreats. The rate of recession increases in a direction from the northwest to southeast (16.6% in the northern part of Sredinny Ridge, 19.2% in the Alney-Chashakonja massif, and 22.9% in Kronotsky Peninsula). In areas

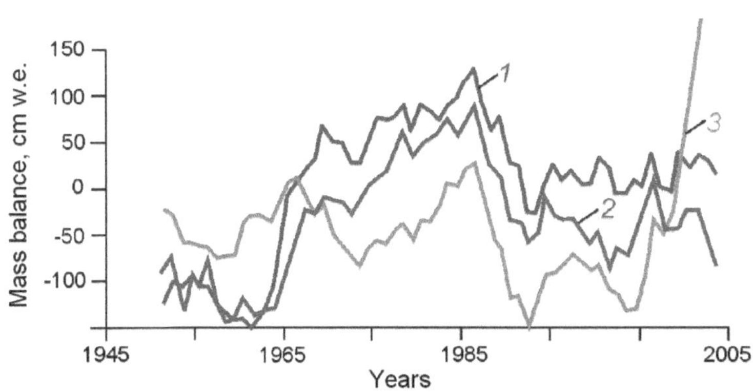

FIGURE 2.46 The glacier mass balance. East coast of Kamchatka Peninsula (running smoothing over 7 years); (1) Kropotkin glacier, (2) Kozelskiy glacier, (3) Koryto glacier.

of contemporary volcanic activity, fluctuations of glaciers are rather diversified. At present, the definition of proportion of volcanic and climatic factors of those fluctuation factors seems to be impossible.

THE TREND IN DYNAMICS OF MOUNTAIN GLACIERS IN RUSSIA

The results of studies of glacier dynamics in the mountain areas of Russia are shown in Table 2.20 and Fig. 2.47. The rate of glacier recession in the Caucasus for a 100-year period is 45.16%; moreover, during the period from the first (Podozersky, 1911) to the second cataloging, based mainly on the data of the middle of the 20th century (Glacier Inventory of the USSR, 1965–1982), the area of glaciers has decreased by 27.46% and during the second part of the 20th century by 17.7% (Khromova et al., 2014). The pattern of glacier retreat in other studied regions is almost the same: during the second half of the 20th century the glaciers of Severo- and Yuzhno-Chuisky Ranges in the Altai have decreased by 24.4% (Shahgedanova et al., 2010), in Katunsky Range by 28.3% (Nosenko et al., 2014), in the Polar Urals by 22.3% (Shahgedanova et al., 2012), in Kamchatka from 16% to 22.9% (Muraviev and Nosenko, 2013; Muraviev, 2014), and in Suntar-Khayata Mountains by 26.8% (Galanin et al., 2013). Obviously, these estimations are not absolute. For example, when using the data of actual aerial photographs from 1945 for comparison, the value of decrease in the glacier area of the Suntar-Khayata Mountains happens to be higher, up to 36%. Similarly for the Kodar glaciers, the value of decrease in glacier area was 35.8% (Stokes et al., 2013). In every case, the evaluation of the accuracy of initial data and the extent of measurement error is important. Our studies have shown that with the use of satellite information, the precision of initial data varies from 1.6% (the Caucasus) to 5.5% (the Altai). This allows the validation of obtained results. They confirm the general trend of glacial retreat on the territory of Russia from the middle of the 20th century to the present that probably has continued the general tendency of deglaciation during the 20th century.

Regular imaging of glacier regions and analysis of satellite imagery within the International Project GLIMS (Global Land Ice Measurements from Space) began in 2000. Thus, it has given an opportunity to assess changes in glaciers during the past 15 years based on compatible data—ASTER images with 15–30 m resolution and in certain instances, satellite images of higher resolution (WorldView, CORONA).

For 60 glaciers in the Altai covered by three satellite images from different time periods, we have estimated glacier areas in 1968, 2008, and 2012 and received 145.43, 122.29, and 118.33 km^2, respectively (Nosenko et al., 2014). In general, during 1968–2008, those glaciers have decreased by 15.9% and during the past 4 years (2008–2012) by an additional 2.7%. Thus, during 40 years (from 1968 to 2008), the area of glaciers was reducing by 0.58 km^2

TABLE 2.20 Glacier Area Shrinkage in the Mountain Regions of Russia[a]

			Data From the Glacier Inventory of the USSR		Glaciers Studied							
							Area Shrinkage					
	No	Glacier System	Number	Area, km^2	Number	Area, km^2	km^2	%	%/yr	Period of Observations, yr	Data	Published Sources
Atlantic Glacier Province	1	Khibiny	4	0.1								
	2	Urals	143	28.66	30	9.17	2.4	22.3	0.47	1953/60–2000	Glacier Inventory of the USSR—ASTER July 14, 2000	Shahgedanova et al. (2012)
	3	Putorana Mts.	22	2.54	67	7.18					Landsat_7 (August 1999)	Sarana (2005)
	4	Byrranga Mts.	96	30.5	56			17	0.47	1967–2003	Glacier Inventory of the USSR—Landsat 2001, 2001, 2003	Ananicheva et al. (2010)
	5	Orulgan ridge	74	18.38	52			60	1.5	1970–2010	Glacier Inventory of the USSR ASTER 2010	Ananicheva and Karpachevsky (2015)
	6	Caucasus	2080	1427.12	1329 (1911)	1967.42 (1911)	540.3	27.46	0.45	1911–1967/79	Podozersky, 1911—The Glacier Inventory of the USSR	Khromova et al. (2014)
					1706 (1999/ 2004)	1174.52 (1999/ 2004)	252.6	17.7	0.49	1967/79–1999/ 2004	Glacier Inventory of the USSR—Landsat ETM+ 12/09/2000, 08/08/1999 ASTER 15/09/2001, 15/08/2004, 25/09/2004	
					478 (2001)	407.3 (2001)	19.2	4.7	0.52	2001–2010	ASTER 15/09/2001- ASTER 29/09/2010	Nosenko et al. (2013)

Continued

TABLE 2.20 Glacier Area Shrinkage in the Mountain Regions of Russia[a]—cont'd

		Data From the Glacier Inventory of the USSR		Glaciers Studied							
						Area Shrinkage					
No	Glacier System	Number	Area, km^2	Number	Area, km^2	km^2	%	%/yr	Period of Observations, yr	Data	Published Sources
7	Altai	1517	923.1	256 Chuya Range	336	82	24.4	0.47	1952–2004	Glacier Inventory of the USSR ASTER 09/10/2004 ISS 09/09/200	Shahgedanova et al. (2010)
				336 Katun Range	289.9	81.9	28.3	0.51	1952–2008	Glacier Inventory of the USSR CORONA 1968, ALOS PRIZM, 2008	Nosenko et al. (2014)
8	Kuznetzk Alatau	91	6.79	6	0.79	0.37	42	1.4	1975–2005	Aerial survey, ground GPS_survey	Kovalenko and Kitov (2011), Shpin (1980), and Kovalenko and Kitov (2011)
9	East Sayan	105	30.3	13	5.1	1.53	36	0.65	1953/58–2001	Glacier Inventory of the USSR Quick Bird 08/2006, WorldView-1 07/2008, 08/2008 Landsat-7ETM+, 08/2001 topographic maps 1:100000 1981 1:50000 1992	Osipov et al. (2013)

	10	Kodar ridge	30	18.8	28	12.6[b]	0.88[b]	17.1[b]	0.53[b]	1963–1995	Glacier Inventory of the USSR—Landsat TM 17/07/1995	Stokes et al. (2013)
					34	11.72[b]	2.19[b]	18.7[b]	3.12[b]	1995–2001	Landsat TM 17/07/1995-Landsat ETM+ 11/07/2001	
							2.52[b]	26.4[b]	2.93[b]	2001–2010	Landsat ETM+ 11/07/2001—Landsat ETM+ 27/07/2010	
	11	Barguzin ridge			7	1.4	0.4	54.4	1.03	1960–2003/2007–2013	Topographic map 1:50000, 1960 Landsat (2003, 2007–2013) WorldView-1 (2008)	Kitov et al. (2014)
	12	Baikal ridge			1	0.45	0.04	9	0.18	1960–2010	Topographic map 1: 50000, 1960 Landsat (2010)	Aleshin (1982) and Kitov et al. (2014)
Pacific Glacier Province	13	Chukotka Upland			47	13.53						Sedov (1988, 1992, 1995, 1996, 1997c)
	14	Koryak Upland	1335	259.5	237	176.6	122.2	69	1.30	1950–2003	Glacier Inventory of the USSR—Landsat 2003	Ananicheva (2009)
		Northeastern Koryak Upland			116	43.96	13.18	30	0.70	1967/85–2008/2010	Sedov' Glacier Inventory Landsat 2008/2010	Sedov (2001) and Ananicheva et al. (2010)
	15	Kolyma Upland			19	3.61						Sedov (1995, 1997c, 2000)

Continued

TABLE 2.20 Glacier Area Shrinkage in the Mountain Regions of Russia[a]—cont'd

		Data From the Glacier Inventory of the USSR		Glaciers Studied							
						Area Shrinkage					
No	Glacier System	Number	Area, km^2	Number	Area, km^2	km^2	%	%/yr	Period of Observations, yr	Data	Published Sources
16	Cherskii Mts.	372	155.4	80	70	14.8	17	0.11	Little Ice Age (ca. AD. 1550–1850)–2001	Landsat 7 ETM+ August, 2001	Gurney et al. (2008)
				186	84.2	23.4	28	0.90	1970–2001	Glacier Inventory of the USSR Landsat	Ananicheva et al. (2006)
17	Suntar–Khayata Mts.	208	201.6	195	199.4	37.2	19.3	0.33	1945–2003	Glacier Inventory of the USSR Landsat	Ananicheva et al. (2006)
				87	84.2[b]	22.56[b]	26.8[b]	0.48[b]	1954–2010	Glacier Inventory of the USSR Landsat 2010	Galanin et al. (2013)
				12	34.66[b]	12.74[b]	36.09[b]	0.55[b]	1945–2011	Arial survey—WorldView 2011	
18	Kamchtka	405	874.0	161 the Northern part of the Sredinny range	347.6	57.7	16.6	0.32	1950–2002	Glacier Inventory of the USSR—ASTER August 18, 2002	Muravyev and Nosenko (2013)
				24 the Alnei–Chasha-kondzha massif	60.5	11.62	19.2	0.32	1950–2010	Glacier Inventory of the USSR—World View July 22, 2010	Muravyev (2014)
				29 the Kronotskii Peninsula	82.3	18.84	22.9	0.36	1950–2013	Glacier Inventory of the USSR—Landsat September 2, 2013	

[a] *No data are available for the changes in the glaciers on Khibiny, the Putorana mountain, the Kolyma Upland, and the Chukotka Upland.*
[b] *The data are presented for open glacier surface.*

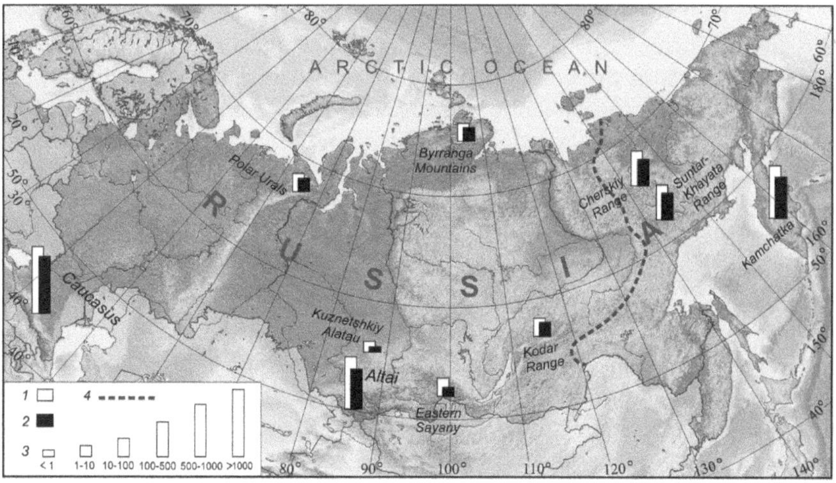

FIGURE 2.47 Glacier area decrease in some mountain regions of Russia from the middle of 20th century (1) to the beginning of the 21st century (2). (3) Glacier area, km^2 and (4) boundaries between the Atlantic and Pacific Glacier provinces.

(0.4%) per year and during the past four years (2008–2012) by 0.99 km^2 (0.68%) per annum. On several glaciers of East Sayan and Kodar (Osipov and Osipova, 2015), an abrupt increase in the rate of glacier area retreat was observed in 2001–2011 in comparison with the previous period of 1850–2001. Until 2002, the rate of the glacial retreat in those regions was almost the same (0.27–0.30% per year), whereas in 2002–2011, it reached 1.27% per year on East Sayan and 3.98% per year in Kodar Ridge. During the past decade, an increase in the rate of reduction of glacier area has been registered in other mountain regions of Eurasia as well as in the Caucasus, Polar Urals, and Suntar-Khayata Mountains.

Thus, the results of our studies demonstrate the reduction of glaciation area in continental Russia that has continued over a 100 years. During the past 50 years, the area of glaciers in different glacier systems has decreased within the range of 17–40% and during the past 10 years from 2.7% to 5%.

CHANGES IN CLIMATIC VARIABLES

The retreat of glacier areas in the territory of Russia took place in consistence with the increase in summer air temperatures (Kotlyakov et al., 2015). In general, during 1950–2010, changes as high as over 2°C were recorded in the south of Europe and in the Aral-Caspian region, and a little lesser change of 1.5–2°C was observed in Mongolia. The peak warming during the past 30 years has been recorded almost in the same regions with over 3°C increase in Mongolia. Summer temperatures in the northern Black Sea region and in the Black Sea coast of the Caucasus have increased by 2–2.5°C. For the most part of the territory—the Far East, East Siberia, and the European part of Russia—an increase in temperature by 1.5°C has been observed, whereas in isolated clusters in the northern Far East, Transbaikal, and in the south and west of the East European Plain, in the Caucasus and Balkans, it was 2–2.5°C. The main regions of glaciation including the Caucasus, Altai, Kamchatka, Polar Urals, and Siberian mountains where significant glacier retreat is being recorded are falling within the areas of temperature rise.

Changes in summer temperature during the past decade (2001–2010), compared to the rather stable period of 1951–1980, in general correspond with the trend during the past 30-years period, although exemptions occur to confirm the features of current trends. First of all, this is a more significant rise in temperature in the south of Western Europe (almost 2°C as in Mongolia) and considerable warming of the north of Western Siberia, up to 1.5°C. These trends in summer temperatures correspond with more intensive glacial area reduction in the Caucasus, Polar Urals, and Altai.

Obviously, winter precipitation that defines glacier accumulation plays a significant role in the life of glaciers (Kotlyakov et al., 2015). Areas where winter precipitation has increased by 20–40% have been identified in Eastern Siberia in the Ob River catchments and in the European territory of Russia in the northern Dvina and Upper Volga catchments. A decrease in precipitation by more than 40% is recorded in the northeast of Eurasia and in the western Black Sea region. In general, during 1989–2001, almost no increase in winter precipitation occurred in glacier areas, and thus winter precipitation did not compensate the impact of summer-temperatures rise in glacier mass balance.

The maximum value of snow thickness can be considered as another index of the level of accumulation (Kotlyakov et al., 2015). In examining the dynamics of winter snow thickness during 1950–2008, it appears that contemporary warming in the greater part of Russia is accompanied by an increase in snow accumulation. It happens because in the greater part of the territory, a rise in winter temperature does not result in significant thaw periods and occurs within negative values of the temperatures. Also, a rise in winter temperatures is accompanied by an intensification of cyclonic circulation in the greater part of Northern Eurasia that results in a recurrence of heavy snowfalls and an increase in the thickness of snow cover that is marked by its highest levels during the winter season.

The highest increase in this parameter is observed in the Sakhalin and Primorye and slightly lower in the east of the European territory of Russia and north of Western Siberia. If the change in maximum thickness of snow cover in 1971–2008 is compared with 1951–1980, it happens to be dramatic—the area with more than a 10% increase stretches from the northeast European territory of Russia to Western Siberia and the north of Eastern Siberia. Biggest changes of over 30% occurred in the Far North and in the south of Western Siberia. In the east of the European territory of Russia and in the Polar Urals, in the south of Khabarovsk Territory, Sakhalin, and Kamchatka, the increase was also significant, by 20%, and it affected the glaciation areas in Kamchatka; however it did not impact the general decrease in glacier area in those regions. Also, the obvious decrease in the highest snow storage has been recorded in glacier areas of the Caucasus, Altai, and Eastern Siberia Mountains that have an apparent influence on increase in the melt rate of glaciers during the past decade.

CONCLUSION

A decrease in the area of contemporary glaciers results in landscape change within a glacial zone, formation of new lakes, and an increase in natural hazards: disastrous mudflows, ice slides, glacier outburst floods, and so forth. The absence or insufficient information on those processes results in socio-economic losses and loss of life. Therefore, it is vital to assess changes in glaciation and their consequences and study how glaciation impacts local and regional water recourses. Change in the area of glaciation, one of the main components of the global climatic system, should be considered while adjusting global climate models and their regional projections.

Remote-sensing data and contemporary GIS technologies allow the definition of glacier boundaries and their usage in subsequent monitoring. The results of glacier research with the use of satellite imagery in different parts of the world, including the territory of Russia, are collected in the database of the International GLIMS Project (https://www.glims.org/).

Our analysis has demonstrated that although the general trend of glacier retreat in mountain regions of Russia is determined by changes in climatic factors, temperature, and precipitation, the diversified rate of glacier retreat in different regions depends upon orographic factors such as elevation of the ranges, roughness, and position of ranges toward moisture-bearing air flows, and many other factors. In a number of regions, the data of annual terrestrial surveys and paleoglaciologic reconstructions indicate a significant mobility of certain glacier fronts when glacier retreats have been interrupted several times within a 100 years with small advances with durations ranging from one to several years (Solomina, 1999; Narozhny and Nikitin, 2003).

Further studies should be focused on improving the methodology of data collection and analysis and creating databases on glacial conditions for modeling and forecasting climate, environmental conditions, and adverse natural events related to the dynamics of glaciers. The actual picture of the size of glaciation would help to adjust parameters of global climatic models and solve the problems of their regional modifications.

ACKNOWLEDGMENT

This paper includes the results of research projects № 01201352478 и № 01201352475 included in the Research Plan of the Institute of Geography, RAS; Programme of Earth Sciences Division, RAS; and International Project GLIMS. Translated by E. Chowdhury.

REFERENCES

Aleshin, G.V., 1982. Modern glaciers and their relief-forming role on the Baikal Range. Geography and Nature Resources 4, 133–136 (in Russian).

Aleshkov, A.N., 1935. Mount Sablya and its glaciers. In: Proceedings of Glacier Expeditions 4. TUEUMS, Leningrad, pp. 56–74 (in Russian).

Ananicheva, M.D., 2009. Glaciological characteristics fields for the ECHAM-4 scenario: the north-eastern Siberia and Kamchatka. Data of Glaciological Studies 107, 81–86 (in Russian).

Ananicheva, M.D., Kapustin, G.A., 2010. Assessment of glaciers change in the Byrranga mountains from satellite images and the Glacier Inventory of the USSR. Ice and Snow 3 (111), 19–26 (in Russian).

Ananicheva, M.D., Kapustin, G.A., Koreysha, M.M., 2006. Glacier changes in Suntar-Khayata mountains and Chersky range from glacier inventory of the USSR and satellite images 2001–2003. Data of Glaciological Studies 101, 163–168 (in Russian).

Ananicheva, M.D., Kapustin, G.A., Mikhailov, A.Y., 2010. Glaciers of the Meynypilgynsky Range: current state and development forecast of glacier systems. Ice and Snow 2 (118), 40–50 (in Russian).

Ananicheva, M.D., Karpachevsky, A., 2015. Glaciers of the Orulgan range: assessment of the current state and possible development for the middle of the 21st century. Environmental Earth Sciences 74 (3), 1985–1995. http://dx.doi.org/10.1007/s12665-015-4605-2.

Ananicheva, M.D., Koreysha, M.M., 2005. Glaciers retreat in north and south of Suntar-Khayata mountains and Chersky Range. Data of Glaciological Studies 99, 18–25 (in Russian).

Ananicheva, M.D., Krenke, A.N., 2007. Mountain glaciation (north-eastern Siberia and Kamchatka as an example). In: Glaciation of North Eurasia in the Recent Past and Nearest Future. Nauka, Moscow, pp. 277–293 (in Russian).

Bashlavin, D.K., 1970. Modern glaciers of Orulgan range. Data of Glaciological Studies 16, 174–178 (in Russian).

Berman, L.L., 1947. Modern glaciation in headwaters of Indigirka river. Geography Questions 4, 33–66 (in Russian).

Chernova, L.P., Khromova, T.E., Zverkova, N.M., Nosenko, G.A., Muraviev, A.Y., 2011. Glaciological provinces of Russia and the belt of minimal snowiness on the continent. Ice and Snow 2 (114), 5–12 (in Russian).

Shanghai. Concise Glacier Inventory, 2008.

Dolgushin, L.D., 1960. Glaciers of the Ural and some features of their evolution. In: Questions of Physical Geography of the Ural, Moscow, pp. 33–60.

Galakhov, V.P., Mukhametov, R.M., 1999. Glaciers of Altai. Nauka, Novosibirsk (in Russian).

Galanin, A.A., Lytkin, V.M., Fedorov, A.N., Kadota, T., 2013. Recession of glaciers in the Suntar-Khayata Mountains and methodological consideration of its assessment. Ice and Snow 53 (4), 30–42. http://dx.doi.org/10.15356/2076-6734-2013-4-30-42 (in Russian).

Ganyushkin, D.A., Chistyakov, K.V., Kunaeva, E.P., Volkov, I.V., Bantsev, D.V., 2016. Current glaciation of the Chikhachev ridge (south-eastern Altai) and its dynamics after maximum of the little ice age. Ice and Snow 56 (1), 29–42. http://dx.doi.org/10.15356/2076-6734-2016-1-29-42 (in Russian).

Glaciation of the North and Central Eurasia in Modern Epoch, 2006. Nauka, Moscow (in Russian).

Glaciation of the Urals, 1966. Nauka, Moscow (in Russian).

Glacier Inventory of the USSR, Vol. 1–20, 1965–1982. Gidrometeoizdat, Moscow, Leningrad (in Russian).

Vinogradov, V.N. (Ed.), 1968. Glacier Inventory of the USSR. Kamchatka, Vol. 20. Chapters 2–4. Gidrometeoizdat, Leningrad (in Russian).
Glazovsky, A.F., Nosenko, G.A., Tsvetkov, D.G., 2005. Glaciers of Urals. Current state and future. Data of Glaciological Studies 98, 207–214 (in Russian).
Glazyrin, G.E., 1985. Distribution and Regime of Mountain Glaciers. Gidrometeoizdat, Leningrad (in Russian).
Golub, N.V., Muraviev, YaD., 2005. Mass balance and Kropotkin glacier fluctuations (Bolshoy Semiachic volcano, eastern Kamchatka) and their relation to climate. Data of Glaciological Studies 99, 26–31 (in Russian).
Govorykha, L.S., 1971. Current state of Byrranga mountain glaciation. Izvestia VGO 103 (6), 510–516 (in Russian).
Grosvald, M.G., 1965. Development of Syan-tuva Upland Relief (Glaciation, Volcanism, Neotectonics). Nauka, Moscow (in Russian).
Gurney, S.D., Popovnin, V.V., Shahgedanova, M., Stokes, C.R., 2008. A glacier inventory for the Buordakh massif, Cherskiy range, northeast Siberia, and evidence for recent glacier recession arctic. Antarctic and Alpine Research 40, 81–88.
Haeberli, W., Huggel, C., Kääb, A., Zgraggen-Oswald, S., Polkvoj, A., Galushkin, I., Zotikov, I., Osokin, N., 2004. The Kolka-Karmadon rock/ice slide of 20 September 2002: an extraordinary event of historical dimensions in North Ossetia, Russian Caucasus. Journal of Glaciology 50 (171), 533–546.
Huggel, C., Zgraggen-Oswald, S., Haeberli, W., Kaab, A., Polkvoi, A., Galushkin, I., Evans, S.G., 2005. The 2002 rock/ice avalanche at Kolka/Karmadon, Russian Caucasus: assessment of extraordinary avalanche formation and mobility, and application of QuickBird satellite imagery. Natural Hazards and Earth System Sciences 5, 173–187.
Kitov, A.D., Kovalenko, S.N., Plyusnin, V.M., 2014. Nival-glacial bodies in the Barguzinsky range. Ice and Snow 54 (1), 48–60. http://dx.doi.org/10.15356/2076-6734-2014-1-48-60 (in Russian).
Khromova, T., Nosenko, G., Kutuzov, S., Muraviev, A., Chernova, L.m, 2014. Glacier area changes in Northern Eurasia. Environmental Research Letters 9 (1), 015003. http://dx.doi.org/10.1088/1748-9326/9/1/015003 (special issue).
Koreysha, M.M., 1963. Modern glaciation of Suntar-Khayata mountains. In:Results of the studies under the IGY program. Glaciology 11. AS USSR, Moscow (in Russian).
Kotlyakov, V.M., Nosenko, G.A., 2001. Progress in the progect GLIMS in Moscow regional center. (RC16). Data of Glaciological Studies 91, 254–256 (in Russian).
Kotlyakov, V.M., Rototaeva, O.V., Nosenko, G.A., Desinov, L.V., Osokin, N.I., Chernov, R.A., 2014. Karmadon Catastrophe: What Happened and What We Should Wait in Future. Codex, Moscow (in Russian).
Kotlyakov, V.M., Khromova, T.E., Nosenko, G.A., Popova, V.V., Chernova, L.P., Murav'ev, A.Ya, Rototaeva, O.V., Nikitin, S.A., Zverkova, N.M., 2015. Recent Glacier Changes in Mountain Regions of Russia. KMK Scientific Press, Moscow (in Russian).
Kotlyakov, V.M., Zi-chu, X., Shu-hong, W., Xin, W., Khromova, T.Y., Nosenko, G.A., 2012. Changing of the Altai glacier system since the mid-twentieth century and its response to the climate warming in future. Ice and Snow 3 (119), 17–24.
Kotlyakov, V.M., Khromova, T.E., Zverkova, N.M., Chernova, L.P., Nosenko, G.A., 2011. Two new glacier systems in Northeastern Eurasia. Doklady Earth Sciences 437, 374–379 (in Russian).
Kovalenko, N.V., 2008. Carrent state of small glaciers in Kuznetsky Alatau. Vestnik MGU. Series Geography 3, 67–71 (in Russian).

Kovalenko, N.V., Kitov, A.D., 2011. Modern glaciers of headwaters of Tompuda river (Barguzinsky range). Vestnik of VSGAO Geography Department 2, 71–88 (in Russian).

Kovalenko, N.V., Kitov, A.D., Sofronov, A.P., 2012. Second glacio-botanic expedition to the North of Barguzin range (headwaters of Verhnaya Akuli and Svetlaya rivers). Vestnik of VSGAO Geography Department 4 (6), 93–97 (in Russian).

Krenke, A.N., 1982. Mass Exchange in Glacier Systems on the Territory of the USSR. Gidrometeoizdat, Leningrad (in Russian).

Muravjev, A.Y., 2014. Glacier size changes in Kronotsky Peninsula and Alney-Chashakondzha Massif, Kamchatka Peninsula in the second half of XX century and the beginning of XXI century. Ice and Snow 54 (2), 22–28. http://dx.doi.org/10.15356/2076-6734-2014-2-22-28 (in Russian).

Muraviev, A.Y., Nosenko, G.A., 2013. Glaciation change in the northern part of the middle range on the Kamchatka Peninsula in the second half of the XX century. Ice and Snow 53 (2), 5–11. http://dx.doi.org/10.15356/2076-6734-2013-2-5-11 (in Russian).

Muraviev, YaD., 1985. Features of crater glaciers regime of Mutnovsky volcano. Volcanic Studies in Kamchatka 1, 74–78 (in Russian).

Muraviev, YaD., Muraviev, A.Y., Osipova, G.B., 2010. Glacier surges in the areas of active volcanism. In: Proceedings of the XV Glaciological Symposia. Kazan, Russia, p. 23 (in Russian).

Muraviev, YaD., Salamatin, A.N., 1993. Projection of glacier ice dynamics in volcano tectonic trenches of Klyuchevskoy volcano. Volcanology and Seismology 4, 43–53 (in Russian).

Muraviev, YaD., Tsvetkov, D.G., Muraviev, A.Y., Osipova, G.B., 2012. Dynamics of the Bilchenok surging glacier in the Klyuchevskaya volcano group. Ice and Snow 2 (118), 31–39. http://dx.doi.org/10.15356/2076-6734-2012-2 (in Russian).

Narozhny, Yu K., Nikitin, S.A., 2003. Recent glaciations of Altay at the beginning of 21th century. Data of Glaciological Studies 95, 93–101 (in Russian).

Nikitin, S.A., 2009. Regularities in the glacial ice distribution in the Russian Altai, storage and dynamics assessment. Data of Glaciological Studies 107, 87–96 (in Russian).

Nikitin, S.A., Vesnin, A.V., Osipov, A.V., 2005. Echo sounding results for surging glaciers in Caucasus and Pamirs. Data of Glaciological Studies 99, 151–153 (in Russian).

Nosenko, G.A., Khromova, T.E., Muraviev, A.Ya, Narozhny, Yu K., Shakhgedanova, M.V., 2010. Application of historical data and modern space images for estimation of glacier size changes in the Altai. Ice and Snow 2 (118), 19–24 (in Russian).

Nosenko, G.A., Khromova, T.E., Rototaeva, O.V., Shakhgedanova, M.V., 2013. Glacier reaction to temperature and precipitation change in Central Caucasus, 2001–2010. Ice and Snow 53 (1), 26–33. http://dx.doi.org/10.15356/2076-6734-2013-1-26-33 (in Russian).

Nosenko, G.A., Nikitin, S.A., Khromova, T.E., 2014. Glacier area and volume changes in the Mountain Altai (Russia) since the mid-twentieth century from space imagery data. Ice and Snow 54 (2), 5–13. http://dx.doi.org/10.15356/2076-6734-2014-2-5-13 (in Russian).

Nosenko, G., Tsvetkov, D., 2003. Assessment of Glaciers Change on Polar Urals from ASTER Imagery. Glaciological Data. NSIDC, pp. 80–82. Report GD-32.

Osipov, E.Y., 2010. Monitoring of contemporary glaciation of Kodar Range using space images Landsat and GIS. In: Proceedings of the International Conference on GIS for Aerospace Monitoring of Natural Hazards, Irkutsk, pp. 15–18 (in Russian).

Osipov, E.Y., Ashmetiev, A.Y., Osipova, O.P., Klevtsov, E.V., 2013. New inventory of glaciers in southeastern part of the Eastern Sayan Mountains. Ice and Snow 53 (3), 45–54. http://dx.doi.org/10.15356/2076-6734-2013-3-45-54 (in Russian).

Osipov, E.Y., Osipova, O.P., 2015. Dynamics of mountain glaciation in the South-Eastern Siberia over the past 160 years. Ice and Snow 55 (2), 33–41. http://dx.doi.org/10.15356/2076-6734-2015-2-33-41 (in Russian).

Panagiotopoulos, F., Shahgedanova, M., Stephenson, D.B., Hannachi, A., 2005. Observed trends and teleconnections of the Siberian high, a recently declining centre of action. Journal of Climate 18, 1411–1422.

Panov, V.D., 1993. Evolution of Modern Glaciation of Caucasus. Gidrometeoizdat, St. Peterburg (in Russian).

Panov, V.D., Ilichev, Y.G., Lurie, P.M., 2002. Ice collapse in the mountains of north Osetia in 2002. Meteorology and Hydrology 12, 94–98 (in Russian).

Perov, V.F., 1968. Snow Packs, Glaciers and Permafrost Relief of Khibiny Mountains. In: Results of Studies under International Geophysical Progects, Glaciology, vol. 22. Nauka, Moscow (in Russian).

Plastinin, L.A., 1998. Remote Sensing and Cartographic Studies of Glacio-nival Complex in Mountain Regions of Siberia (Morphology and Dynamics of Glaciers, Snow Packs and Aufeises of the Kodar Range in Zabaikalye). Irkutsk (in Russian).

Plusnin, V.M., 2003. Landscape Analysis of Mountain Areas. Institute of Geography SBRAS, Irkutsk (in Russian).

Plusnin, V.M., Drozdova, O.V., Kitov, A.D., Kovalenko, S.N., 2008. Dynamics of mountain geosystems of southern Siberia. Geography and Natural Resources 2, 5–13 (in Russian).

Podozersky, K., 1911. Glaciers of Caucasus (Inventory of Caucasus Glaciers). Zap. KORGO, Tiflis (in Russian).

Popov, Y.N., 1948. Area of contemporary glaciation on the North-East of USSR. Izvestiya VGO 80 (2), 182–183 (in Russian).

Popov, Y.N., 1947. About modern glaciation of north-east of Asia in connection with former glaciation. Izvestiya VGO 79 (3), 280–287 (in Russian).

Rototaeva, O.V., Kotlyakov, V.M., Nosenko, G.A., Khmelevskoy, I.F., Chernov, P.A., 2005. Historical information on surges of pulsing glaciers in the Northern Caucasus and Karmadon catastrophe. 2002y. Data of Glaciological Studies 98, 136–145 (in Russian).

Rototaeva, O.V., Nosenko, G.A., Khmelevskoy, I.F., 2009. Variation of the factors that determine the dynamics of Elbrus glaciers. Data of Glaciological Studies 107, 57–67 (in Russian).

Rototaeva, O.V., Nosenko, G.A., Khmelevskoy, I.F., Tarasova, L.N., 2003. Balance state of Garabashi Glacier (Elbrus) in the eighties and nineties of the 20th century. Data of Glaciological Studies 95, 111–121 (in Russian).

Sarana, V.A. Glaciological notes on the data of eastern Taimyr expedition in 2007y. Field studies in the period of International Polar Year. Arctic. Cryosphere. (in Russian). http://www.polaruni.ru/assets/files/files/1_121.pdf.

Sarana, V.A., 2005. Glaciers of Putorana plato. Data of Glaciological Studies 98, 19–29 (in Russian).

Sarana, V.A., 2004. Current state of glaciers in Putorana plato. Data of Glaciological Studies 96, 218–224 (in Russian).

Sedov, R.V., 2000. Mass balance of glacier accumulation area in the mountains of Okhotsk coast. Data of Glaciological studies 89, 206–208.

Sedov, R.V., 2001. Glacier inventory of the North-East part of Koryak upland. Data of Glaciological Studies 91, 195–224 (in Russian).

Sedov, R.V., 1997a. Glaciers of Chukotka. Data of Glaciological Studies 82, 213–217 (in Russian).

Sedov, R.V., 1997b. Snow-ice formations of the south Kharaulach and northern Orulgan, Siberia. Data of Glaciological Studies 83, 172–176 (in Russian).

Sedov, R.V., 1997c. Glaciers of the Taigonos peninsula. Data of Glaciological Studies 82, 218–221.

Sedov, R.V., 1996. Glaciers of providence mountain massif. Data of Glaciological Studies 80, 142–145 (in Russian).

Sedov, R.V., 1995. Glaciers and snow patches of the Okhotsk shore mountains. Data of Glaciological Studies 79, 139–144 (in Russian).

Sedov, R.V., 1992. Glaciers of Chantalski range. Data of Glaciological Studies 75, 102–107 (in Russian).

Sedov, R.V., 1990. Corries and glaciers of Tenianyi range. Data of Glaciological Studies 70, 74–78 (in Russian).

Sedov, R.V., 1988. Glaciers of Iskaten' range. Data of Glaciological Studies 62, 129–133 (in Russian).

Shahgedanova, M., Nosenko, G., Bushueva, I., Ivanov, M., 2012. Changes in area and geodetic mass balance of small glaciers, Polar Urals, Russia, 1950–2008. Journal of Glaciology 58 (211), 953–964.

Shahgedanova, M., Nosenko, G., Khromova, T., Muraveyev, A., 2010. Glacier shrinkage and climatic change in the Russian Altai from the mid-20th century: an assessment using remote sensing and PRECIS regional climate model. Journal of Geophysical Research 115, D16107. http://dx.doi.org/10.1029/2009JD012976.

Shahgedanova, M., Nosenko, G., Kutuzov, C., Rototaeva, O., Khromova, T., 2014. Deglaciation of the Caucasus Mountains, Russia/Georgia, in the 21st century observed with ASTER satellite imagery and aerial photography. The Cryosphere 8, 2367–2379. http://dx.doi.org/10.5194/tc-8-2367-2014.

Shahgedanova, M., Popovnin, V., Aleinikov, A., Stokes, C.R., 2011. Geodetic mass balance of Azarova glacier Kodar mountains, Eastern Siberia and its links to observed climatic change. Annals of Glaciology 52 (58), 129–137.

Shpin, P.S., 1980. Glaciation of Kyznetsky Alatau. Nauka, Moscow (in Russian).

Solomina, O.N., 1999. Mountain Glaciations of the Northern Eurasia during the Holocene. Nauchnyi Mir, Moscow (in Russian).

Solomina, O.N., Filatov, E.S., 1998. Changes of mountain glaciers in north-east Russia from maximum of "Little ice age" to the middle of XX century. Izvestia RAS Series Geography 2, 53–63 (in Russian).

Stokes, C.R., Shahgedanova, M., Evans, I.S., Popovnin, V.V., 2013. Accelerated loss of alpine glaciers in the Kodar Mountains, south-eastern Siberia. Global and Planetary Change 101, 82–96.

Surazakov, A.B., Aizen, V.B., Aizen, E.M., Nikitin, S.A., 2007. Glacier changes in the Siberian Altai Mountains, Ob river basin (1952–2006) estimated with high resolution imagery. Environmental Research Letters 2, 045017.

Tielidze, L.G., Gadrani, L., Kumladze, R., 2015. One century record of changes at Nenskra and Nakra river basins glaciers, Causasus Mountains, Georgia. Natural Science 7, 151–157. http://dx.doi.org/10.4236/ns.2015.73017.

Troitsky, L.S., Kemmerikh, A.O., 1966. Glacier Inventory of the USSR. In: North area. Chapter 3. Urals, vol. 3. Gidrometeoizdat, Leningrad, pp. 1–42 (in Russian).

Tronov, B.V., 1925. Glacier Inventory of Altai. Izvestia RGO 57 (2), 107–159 (in Russian).

Tsvetkov, D.G., 1970. 10 years of photo geodetic survey on the Polar Urals glaciers. (Experience of land surface survey and creation of topographic plans of small glaciers with topo maps of glaciers IGRAN and Obrucheva, 1:5000 scale as the appendix). Data of Glaciological Studies 16, 245–257 (in Russian).

Tutubalina, O.V., Chernomorets, S.S., Petrakov, D.A., 2005. Kolka glacier before the crash of 2002: new data. Earth Cryosphere 9 (4), 62–71 (in Russian).

Vaskovsky, A.P., 1955. Modern glaciation of North-East of the USSR. Data of geology and mineral resources of the USSR North-East. Magadan 9, 71–91 (in Russian).

Vinogradov, V.N., Ivanov, B.V., Chirkov, A.M., 1977. Breakthrough side craters in the glacier body. Geography Questions of Kamchatka 7, 31–37 (in Russian).

Vinogradov, V.N., Muraviev, YaD., 1992. Kozelsky Glacier. (Avacha Volcanoes Group). Gidrometeoizdat, Leningrad (in Russian).

Vinogradov, V.N., Muraviev, YaD., 1989. Glaciers development features in the conditions of active volcanism. Data of Glaciological Studies 66, 93–99 (in Russian).

Vinogradov, V.N., Muraviev, YaD., 1982. Variability of modern glaciers in volcanic area of Kamchatka. Data of Glaciological Studies 42, 164–170 (in Russian).

World Atlas of Snow and Ice Resources, 1997 (Moscow).

Yamaguchi, S., Naruse, R., Sugiyama, S., Matsumoto, T., Murav'yev, Y.D., 2003. Initial investigations of dynamics of the maritime Koryto glacier, Kamchatka, Russia. Journal of Glaciology 49 (165), 173–178.

Chapter 3

Inorganic Chemistry in the Mountain Critical Zone: Are the Mountain Water Towers of Contemporary Society Under Threat by Trace Contaminants?

G. Le Roux,[*,§,1] S.V. Hansson[*,§] and A. Claustres[*,§]
*Université de Toulouse; INP, UPS, EcoLab/ENSAT, Castanet Tolosan, France; [§]CNRS, EcoLab, Castanet Tolosan, France
[1]Corresponding author: E-mail: gael.leroux@ensat.fr

Chapter Outline

Mountain Ice and Water. http://dx.doi.org/10.1016/B978-0-444-63787-1.00003-2

INTRODUCTION

It has been established that the water cycle in mountain areas is likely to be impacted by climate change and by local pressure from humans, and if so, the same impacts will apply to the cycles of inorganic contaminants that are intrinsically associated with human activities. The current export and future of inorganic contaminants is related to future changes in hydrology as well as to geochemical processes directed by natural processes and mankind currently at work on mountain watersheds.

Mountains are not only impacted by enhanced deposition of inorganic contaminants but many of them are also the place of past and present mining activities that have increased the level of available contaminants in soils, sediments, and watersheds. Despite environmental improvements such as protocols of the "Clean Air Act," or prohibition of the use of leaded petrol in most countries of the world, pollution from potentially harmful trace elements (PHTEs), such as trace metals, is still an environmental problem. In particular, the energy demand in emerging countries such as China is inducing higher emissions of trace metals due to coal power plants. Other activities, such as the dysfunctional recycling of household electronics or the use of PHTE in nanotechnologies, also induce PHTE environmental emissions. Biogeochemical cycles of metals are still highly disturbed by humans. This has an impact, not only on ecosystem health and human health, but also unexpected effects in the dispersal of these metals. For example, Cziczo et al. (2009) showed that aerosols enriched in anthropogenic lead were better nucleation nuclei than aerosols containing no lead.

The critical zone is defined as the outer Earth's skin, from the treetops down to the maximal extent of groundwater (Brantley et al., 2007). The *mountain critical zone* has specificities including first its high topographic variability. Other particularities (Table 3.1), such a short vegetation growing season or a high snow cover have an impact in mountainous areas, not only on the water cycle but also on the fate of the PHTEs.

WHY GOING INTO THE FIELD TO INVESTIGATE MOUNTAINS

If remote sensing or real-time monitoring tools have been rising sharply for several years and allow for a nearly continuous observation of basic watershed parameters, the study in the field of the atmosphere—soil interface (deposition) and further export of PHTEs in the mountain critical zone is still needed to feed the models and improve spatial and temporal predictions of contaminant availability. This is especially true in remote areas of major research centers and in hard-to-reach areas such as mountains. These are crucial because they are extremely sensitive to environmental change zones: they are anthropically limited, with a short growing season and low warmth. These areas are also

TABLE 3.1 Main Characteristics of the Mountain Temperate Critical Zone for Potentially Harmful Trace Elements (PHTE)

Mountain Critical Zone	Features	Impacts on the Ecogeochemistry
Atmosphere	Orographic phenomena	Higher atmospheric deposition than expected
Weather	High seasonality	Snow deposition, snowmelt, heavy storm rains, and spring floods
Geology	Potential complexity of the geological features, discrete mineralizations, "young terranes"	Potential microscale natural sources of PHTE for the watershed, erodible mining wastes from abandoned ancient mines and potential mines in the future
Topography	High variability, large slopes	Increasing the risk of extreme hydrogeochemical events
Soils	Young soils, thin and fragile soils, presence of potential deep histosoils	Easily erodible, presence of organic matter
Vegetation	Specific mountain-boreal vegetation	Highly impacted by climate variability ("frontier ecosystem")
Forests	A mix between human and natural management – presence of invasive species (spruces)	Soil acidification and impoverishment in limited-resource soils
Grasslands	Grasslands: altitude dependent	Impacted by mining and pastoralism
Headwater catchments	Large incisions, snow dependent, highly seasonal	Erosion and transport of PHTE with suspended matter
Aquatic mountain ecology	High-altitude "white" lakes in the past	Introduction of invasive species, modification of the biogeochemical cycles of PHTEs

subject to specific atmospheric mechanisms compared to the temperate plains. They induce higher atmospheric deposition, but it is more difficult to quantify because few are localized in easily accessible areas (Lovett, 1994; Weathers et al., 2000). The ecological functions and services of these areas are as important as freshwater reserves, as carbon stocks, or by their biological diversity. Up to now, remote sensing and real-time monitoring tools are limited due to the harsh climatic conditions in high mountain areas. The winter season is quite often less monitored due the unstable climatic conditions, with ice, fog, and snow making a hard life for scientific instruments. Shadow areas due to the topography are limiting remote sensing, and also wireless signal transmission.

A need thus exists to go in the field (Fig. 3.1) to verify the validity of online measurements if there are some, or to complete lower altitude measurements by highest points of measurements.

POTENTIALLY HARMFUL TRACE ELEMENTS

At present, anthropogenic fluxes of up to 62 chemical elements surpass their corresponding natural fluxes, i.e., the contribution of human activities to the atmospheric cycles of these elements far exceeds that of natural contributions (Sen and Peucker-Ehrenbrink, 2012). Among these elements many are considered as PHTEs, which hold limited to no known biological function, such as lead (Pb), arsenic (As), antimony (Sb), or mercury (Hg). Combined, the concentration of these elements constitutes less than 0.1% of the Earth's crust. Several studies have clearly shown that the contamination of PHTEs is widespread and that PHTEs can be found even in remote areas that are far distant from contamination sources. For instance, Rosman et al. (1997) showed that Pb contamination, as recorded in Greenland ice cores, can be dated as far back as 500 BC, which is further supported by numerous studies on lake-sediment cores (Cooke and Bindler, 2015). Even in such a remote area as the Arctic, contamination of Pb and Sb has been shown to occur as far back as

FIGURE 3.1 Photo of a winter field trip in the Bassies area, Central Pyrenees, France to investigate trace metal deposition in snow.

Roman times (Gabrielli and Vallelonga, 2015 and references therein). The occurrence and distribution of PHTEs on both geographical and temporal scales are thus widespread, and mountain areas are no exception.

Sources

Natural Sources

PHTEs are naturally present at low concentrations in rocks and in bedrock, and as a result of physical and chemical weathering, they are thus also naturally present in soil and surface waters. Various natural processes therefore enable their dispersal throughout the environment (Colbeck, 2008; Nriagu et al., 1988; Pacyna and Pacyna, 2001), and from a global point of view the following sources can be distinguished:

- *Terrigenous or lithogenic sources*: dispersal from wind erosion of rocks and soils (i.e., >20% of natural derived Cu, Ni, Pb, Sb, and Zn in the environment).
- *Volcanic sources*: dispersal through volcanic activities discharging emission of a significant amount of PHTEs (i.e., ~20% of As, Cd, Cr, Cu, Hg, Ni, Pb, and Sb) up to the stratosphere.
- *Sea spray*: dispersal through suspending marine water droplets that contributes to about 10% of total PHTEs emissions.
- *Biogenic sources*: biomass fires driven either by natural or anthropogenic processes acts as point sources of PHTEs.

Anthropogenic Sources

Anthropogenic sources of PHTEs are mainly due to high-temperature combustion activities resulting in volatilization of trace elements or their release in the form of very fine aerosols (<μm). In the case of erosion or dust emission, without any underlying high-temperature process, emissions tend to be much more localized (i.e., mining activities). Broadly speaking, the activities can be categorized into the following different sources/activities (De Traubenberg et al., 2013; Pacyna and Pacyna, 2001):

- *Energy production by combustion*: the dominating anthropogenic source of PHTEs emission, entailing burning of wood, coal, and oil (As, Cd, Cu, Hg, Ni, Pb, Sb, Se, V, and Zn),
- *The metallurgical industry*: emission of dust near the extraction and point of exploitation, high-temperature processing of ores emit aerosols rich in trace elements (Cd, Cu, Ni, Pb, V, and Zn). The proportion of individual elements emitted in the aerosols depends on the type of ore processed.
- *Other industrial processes*: high-temperature processing and manufacturing (As, Cr, Cu, Ni, Pb, and Zn)
- *Transport*: road traffic (Cd, Cu, Fe, Ni, Pb, and Zn), erosion of brake pads (Cu, and Sb), erosion of train rails (Cu)

- *Waste treatment*: incineration of household waste (As, Cr, Cu, Ni, Pb, Sb, Se, V, and Zn)

The responsibility of humans regarding the dispersion of aerosols and PHTEs by anthropogenic activities is undeniable, but mankind also holds a strong responsibility in the intensification of natural dispersal processes. For example, the burning of biomass, i.e., forest fires, have been naturally occurring as a part of plant development and ecological renewal for more than 400 million years, yet these fires are now mainly driven and controlled by humans (Andreae and Rosenfeld, 2008). Similarly, desertification of semiarid lands and the intensification of agriculture on bare soil also helps to accelerate the natural suspension of dust (Rauch and Pacyna, 2009) and thus also the dispersal of PHTEs.

Transport

The atmospheric boundary layer is located between 0 and 1−2 km altitude above ground level (Fig. 3.2). Below the boundary layer, contaminant transport is dominated by vertical transport and PHTEs bound to fine particles are subjected to turbulence transport. PHTEs could accumulate in this boundary layer if not washed out by rainfall. They could also be transported in the free troposphere where lateral transport is more dominant. Aerosols and related inorganic contaminants can therefore be transported long distances.

The atmosphere thus has a significant role in the dispersion of PHTEs in the environment, whether they occur in particulate form (solid or liquid aerosols), or as gas (e.g., certain forms of Hg). Because long-distance transport is borderless, contaminants can be transported thousands of kilometers before being deposited, sometimes reaching the most remote areas of the globe (Gabrielli and Vallelonga, 2015; Lee et al., 2008b). The transport of particles, and the distance they may travel, depends not only on their size but also on their reactivity.

The reactivity and particle size determine the residence time (τ) of the compounds in the atmosphere and can be divided into:

- *Compounds and chemical speciation with a short residence time ($\tau < 100$ s).* These are commonly the most reactive form of chemical speciation, which tend to deposit fairly quickly after suspension in the air masses. As these particles do not remain airborne for long, deposition therefore occurs in close proximity to the source of emission.
- *Compounds and chemical speciation with a mean residence time (100 s $< \tau < 1$ year).* If hemispherical mixing occurs, these compounds may remain suspended in the air on a timescale of months. This is, for example, the case for very fine aerosols for which sedimentation is not possible, or chemicals of highly reactive speciation (e.g., gaseous oxidized Hg; Holmes et al., 2010).

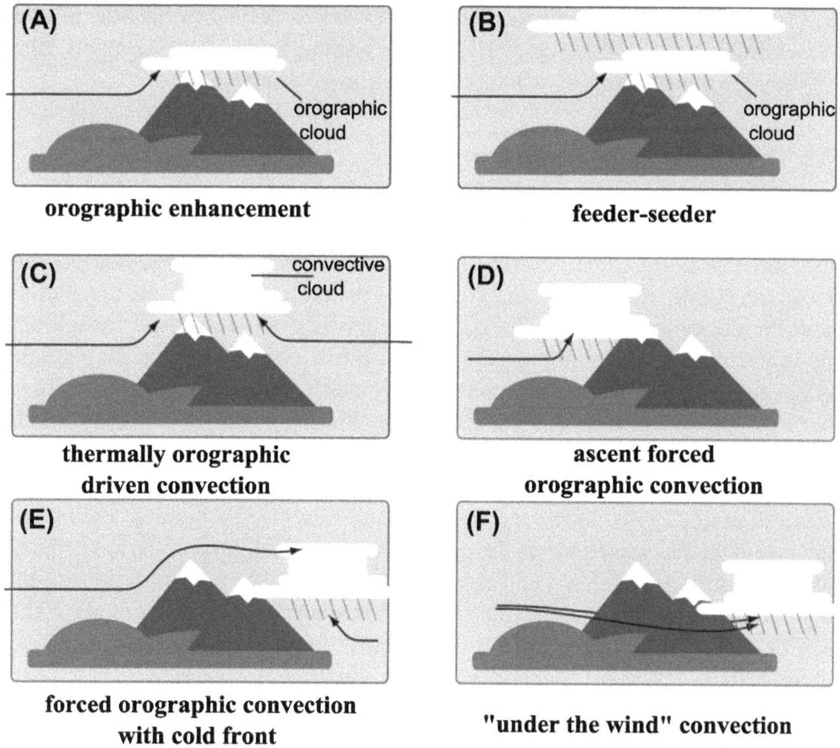

FIGURE 3.2 Schematic representation of different modes of wet precipitation causing differences in precipitation amount and mechanisms of scavenging in mountain areas. Orographic enhancement (A), with "feeder-seeder" effect (B), as well as thermic (C) and forced (D–F) convection leading to rainstorms, are represented.

- *Compounds and chemical speciation with a long residence time ($\tau > 1$ year).* In this case, the interhemispheric mixing becomes possible, as in, for example, the case of gaseous elemental Hg (Holmes et al., 2010), which can thus remain in the atmosphere on a timescale of years.

Deposition

The transport and circulation of PHTEs in the atmosphere, and thus the concentration, can be reduced by deposition which varies depending on weather parameters. Atmospheric deposition that controls the deposited metals onto the ground occurs in the form of:

- *Dry deposition*: aerosols and particle fallout in the absence of precipitation (e.g., rain and snow).
- *Wet deposition*: aerosols and particle fallout driven by precipitation events, either as dissolved in the aqueous phase or in particulate form, physically brought to the ground by the wet precipitation (e.g., rain and snow).

- *Occult deposition*: aerosols and particles are dissolved in aquatic solution with dew and/or mist, i.e., aerosols are present in the water droplets when they are intercepted by a barrier (commonly vegetation).

Dry Deposition

Dry deposition is the process whereby particles are removed from the atmosphere over time due to gravity, and where deposition onto surfaces occurs without involving precipitation (snow, rain, or mist spray; Lovett and Kinsman, 1990; Ruijrok et al., 1995). It essentially depends on the particle size, the concentration, and the reactivity between the deposited particle and the impacted surface (Lindberg et al., 1982). It is considered that the particles with a size of >0.8 μm can easily settle, whereas those with a diameter <0.08 μm tend to track the movement of air masses, and therefore be transported over long distances. The particles of intermediate size (i.e., 0.08–0.8 μm) thus have a behavior influenced by the two deposition processes (Ruijrok et al., 1995). However, due to changes in physiochemical parameters that occur at the interface of the troposphere and the atmospheric boundary layer, the particles with a size between 0.1 and 1 μm do not easily breach this layer and thus have a tendency to remain in suspension (Weathers et al., 2000).

PHTEs are mainly transported in the form of fine particles (Colbeck, 2008) or, in certain cases, as coarse particles originating from desert or local dust sources, and thus not commonly deposited in the form of dry particulates as wet deposition is the dominating process of removal from atmospheric air masses. The exception to this is mercury in its elemental gaseous form (Hg, gaseous elemental mercury) which may correspond to 50% of the deposit onto continental surfaces (Holmes et al., 2010).

Wet Deposition

As opposed to dry deposition, wet deposition is entirely connected to, and controlled by, wet precipitation events (i.e., rain, snow, and hail). Due to their size, which limits their suspension in the air masses and therefore their atmospheric residence time, the majority of aerosols are primarily deposited through wet deposition. However, fine size particles and aerosols that normally would be deposited through dry deposition can also be effectively leached and/or washed out of the atmosphere during wet precipitation events (Montoya-Mayor et al., 2013), and thereby reducing their atmospheric concentration. Such leaching can be divided in two separate phenomena (Bourcier, 2009; Engelmann, 1965; Ishikawa et al., 1995), which include the following:

- *Scavenging in the cloud (within cloud scavenging and/or rainout)* where condensation forms around one or more aerosols that serve as a core (nucleation).

- *Scavenging below the cloud (below cloud scavenging and/or washout)* where particles in the air column below the cloud are incorporated into droplets due to impaction.

These two phenomena may occur either separately or together, as a function of the concentration of aerosol in and under the cloud at the time of precipitation.

The striking topographical changes that occur in mountains interact with precipitation processes, often resulting in precipitation events, thus leading to a higher degree of wet deposition rather than dry. Roe (2005) described the phenomena specific to these environments with pronounced reliefs, which may cause peculiarities in precipitation patterns, as "orographic enhancement" (or orographic increase).

However, the phenomenon of orographic increase can be more complex locally and result in inhomogeneous deposition on a local level. Examples of precipitation phenomena in Fig. 3.2 show the process that may lead to differences in precipitation patterns in mountain areas. The topography can cause the formation of an orographic cloud as pressure decreases when rising air masses are gaining altitude. The formation of an orographic cloud by rising air masses can lead to the following two phenomena (Fig. 3.2):

- Droplets are formed around a condensation nuclei until they reach a level of saturation and precipitate (Fig. 3.2A).
- The orographic cloud droplets are impacted by rainwater from a higher altitude cloud (Fig. 3.2B). This is the "feeder-seeder" effect, which enhances below cloud scavenging and/or washout under the orographic cloud.

In both cases, the deposition will be more intense at altitude than in the adjacent valley.

Other phenomena, although less frequent, can also occur. During calm weather conditions, air masses heated by the sun can rise up the mountainsides from the valley by means of convection and cause localized thunderstorms and rain (Fig. 3.2C). During these events, the cloud is located as in Fig. 3.2A but it is a high-altitude cloud that can wash out several atmospheric layers at the same time. In the case of side convection alone, the stormy cloud can be on the same side (Fig. 3.2D). In the case of winds being lower on the opposite slope, either due to a cold air front (Fig. 3.2E) or due to topographic circumvention (Fig. 3.2F), the cloud could form on the opposite slope as well. More details on orographic processes and modeling are given in Roe (2005).

In these different cases of enhanced deposition, aerosols and the fallout of bound PHTEs are increased locally. Some examples of this, with, PHTE or ^{210}Pb (a natural radionuclide tracer of aerosol), are discussed in the following text.

On the scale of several mountain ranges in France (Alps, Puy-de-Dôme, Corsica), a relationship exists between annual rainfall, altitude, and the amount of fine aerosols that are washed out and deposited. The results obtained by Le Roux et al. (2008) show that the amount of ^{210}Pb (proxy of fine particles) stored in mountain soils increases with altitude and annual rainfall. By using a mass-balance approach, they conclude that 50% of ^{210}Pb is derived due to the leaching from low atmospheric layers, or as occult deposition of aerosols. Measures of bioaccumulated PHTEs in moss show a clear increase with altitude for Pb, Zn, and Cd bioaccumulation along several altitudinal transects in the Alps (Zechmeister, 1995).

Precipitation models predict an increase of 0.4–0.9 mm/m year based on an annual precipitation of 1000 mm (Lovett and Kinsman, 1990; Ollinger et al., 1993). This would lead to a contaminant deposition proportionate to precipitation. However, this is not the case. Examples in Fowler et al. (1998) show that for a 30% increase in rainfall, increased aerosol deposition may be as high as 150% (based on inventories of ^{210}Pb) within an 800 m difference in altitude, which is well above forecasts. The work of Le Roux et al. (2008) highlights the scavenging of aerosols from low atmospheric layers, which potentially could have a higher concentration of aerosols, as an explanation of the observed phenomenon.

In cases where significant sources of PHTEs emissions are local (e.g., occurring only on one side of the mountain) and weather conditions are stable, aerosols can accumulate at the top of the upper level of the atmospheric boundary layer. If the boundary layer is located below the highest levels of the surrounding mountains, the deposition of aerosols and contaminants will not be more intense at higher altitude, due to scavenging from the lower layers of the atmosphere. This is a common occurrence in, for example, California (Munger et al., 1983).

Deposition by Cloud or Fog Interception: Occult Deposition

Another deposition, such as cloud or fog interception, is generally called *occult deposition* because of the difficulties in the ability to accurately measure it using classical instruments such as rain gauges. This type of deposition is intermediate between the dry and the wet deposition. As wet deposition, aerosols are integrated into water droplets, and as dry deposition, the main mechanisms of deposition are driven by turbulent transport, impaction onto surfaces, and sedimentation due to gravity (Lovett and Kinsman, 1990). Cloud or fog interception falls within both of these categories. For instance, studies in different ecosystems of the world have shown the importance of such deposition by cloud or fog interception (Lovett, 1994), where concentrations of ions in rainwater deposited in mountains (Weathers et al., 1988) are one to two degrees of magnitude greater than the ion concentration in rainwater deposited in the forest (Lindberg et al., 1982).

PHTEs Deposition and Accumulation Along Altitudinal Gradients

Artificial Radionuclides

Artificial radionuclides, for example ^{137}Cs, plutonium isotopes, or ^{241}Am, are specific contaminants as they are radioactive and emitters of alpha, beta, and gamma particles that all could affect biological processes (Fig. 3.3). Artificial radionuclides are produced by various human activities, including the following:

- Locally around nuclear reactors and nuclear waste-treatment plants,
- Broadly in the past by the nuclear weapon tests (NWT) and the fallout of a Soviet satellite,
- Point sources in the form of accidents in nuclear plants.

Among the radionuclides that are important from an environmental point of view, ^{137}Cs is the one that has a very similar behavior to many inorganic contaminants: It is associated with aerosols in the atmosphere and its greater fallout in mountain areas is due to similar air mechanisms as inorganic contaminants. Because of the low amount of potential anthropogenic sources, ^{137}Cs behavior at the atmosphere–soil interface in mountains is easier to interpret than other PHTEs. ^{137}Cs has also a long half-life (30 years) that makes its impact measurable several years after its deposition.

It has been shown that for NWT and Chernobyl-derived ^{137}Cs, this radionuclide is enhanced by different atmospheric processes:

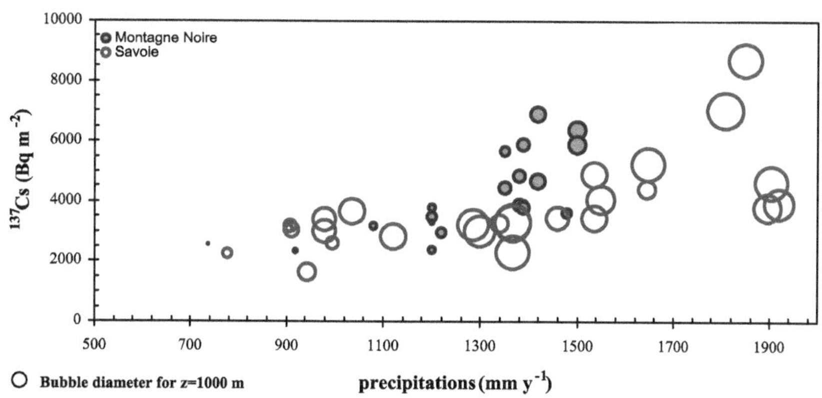

FIGURE 3.3 Bubble chart showing the relationship between ^{137}Cs total soil inventories and annual precipitations in Savoie (*empty red circles* (gray in print versions)), in Montagne Noire (*full blue circles* (dark gray in print versions)) with altitude as a third parameter corresponding to the diameter of the bubble. *Data from Le Roux, G., Duffa, C., Vray, F., Renaud, P., 2010. Deposition of artificial radionuclides from atmospheric Nuclear Weapon Tests estimated by soil inventories in French areas low-impacted by Chernobyl. Journal of Environmental Radioactivity 101, 211–218. http://dx.doi.org/10.1016/j.jenvrad.2009.10.010.*

1. A larger probability that radionuclides transported by air mass will be washed out by rain during orographic events,
2. Orographic deposition is enhanced by processes such as the feeder-seeder effect or fog interception (Le Roux et al., 2010; Fig. 3.4),
3. Enhanced in the watershed by retention in different mountain substrates such as wetlands or forest soils (Pourcelot et al., 2003). Similar conclusions can also be drawn for americium and plutonium isotopes that were transported in the atmosphere.

The accident at Fukushima Daiichi Nuclear Power Plant (FDNPP), Japan, released micron-sized, radioactive particles to the atmosphere. These particles contain different artificial radionuclides including ^{137}Cs. Hososhima and Kaneyasu (2015) showed that enhanced radioactive deposition in the Nikko Mountain National Park, connected to the nuclear plant accident at FDNPP, is related to the altitude and that it is surely also caused by orographic-deposition phenomena. A similar deposition pattern, connected to altitude, has also been shown in the United States. For example, an enhanced transfer of ^{137}Cs and ^{90}Sr from NWT to fish (i.e., trout) at high-altitude (>2500 m) Colorado sites compared to low-altitude sites was shown by Whicker et al. (1972). Persistency of ^{137}Cs can be relatively high as demonstrated by a long-term study of a Norwegian mountain arctic lake (Brittain and Gjerseth, 2010).

PHTEs and Altitudinal Dependency

The different orographic deposition processes can enhance long-range transported contaminant deposition in mountains as previously seen for artificial radionuclides or as evidenced using ^{210}Pb inventories in altitudinal gradients (Fowler et al., 1998; Le Roux et al., 2008). Zechmeister (1995) showed that PHTE concentrations (Pb, Cd, and Zn) in lichens in the Alps are correlated

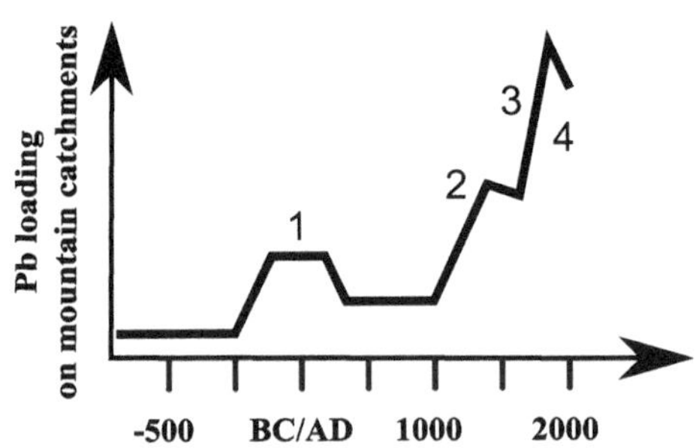

FIGURE 3.4 Typical Pb loading on mountainous environment that occurs in Western European and Mediterranean mountains. (1) Antiquity, (2) Middle Ages, (3) Industrial Revolutions, and (4) use of leaded gasoline and ban of its use in the 1990s.

with altitude. Based on a similar approach using mosses, Gerdol and Bragazza (2006) found a different pattern with higher concentrations at mid-altitude that could be related to higher occult deposition. Altitudinal dependency of PHTE inventories in soils is not obligatorily found in snow precipitations (Bacardit and Camarero, 2010; Yeo and Langley-Turnbaugh, 2010). This can reflect the influence of other parameters (air mass origin, vegetation interception, etc.) on short-time deposition over long-term orographic phenomenon. This emphasizes the need of long-term studies of precipitation along altitudinal transects to reconcile results from soils and mosses with results from orographic precipitations and occult deposition. Orographic phenomenon, such as the feeder-seeder effect or occult deposition, could enhance PHTE deposition, stock and release in mountain soils, wetlands, and sediments due to higher scavenging of aerosols-carrying anthropogenic PHTEs. It has been suggested that additionally to vegetation interception and enhanced orographic aerosol deposition, PHTE altitudinal soil accumulation (Ag, Cd, Pb, and Sb) could also be due to cold trapping on the higher Tibetan plateau (Bing et al., 2016).

Mercury is one of the most investigated PHTEs, but a fraction of Hg is present in a gaseous form in the atmosphere making this PHTE different from metals and metalloids such as Pb or Sb. It has been largely investigated in Arctic areas, and the mercury biogeochemical cycle in the Arctic shares some similarities with the Hg cycle in mountain areas. Hg soil accumulation increases with the elevation (and decreasing temperature) in mountainous areas (Zhang et al., 2013) due to cold trapping of gaseous mercury. Another driver of terrestrial retention of Hg includes air–vegetation gas exchanges (Demers et al., 2013; Enrico et al., 2016). These drivers are interrelated and could be difficult to distinguish, especially in an elevation gradient followed by a vegetation change (Blackwell and Driscoll, 2015).

LEGACY POLLUTION IN MOUNTAIN ENVIRONMENTS

Although we largely associate impacts from anthropogenic activities on the surrounding environment with the industrial era and its advancements in technology, the emission and spread of pollutants has a millennial-scale history in many parts of the globe, characteristically leaving a lasting imprint on the environment. Pollution from preindustrial times caused by mining activities such as ore exploitation and processing, when there were few if any environmental controls, could be substantial (Brännvall et al., 1999; Cortizas et al., 2002; Monna et al., 2004). As mining also entailed widespread land use that included forest disturbance and agriculture, the small-scale but generally widespread mining activities thus amounted to a geographically large-scale impact on the landscape (Bindler et al., 2012; Jouffroy-Bapicot et al., 2007; Monna et al., 2004). It is especially the case in mountain areas where geological features made metallic ores easily available for preindustrial miners.

Environmental Archives: Example of Pb Pollution

Cores derived from natural archives, such as ice cores, lake-sediment cores, and peat cores, allow us to study the aspects of past environmental conditions, such as the net accumulation of elements, deposition, and atmospheric cycling, on time-scales much longer than the information available from contemporary monitoring programs or written records. The magnitude of historical PHTE deposition, and its changes over time, can therefore be reconstructed using such natural archives.

An example of this is the reconstruction of Pb pollution and its global dimension through history. Initially shown by the groundbreaking work of Clair Patterson et al. (Murozumi et al., 1969; Settle and Patterson, 1980; Shirahata et al., 1980), then further refined by Jerome Nriagu (1996), the pollution legacy of Pb has now been reproduced in numerous publications on a world wide scale using both ice and snow records (Gabrielli and Vallelonga, 2015; Hong et al., 1994; Rosman et al., 1997), lake-sediment records (Abbott and Wolfe, 2003; Cooke and Bindler, 2015; Lee et al., 2008a; Renberg et al., 1994), and peat records (Aaby and Jacobsen, 1978; Bao et al., 2015; De Vleeschouwer et al., 2010 and references therein; Jensen, 1997; Lee and Tallis, 1973; Shotyk et al., 1998; Weiss et al., 1999).

Studies of trace metal pollution using natural archives have not only reconstructed contamination history but, more importantly, also have been able to shown that preindustrial pollution and its impact on the environment on a local-scale commonly surpasses the impact of contemporary pollution (Forel et al., 2010; Le Roux et al., 2005; Monna et al., 2004; Weiss et al., 1999). Specifically, reconstructed contamination records obtained in mountain areas, linked to nearby ancient metallurgical activities, show that the preindustrial contribution could exceed that of the last 200 years, especially for Pb and Sb (Camarero et al., 1998; Forel et al., 2010; Le Roux et al., 2005).

Fig. 3.4 shows the general chronological pattern of past Pb loads in Western/Southern Europe. This pattern is highly variable, and especially so in mountain areas, due to local activities such as mining or metallurgical workshops; yet it is commonly found in environmental records from, for example, the Pyrenees and the Alps. On other continents, the PHTE load in the past is less investigated (Marx et al., 2016); however, some chronologies of pollution patterns that follow the developments of large civilizations can also be found. Some examples of this are the increasing pollution in the Andes during the Incas period (Cooke and Bindler, 2015; De Vleeschouwer et al., 2014) or in Central China during the early Han dynasty (Lee et al., 2008a) or Southern China during the Yuan period(Hillman et al., 2015).

Reconstructions of Other Trace Elements

Publication records show that in terms of pollution reconstructions, Pb is by far the most well-documented trace element, much ado to the assumption of Pb being relatively stable (Shotyk et al., 1997) and thus yielding a well-preserved signal stored in natural archives (Catalan, 2015; Catalan et al., 2013; Hansson et al., 2015; Shotyk and Le Roux, 2005). However, Pb is not the only trace element that has been successfully reconstructed using environmental archives. For example, other studies using peat and lakes as archives have shown the pollution record of As, Cu, Cr, Hg, Ni, Sb, and Zn (Catalan, 2015; Hansson et al., 2015 and references therein). Yet a comprehensive reconstruction of trace elements other than Pb can still be considered restricted, based on the limited geographical coverage. The majority of studies on pollution history have focused on European sites, and even fewer on mountainous areas. A need for further investigations of PHTEs on both spatial and temporal scale, and in mountain areas specifically, still resides (Catalan and Rondón, 2016).

ECOLOGICAL IMPACT OF PHTES IN THE MOUNTAIN ENVIRONMENT

Release of Inorganic Pollutants Into the Mountain Watersheds

One key aspect of mountains is the presence of buffer zones that include organic soils, forests, and wetlands (Bacardit et al., 2012; Gandois et al., 2010; González et al., 2006; Fig. 3.5). Not only do these buffer zones retain and store contaminants, but their alteration or destruction can lead to profound disturbances in the biogeochemical functioning of the mountain critical zone. As identified by Gurung et al. (2012), the current global change research in mountains focus mainly on studying climate change and its impact on mountain ecosystems. Yet, there is an urgent need to understand additional drivers (such as local human activities like mining or hydroelectricity) and disentangle their cascading impacts on, and disturbances of, mountain ecosystems. Gurung et al. also stated that to detect such impacts and attribute them to specific drivers of change, suitable indicators need to be identified. Long-term monitoring is therefore necessary both to understand which drivers are concretely involved and to quantify the associated consequences on biogeochemical processes in the mountain critical zone. Few critical zone observatories (CZO) in Northern America and Europe are located in mountain environments (i.e., Southern Sierra CZO in Unites States and Strengbach-Vosges Mountains in France) and even fewer are actively monitoring and investigating ecogeochemical processes in mountain watersheds. This is despite the fact that many studies have demonstrated a mining PHTE legacy in remote areas and their resulting current impact on the environment (Bindler et al., 2012; Camizuli et al., 2014; Mariet et al., 2016; Monna et al., 2011). Additionally, mountain areas can be geologically enriched in PHTEs (i.e., As; González et al., 2006; Zaharescu et al., 2009), and these PHTEs can be remobilized and released in the watershed due to environmental modifications such as melting of glaciers or dehydration of wetlands.

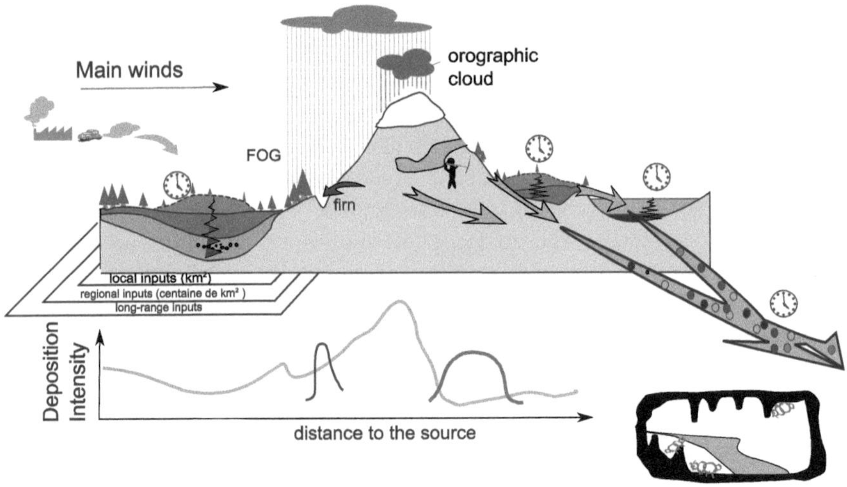

FIGURE 3.5 Atmosphere–soil–water in mountain ecosystems: the mountain critical zone. *Clocks* indicate environmental archives but also potential reservoir compartments for anthropogenic potentially harmful trace element (PHTE). The *arrows with dots* indicate that different PHTE will have different chemical fates, depending on their time and manner of deposition. The inset shows an example of PHTE deposition (accumulation intensity) in soil with deposition enhanced by canopy edge interception and orographic deposition (blue (light gray in print versions)) but also local activities such as mining (gray). Microtopography plays also a role in PHTE accumulation by concentrating snow in winter. A key compartment not discussed in this chapter is caves that are highly sensitive to environmental changes and that are frequently present in mountainous environments.

Potential Impacts

Because of higher concentrations in a localized area, it has been shown that former and present mining mountain areas have a clear impact on the transfer and effect of PHTEs in mountain-living organisms (Camizuli et al., 2014; Mariet et al., 2016; Qu et al., 2010). A key impact is that PHTE bioavailability is enhanced in high mountain areas where the chemical buffering capacity is limited, such as in clear soft water lakes (Köck and Hofer, 1998). Altitudinal ecological impact of enhanced deposition of PHTE is less clear since not only trace metal accumulation drives their transfer to the living organisms. Blais et al. (2006) observed an increase of Hg in fish with altitude in the Pyrenees as well as some organic contaminants. This is similar to a pattern that was observed for Pb in chamois from the Tatras mountains by Janiga (2008). On the other hand, no clear evidence exists of a gradient impact of Hg transfer in amphibians from the Sierra Nevada. Also no evidence occurs of hot spots of mercury concentrations in tadpoles in the Sierra Nevada due to the transportation of atmospheric Hg from the nearby San Joachim valley (Bradford et al., 2012).

Although the ecological impacts and effects of PHTEs on biota could constitute an entire chapter of its own, we believe that a brief summary example for one PHTE: Hg is still beneficial with specific references to mountain areas and therefore it is included below. We would like to remark however, that this summary is not conclusive, and we therefore encourage readers to turn to the literature listed below, and references therein for further information on the matter. For Cd and Pb, the readers are encouraged to consult Köck and Hofer (1998).

As mentioned previously, many PHTEs hold few to no biological functions and are often considered highly toxic. This applies not only to humans but to other biota as well. In the case of, for instance Hg, many studies have focused on the connection between Hg contamination in food webs and the potential effect on fish (Gantner et al., 2009; Mason et al., 2000; Tsui et al., 2014, 2012). Unlike mammals and birds, that can to some extent excrete Hg and thus reduce the internal concentration of Hg through either digestion or shedding of hair, fur, or feathers, fish have no clear possibility to reduce their internal concentration of Hg. Instead, the Hg is bioaccumulated within the individual thus leading to a biomagnification up the food chain (Mason et al., 1996; Power et al., 2002). Some of the effects as seen by an increased concentration of PHTEs in fish are: reproductive difficulties such a reduced sperm mobility (Dietrich et al., 2010) or maternal transfer from female to egg (Hammerschmidt et al., 1999; Hammerschmidt and Sandheinrich, 2005; Sackett et al., 2013), reduced growth length (Simoneau et al., 2005), impairment in the feeding behavior (Fjeld et al., 1998), and histological changes in livers and kidneys (Rhea et al., 2013) to name a few. Sanches-Galan et al. (1998) even showed that increased concentrations of metals (e.g., Hg) could lead to a higher abundance of micronuclei in fish blood, that is a mutation on the cellular level where the cell division is not functioning properly and an "extra" micronuclei is formed within the blood cell (Blais et al., 2006).

CONCLUSION AND FUTURE OUTLOOK

Despite several decades of research on PHTEs in mountains, many questions remains due to the complexity of mountainous areas. PHTE dispersion has varied in its form through time and in the past it was dominated by local human activities but it has been influenced by long-range transportation of pollution from urban and industrial centers. Additionally to legacy pollution, mountain environments are still impacted by PHTE atmospheric deposition. Higher inputs in the environment could be expected in the future for some PHTE such as Hg emitted by coal combustion. New forms and pathways of PHTE dispersion can also be expected because of the use of nanomaterials and emerging trace elements. The quote written by Weathers et al. (2000) is thus still relevant: "Our results suggest that consideration of variation in deposition across such landscape features as forest edges and gaps, elevation, aspect, and vegetation type should be considered in future modeling efforts. In areas with a long mining history, the pollution legacy should also be considered."

Environmental changes, including climate change (Janiga, 2008), are likely to produce unexpected cascading impacts between PHTE biogeochemical cycles and mountainous ecosystems (including water quality). It is therefore urgent to further investigate the feedbacks between metals and impaired functioning of the mountain critical zone.

ACKNOWLEDGMENTS

This research is supported by a Young Researcher Grant of the Agence National de la Recherche (ANR), project ANR-15-CE01-0008 "TRAM." Adrien Claustres' PhD was funded by a University of Toulouse grant. Sophia Hansson is funded by AXA Research Fund and Prestige/Campus France—cofunded by Marie Curie. Generous support by Labex DRIIHM and Observatoire Hommes et Milieux Haut Vicdessos, as well as ADEME is also acknowledged.

REFERENCES

Aaby, B., Jacobsen, J., 1978. Changes in biotic conditions and metal deposition in the last millennium as reflected in ombotrophic peat in Draved Mose. Aarbog-Danmarks Geologiske Undersoegelse 5–43.

Abbott, M.B., Wolfe, A.P., 2003. Intensive pre-Incan metallurgy recorded by lake sediments from the Bolivian Andes. Science 301, 1893–1895.

Andreae, M., Rosenfeld, D., 2008. Aerosol–cloud–precipitation interactions. Part 1. The nature and sources of cloud-active aerosols. Earth-Science Reviews 89, 13–41.

Bacardit, M., Camarero, L., 2010. Atmospherically deposited major and trace elements in the winter snowpack along a gradient of altitude in the Central Pyrenees: the seasonal record of long-range fluxes over SW Europe. Atmospheric Environment 44, 582–595. http://dx.doi.org/10.1016/j.atmosenv.2009.06.022.

Bacardit, M., Krachler, M., Camarero, L., 2012. Whole-catchment inventories of trace metals in soils and sediments in mountain lake catchments in the Central Pyrenees: apportioning the anthropogenic and natural contributions. Geochimica et Cosmochimica Acta 82, 52–67. http://dx.doi.org/10.1016/j.gca.2010.10.030.

Bao, K., Shen, S., Wang, G., Le Roux, G., 2015. Atmospheric deposition history of trace metals and metalloids for the last 200 years recorded by three peat cores in Great Hinggan mountain, Northeast China. Atmosphere 6.

Bindler, R., Yu, R., Hansson, S., Claßen, N., Karlsson, J., 2012. Mining, metallurgy and the historical origin of mercury pollution in lakes and watercourses in Central Sweden. Environmental Science & Technology 46, 7984–7991. http://dx.doi.org/10.1021/es300789q.

Bing, H., Wu, Y., Zhou, J., Li, R., Luo, J., Yu, D., 2016. Vegetation and cold trapping modulating elevation-dependent distribution of trace metals in soils of a high mountain in eastern Tibetan Plateau. Scientific Reports 6.

Blackwell, B.D., Driscoll, C.T., 2015. Deposition of mercury in forests along a montane elevation gradient. Environmental Science & Technology. http://dx.doi.org/10.1021/es505928w, 150330144737006.

Blais, J.M., Charpentié, S., Pick, F., Kimpe, L.E., Amand, A.S., Regnault-Roger, C., 2006. Mercury, polybrominated diphenyl ether, organochlorine pesticide, and polychlorinated biphenyl concentrations in fish from lakes along an elevation transect in the French Pyrénées. Ecotoxicology and Environmental Safety 63, 91–99. http://dx.doi.org/10.1016/j.ecoenv.2005.08.008.

Bourcier, L., 2009. Transport et dépôt de particules et radionucléides au puy de Dôme, France. Université Blaise Pascal-Clermont-Ferrand II.

Bradford, D.F., Kramer, J.L., Gerstenberger, S.L., Tallent-Halsell, N.G., Nash, M.S., 2012. Mercury in tadpoles collected from remote alpine sites in the southern Sierra Nevada mountains, California, USA. Archives of Environmental Contamination and Toxicology 62, 135–140.

Brännvall, M.-L., Bindler, R., Renberg, I., Emteryd, O., Bartnicki, J., Billström, K., 1999. The Medieval metal industry was the cradle of modern large-scale atmospheric lead pollution in northern Europe. Environmental Science & Technology 33, 4391–4395.

Brantley, S.L., Goldhaber, M.B., Ragnarsdottir, K.V., 2007. Crossing disciplines and scales to understand the critical zone. Elements 3, 307–314.

Brittain, J.E., Gjerseth, J.E., 2010. Long-term trends and variation in 137Cs activity concentrations in brown trout (*Salmo trutta*) from Øvre Heimdalsvatn, a Norwegian subalpine lake. Hydrobiologia 642, 107–113.

Camarero, L., Masqué, P., Devos, W., Ani-Ragolta, I., Catalan, J., Moor, H.C., Pla, S., Sanchez-Cabeza, J.A., 1998. Historical variations in lead fluxes in the Pyrenees (northeast Spain) from a dated lake sediment core. Water, Air, and Soil Pollution 105, 439–449.

Camizuli, E., Monna, F., Scheifler, R., Amiotte-Suchet, P., Losno, R., Beis, P., Bohard, B., Chateau, C., Alibert, P., 2014. Impact of trace metals from past mining on the aquatic ecosystem: a multi-proxy approach in the Morvan (France). Environmental Research 134, 410–419. http://dx.doi.org/10.1016/j.envres.2014.07.008.

Catalan, J., Rondón, J.C.D., 2016. Perspectives for an integrated understanding of tropical and temperate high-mountain lakes. Journal of Limnology 75.

Catalan, J., Pla-Rabés, S., Wolfe, A.P., Smol, J.P., Rühland, K.M., Anderson, N.J., Kopáček, J., Stuchlík, E., Schmidt, R., Koinig, K.A., Camarero, L., Flower, R.J., Heiri, O., Kamenik, C., Korhola, A., Leavitt, P.R., Psenner, R., Renberg, I., 2013. Global change revealed by palaeolimnological records from remote lakes: a review. Journal of Paleolimnology 49, 513–535. http://dx.doi.org/10.1007/s10933-013-9681-2.

Catalan, J., 2015. Tracking long-range atmospheric transport of trace metals, polycyclic aromatic hydrocarbons, and organohalogen compounds using lake sediments of mountain regions. In: Environmental Contaminants. Springer, pp. 263–322.

Colbeck, I., 2008. Environmental Chemistry of Aerosols. Wiley Online Library.

Cooke, C.A., Bindler, R., 2015. Lake sediment records of preindustrial metal pollution. In: Environmental Contaminants. Springer, pp. 101–119.

Cortizas, A.M., Garcıa-Rodeja, E., Pombal, X.P., Muñoz, J.N., Weiss, D., Cheburkin, A., 2002. Atmospheric Pb deposition in Spain during the last 4600 years recorded by two ombrotrophic peat bogs and implications for the use of peat as archive. The Science of the Total Environment 292, 33–44.

Cziczo, D.J., Stetzer, O., Worringen, A., Ebert, M., Weinbruch, S., Kamphus, M., Gallavardin, S.J., Curtius, J., Borrmann, S., Froyd, K.D., et al., 2009. Inadvertent climate modification due to anthropogenic lead. Nature Geoscience 2, 333–336.

De Traubenberg, C.R., Galsomiès, L., Martinet, Y., 2013. Pollution atmosphérique par les métaux en France: 10 ans de biosurveillance des retombées. EDP Sciences.

De Vleeschouwer, F., Le Roux, G., Shotyk, W., 2010. Peat as an archive of atmospheric pollution and environmental change: a case study of lead in Europe. PAGES Magazine 18, 20–22.

De Vleeschouwer, F., Vanneste, H., Mauquoy, D., Piotrowska, N., Torrejón, F., Roland, T., Stein, A., Le Roux, G., 2014. Emissions from pre-Hispanic metallurgy in the South American atmosphere. PLoS One 9, e111315.

Demers, J.D., Blum, J.D., Zak, D.R., 2013. Mercury isotopes in a forested ecosystem: implications for air-surface exchange dynamics and the global mercury cycle. Global Biogeochemical Cycles 27, 222–238.

Dietrich, G.J., Dietrich, M., Kowalski, R., Dobosz, S., Karol, H., Demianowicz, W., Glogowski, J., 2010. Exposure of rainbow trout milt to mercury and cadmium alters sperm motility parameters and reproductive success. Aquatic Toxicology 97, 277–284.

Engelmann, R.J., 1965. The Calculation of Precipitation Scavenging. Pacific Northwest Lab, Battelle-Northwest, Richland, Washington.

Enrico, M., Roux, G.L., Marusczak, N., Heimbürger, L.-E., Claustres, A., Fu, X., Sun, R., Sonke, J.E., 2016. Atmospheric mercury transfer to peat bogs dominated by gaseous elemental mercury dry deposition. Environmental Science & Technology 50, 2405–2412.

Fjeld, E., Haugen, T., Vøllestad, L., 1998. Permanent impairment in the feeding behavior of grayling (*Thymallus thymallus*) exposed to methylmercury during embryogenesis. The Science of the Total Environment 213, 247–254.

Forel, B., Monna, F., Petit, C., Bruguier, O., Losno, R., Fluck, P., Begeot, C., Richard, H., Bichet, V., Chateau, C., 2010. Historical mining and smelting in the Vosges Mountains (France) recorded in two ombrotrophic peat bogs. Journal of Geochemical Exploration 107, 9–20.

Fowler, D., Smith, R., Leith, I., Crossley, A., Mourne, R., Branford, D., Moghaddam, M., 1998. Quantifying Fine-Scale Variability in Pollutant Deposition in Complex Terrain Using 210Pb Inventories in Soil. In: Biogeochemical Investigations at Watershed, Landscape, and Regional Scales. Springer, pp. 459–470.

Gabrielli, P., Vallelonga, P., 2015. Contaminant records in ice cores. In: Environmental Contaminants. Springer, pp. 393–430.

Gandois, L., Tipping, E., Dumat, C., Probst, A., 2010. Canopy influence on trace metal atmospheric inputs on forest ecosystems: speciation in throughfall. Atmospheric Environment 44, 824–833. http://dx.doi.org/10.1016/j.atmosenv.2009.11.028.

Gantner, N., Hintelmann, H., Zheng, W., Muir, D.C., 2009. Variations in stable isotope fractionation of Hg in food webs of Arctic lakes. Environmental Science & Technology 43, 9148–9154.

Gerdol, R., Bragazza, L., 2006. Effects of altitude on element accumulation in alpine moss. Chemosphere 64, 810–816.

González, Z.I., Krachler, M., Cheburkin, A.K., Shotyk, W., 2006. Spatial distribution of natural enrichments of arsenic, selenium, and uranium in a minerotrophic peatland, Gola di Lago, Canton Ticino, Switzerland. Environmental Science & Technology 40, 6568–6574.

Gurung, A.B., von Dach, S.W., Price, M.F., Aspinall, R., Balsiger, J., Baron, J.S., Sharma, E., Greenwood, G., Kohler, T., 2012. Global change and the World's mountains—research needs and emerging themes for sustainable development: a synthesis from the 2010 Perth II conference. Mountain Research and Development 32, S47–S54. http://dx.doi.org/10.1659/MRD-JOURNAL-D-11-00084.S1.

Hammerschmidt, C.R., Sandheinrich, M.B., 2005. Maternal diet during oogenesis is the major source of methylmercury in fish embryos. Environmental Science & Technology 39, 3580–3584.

Hammerschmidt, C.R., Wiener, J.G., Frazier, B.E., Rada, R.G., 1999. Methylmercury content of eggs in yellow perch related to maternal exposure in four Wisconsin lakes. Environmental Science & Technology 33, 999–1003.

Hansson, S.V., Bindler, R., De Vleeschouwer, F., 2015. Using peat records as natural archives of past atmospheric metal deposition. In: Environmental Contaminants. Springer, pp. 323–354.

Hillman, A.L., Abbott, M.B., Yu, J., Bain, D.J., Chiou-Peng, T., 2015. Environmental legacy of copper metallurgy and Mongol silver smelting recorded in Yunnan lake sediments. Environmental Science & Technology 49, 3349–3357.

Holmes, C.D., Jacob, D.J., Corbitt, E.S., Mao, J., Yang, X., Talbot, R., Slemr, F., 2010. Global atmospheric model for mercury including oxidation by bromine atoms. Atmospheric Chemistry and Physics 10, 12037–12057.

Hong, S., Candelone, J.-P., Patterson, C.C., Boutron, C.F., 1994. Greenland ice evidence of hemispheric lead pollution two millennia ago by Greek and Roman civilizations. Science 265, 1841–1843.

Hososhima, M., Kaneyasu, N., 2015. Altitude-dependent distribution of ambient gamma dose rates in a mountainous area of Japan caused by the fukushima nuclear accident. Environmental Science & Technology 49, 3341–3348.

Ishikawa, Y., Murakami, H., Sekine, T., Yoshihara, K., 1995. Precipitation scavenging studies of radionuclides in air using cosmogenic ^{7}Be. Journal of Environmental Radioactivity 26, 19–36.

Janiga, M., 2008. Potential effects of global warming on atmospheric lead contamination in the mountains. In: The Socio-economic Causes and Consequences of Desertification in Central Asia. Springer, pp. 231–247.

Jensen, A., 1997. Historical deposition rates of Cd, Cu, Pb, and Zn in Norway and Sweden estimated by ^{210}Pb dating and measurement of trace elements in cores of peat bogs. Water, Air, and Soil Pollution 95, 205–220.

Jouffroy-Bapicot, I., Pulido, M., Baron, S., Galop, D., Monna, F., Lavoie, M., Ploquin, A., Petit, C., de Beaulieu, J.-L., Richard, H., 2007. Environmental impact of early palaeometallurgy: pollen and geochemical analysis. Vegetation History and Archaeobotany 16, 251–258.

Köck, G., Hofer, R., 1998. Origin of cadmium and lead in clear softwater lakes of high-altitude and high-latitude, and their bioavailability and toxicity to fish. In: Fish Ecotoxicology. Springer, pp. 225–257.

Le Roux, G., Aubert, D., Stille, P., Krachler, M., Kober, B., Cheburkin, A., Bonani, G., Shotyk, W., 2005. Recent atmospheric Pb deposition at a rural site in southern Germany assessed using a peat core and snowpack, and comparison with other archives. Atmospheric Environment 39, 6790–6801.

Le Roux, G., Pourcelot, L., Masson, O., Duffa, C., Vray, F., Renaud, P., 2008. Aerosol deposition and origin in French mountains estimated with soil inventories of ^{210}Pb and artificial radionuclides. Atmospheric Environment 42, 1517–1524. http://dx.doi.org/10.1016/j.atmosenv.2007.10.083.

Le Roux, G., Duffa, C., Vray, F., Renaud, P., 2010. Deposition of artificial radionuclides from atmospheric Nuclear Weapon Tests estimated by soil inventories in French areas low-impacted by Chernobyl. Journal of Environmental Radioactivity 101, 211–218. http://dx.doi.org/10.1016/j.jenvrad.2009.10.010.

Lee, J., Tallis, J., 1973. Regional and historical aspects of lead pollution in Britain. Nature 245, 216.

Lee, C.S., Qi, S., Zhang, G., Luo, C., Zhao, L.Y., Li, X., 2008a. Seven thousand years of records on the mining and utilization of metals from lake sediments in central China. Environmental Science & Technology 42, 4732–4738.

Lee, K., Do Hur, S., Hou, S., Hong, S., Qin, X., Ren, J., Liu, Y., Rosman, K.J., Barbante, C., Boutron, C.F., 2008b. Atmospheric pollution for trace elements in the remote high-altitude atmosphere in central Asia as recorded in snow from Mt. Qomolangma (Everest) of the Himalayas. The Science of the Total Environment 404, 171–181.

Lindberg, S.E., Harriss, R.C., Turner, R.R., 1982. Atmospheric deposition of metals to forest vegetation. Science 215, 1609–1611.

Lovett, G.M., Kinsman, J.D., 1990. Atmospheric pollutant deposition to high-elevation ecosystems. Atmospheric Environment. Part A. General Topics 24, 2767–2786.

Lovett, G.M., 1994. Atmospheric deposition of nutrients and pollutants in North America: an ecological perspective. Ecological Applications 4, 629–650.

Mariet, A.-L., de Vaufleury, A., Bégeot, C., Walter-Simonnet, A.-V., Gimbert, F., 2016. Palaeopollution from mining activities in the Vosges mountains: 1000 years and still bioavailable. Environmental Pollution 214, 575–584.

Marx, S.K., Rashid, S., Stromsoe, N., 2016. Global-scale patterns in anthropogenic Pb contamination reconstructed from natural archives. Environmental Pollution 213, 283–298.

Mason, R.P., Reinfelder, J.R., Morel, F.M., 1996. Uptake, toxicity, and trophic transfer of mercury in a coastal diatom. Environmental Science & Technology 30, 1835–1845.

Mason, R., Laporte, J.-M., Andres, S., 2000. Factors controlling the bioaccumulation of mercury, methylmercury, arsenic, selenium, and cadmium by freshwater invertebrates and fish. Archives of Environmental Contamination and Toxicology 38, 283–297.

Monna, F., Galop, D., Carozza, L., Tual, M., Beyrie, A., Marembert, F., Chateau, C., Dominik, J., Grousset, F.E., 2004. Environmental impact of early Basque mining and smelting recorded in a high ash minerogenic peat deposit. The Science of the Total Environment 327, 197–214. http://dx.doi.org/10.1016/j.scitotenv.2004.01.010.

Monna, F., Camizuli, E., Revelli, P., Biville, C., Thomas, C., Losno, R., Scheifler, R., Bruguier, O., Baron, S., Chateau, C., Ploquin, A., Alibert, P., 2011. Wild brown trout affected by historical mining in the Cévennes National Park, France. Environmental Science & Technology 45, 6823–6830. http://dx.doi.org/10.1021/es200755n.

Montoya-Mayor, R., Fernández-Espinosa, A., Seijo-Delgado, I., Ternero-Rodríguez, M., 2013. Determination of soluble ultra-trace metals and metalloids in rainwater and atmospheric deposition fluxes: a 2-year survey and assessment. Chemosphere 92, 882–891.

Munger, J., Waldman, J., Jacob, D., Hoffman, M., 1983. Vertical variability and short term temporal trends in precipitation chemistry. Precipitation Scavenging, Dry Deposition and Resuspension 1, 275–281.

Murozumi, M., Chow, T.J., Patterson, C., 1969. Chemical concentrations of pollutant lead aerosols, terrestrial dusts and sea salts in Greenland and Antarctic snow strata. Geochimica et Cosmochimica Acta 33, 1247–1294.

Nriagu, J.O., Pacyna, J.M., et al., 1988. Quantitative assessment of worldwide contamination of air, water and soils by trace metals. Nature 333, 134–139.

Nriagu, J.O., 1996. A history of global metal pollution. Science 272, 223.

Ollinger, S.V., Aber, J.D., Lovett, G.M., Millham, S.E., Lathrop, R.G., Ellis, J.M., 1993. A spatial model of atmospheric deposition for the northeastern US. Ecological Applications 3, 459–472.

Pacyna, J.M., Pacyna, E.G., 2001. An assessment of global and regional emissions of trace metals to the atmosphere from anthropogenic sources worldwide. Environmental Reviews 9, 269–298.

Pourcelot, L., Louvat, D., Gauthier-Lafaye, F., Stille, P., 2003. Formation of radioactivity enriched soils in mountain areas. Journal of Environmental Radioactivity 68, 215–233.

Power, M., Klein, G., Guiguer, K., Kwan, M., 2002. Mercury accumulation in the fish community of a sub-Arctic lake in relation to trophic position and carbon sources. Journal of Applied Ecology 39, 819–830.

Qu, X., Wu, N., Tang, T., Cai, Q., Park, Y.-S., 2010. Effects of heavy metals on benthic macroinvertebrate communities in high mountain streams. In: Annales de Limnologie-International Journal of Limnology. EDP Sciences, pp. 291–302.

Rauch, J.N., Pacyna, J.M., 2009. Earth's global Ag, Al, Cr, Cu, Fe, Ni, Pb, and Zn cycles. Global Biogeochemical Cycles 23.

Renberg, I., Persson, M.W., Emteryd, O., 1994. Pre-industrial Atmospheric Lead Contamination Detected in Swedish Lake Sediments.

Rhea, D.T., Farag, A.M., Harper, D.D., McConnell, E., Brumbaugh, W.G., 2013. Mercury and selenium concentrations in biofilm, macroinvertebrates, and fish collected in the Yankee Fork of the Salmon River, Idaho, USA, and their potential effects on fish health. Archives of Environmental Contamination and Toxicology 64, 130–139.

Roe, G.H., 2005. Orographic Precipitation. Annual Review of Earth and Planetry Sciences 33, 645–671. http://dx.doi.org/10.1146/annurev.earth.33.092203.122541.

Rosman, K.J., Chisholm, W., Hong, S., Candelone, J.-P., Boutron, C.F., 1997. Lead from Carthaginian and Roman Spanish mines isotopically identified in Greenland ice dated from 600 BC to 300 AD. Environmental Science & Technology 31, 3413–3416.

Ruijrok, W., Davidson, C.I., Nicholson, W., et al., 1995. Dry deposition of particles. Tellus B 47, 587–601.

Sackett, D.K., Aday, D.D., Rice, J.A., Cope, W.G., 2013. Maternally transferred mercury in wild largemouth bass, *Micropterus salmoides*. Environmental Pollution 178, 493–497.

Sanchez-Galan, S., Linde, A., Izquierdo, J., García-Vázquez, E., 1998. Micronuclei and fluctuating asymmetry in brown trout (*Salmo trutta*): complementary methods to biomonitor freshwater ecosystems. Mutation Research/Genetic Toxicology and Environmental Mutagenesis 412, 219–225.

Sen, I.S., Peucker-Ehrenbrink, B., 2012. Anthropogenic disturbance of element cycles at the Earth's surface. Environmental Science & Technology 46, 8601–8609. http://dx.doi.org/10.1021/es301261x.

Settle, D.M., Patterson, C.C., 1980. Lead in albacore: guide to lead pollution in Americans. Science 207, 1167–1176.

Shirahata, H., Elias, R.W., Patterson, C., Koide, M., 1980. Chronological variations in concentrations and isotopic compositions of anthropogenic atmospheric lead in sediments of a remote subalpine pond. Geochimica et Cosmochimica Acta 44, 149–162.

Shotyk, W., Le Roux, G., 2005. Biogeochemistry and cycling of lead. Metal Ions in Biological Systems 43, 239–275.

Shotyk, W., Cheburkin, A., Appleby, P., Fankhauser, A., Kramers, J., 1997. Lead in three peat bog profiles, Jura mountains, Switzerland: enrichment factors, isotopic composition, and chronology of atmospheric deposition. Water, Air, and Soil Pollution 100, 297–310.

Shotyk, W., Weiss, D., Appleby, P., Cheburkin, A., Frei, R., Gloor, M., Kramers, J., Reese, S., Van Der Knaap, W., 1998. History of atmospheric lead deposition since 12,370 ^{14}C yr BP from a peat bog, Jura mountains, Switzerland. Science 281, 1635–1640.

Simoneau, M., Lucotte, M., Garceau, S., Laliberté, D., 2005. Fish growth rates modulate mercury concentrations in walleye (*Sander vitreus*) from eastern Canadian lakes. Environmental Research 98, 73–82.

Tsui, M.T.K., Blum, J.D., Kwon, S.Y., Finlay, J.C., Balogh, S.J., Nollet, Y.H., 2012. Sources and transfers of methylmercury in adjacent river and forest food webs. Environmental Science & Technology 46, 10957–10964.

Tsui, M.T.K., Blum, J.D., Finlay, J.C., Balogh, S.J., Nollet, Y.H., Palen, W.J., Power, M.E., 2014. Variation in terrestrial and aquatic sources of methylmercury in stream predators as revealed by stable mercury isotopes. Environmental Science & Technology 48, 10128–10135. http://dx.doi.org/10.1021/es500517s.

Weathers, K.C., Likens, G.E., Bormann, F.H., Bicknell, S.H., Bormann, B.T., Daube, B.C., Eaton, J.S., Galloway, J.N., Keene, W.C., et al., 1988. Cloudwater chemistry from ten sites in North America. Environmental Science & Technology 22, 1018–1026.

Weathers, K., Lovett, G., Likens, G., Lathrop, R., 2000. The effect of landscape features on deposition to Hunter mountain, Catskill mountains, New York. Ecological Applications 10, 528–540.

Weiss, D., Shotyk, W., Appleby, P.G., Kramers, J.D., Cheburkin, A.K., 1999. Atmospheric Pb deposition since the industrial revolution recorded by five Swiss peat profiles: enrichment factors, fluxes, isotopic composition, and sources. Environmental Science & Technology 33, 1340–1352.

Whicker, F.W., Nelson, W.C., Gallegos, A.F., 1972. Fallout ^{137}Cs and ^{90}Sr in trout from mountain lakes in Colorado. Health Physics 23, 519–527.

Yeo, B., Langley-Turnbaugh, S., 2010. Trace element deposition on mount Everest. Soil Survey Horizons 51, 72–78.

Zaharescu, D.G., Hooda, P.S., Fernandez, J., Soler, A.P., Burghelea, C.I., 2009. On the arsenic source mobilisation and its natural enrichment in the sediments of a high mountain cirque in the Pyrenees. Journal of Environmental Monitoring 11, 1973–1981.

Zechmeister, H.G., 1995. Correlation between altitude and heavy metal deposition in the Alps. Environtal Pollution 89, 73–80.

Zhang, H., Yin, R., Feng, X., Sommar, J., Anderson, C.W.N., Sapkota, A., Fu, X., Larssen, T., 2013. Atmospheric mercury inputs in montane soils increase with elevation: evidence from mercury isotope signatures. Scientific Reports 3. http://dx.doi.org/10.1038/srep03322.

Chapter 4

Water and Sustainability in the Lake Mývatn Region of Iceland: Historical Perspectives and Current Concerns

R. Sigurðardóttir,* A.E.J. Ogilvie,[§,1] Á.D. Júlíusson,[¶] V. Hreinsson* and M.T. Hicks[||]

*Reykjavik Academy, Reykjavik, Iceland; [§]Stefansson Arctic Institute, Akureyri, Iceland and Institute of Arctic and Alpine Research, Boulder, Colorado, United States; [¶]Reykjavik Academy and National Museum of Iceland, Reykjavik, Iceland; [||]City University of New York, New York, United States

[1]Corresponding author: E-mail: astrid.ogilvie@unak.is

Chapter Outline

Mountain Ice and Water. http://dx.doi.org/10.1016/B978-0-444-63787-1.00004-4

INTRODUCTION

The geographical focus of this chapter is the highland wetland region of Mývatn in the northeast of Iceland in the county of Suður ("southern") Thingeyjarsýsla, in the municipality of Skútustaðir (see Fig. 4.1. for an image of the striking landscape). It is likely that the area was one of the first to be settled in Iceland in the wake of the late-9th century "Viking" expansion westwards. The historical period under consideration is primarily the mid-19th century to the present day, and thus includes the period of Danish rule (from c.1523 to 1944). The key aspect that is considered is the way in which sustainable economic practices were linked to the dominant feature of the local environment, its water, in particular the lake from which the region takes its name, as well as current concerns and issues during the present day. The following is an extract from a description of the Mývatn district written in 1849, and has been chosen because it serves to highlight a number of issues relevant to this chapter, such as the geological features, the wetland environment, and the remoteness of the region from other inhabited parts. The original Icelandic is given first and is followed by an English translation.

> *Um þessa hrepps afstöðu og ásigkomulag yfir höfuð, er þess fyrst að geta: að Sveitin er hálend og liggur langt upp til fjalla, hérumbil í 6 til 8 mílna fjárlægð frá næsta verslunarstað, er hún því erfið mjög hvað alla aðdrætti snertir á höndlunarvöru, trjávið og sjávarafla; hún er og ill og erfið yfirferðar vegna hrauns og gjáa. Í vatni því er í Sveitinni liggur og kallað er Mývatn eru margar eyjar og hólmar og í sumum þeirra allgóður heyskapur og nokkurt andarvarp. Silungsveiði er og í vatninu á flestum tímum ársins. Á sumum stöðum er landrými nokkurt og landkostir allgóðir; hér og hvar eru dálitlar skógarleifar og grasatekja; einnig má telja afréttarlönd er sveitinni tilheyra. Yfirhöfuð eru landkostir töluverðir, en þó mjög misjafnir og heyskapur eins. Öllum þessum Sveitarinnar helstu gæðum er þannig varið, að þau útheimta sérlega atorku sem því aðeins geta komið að liði, að sveitarmenn hafi bæði nægilega krafta og gott samheldi, og er það í hinu fyrra tilliti ekki lítill styrkur, að flestir innansveitarmenn eiga jarðir þær er í heinni liggja, þó að því minna þarf út að gjalda, og nýtur hún sín því að mestu frjáls.*
>
> Jarðamat (1849)

> *With regard to the general position and condition of this parish, this may be mentioned first: the district consists of highland, and is situated far up in the mountains, approximately 6 to 8 miles from the nearest trading station. Thus it is difficult to obtain supplies, including trade goods, timber, and products from the sea. It is also difficult to travel about the district because of the lava formations and chasms. There is a lake in the community called Mývatn where there are many islands and islets. On some of them excellent hay is produced, and ducks also nest there. There is year-round Arctic char fishing in the lake. In some places there is much habitable land and fertile ground. Here and there are small*

wooded places and areas where Iceland moss may be found. There are also outlying pastures belonging to the district. On the whole, the habitable land is reasonably fertile, but it is variable, and the hay production also. The utilization of the district's resources requires the population of the area to be exceptionally hardy. However, this would be to little avail if the inhabitants did not have strength enough, and the ability to cooperate well. In particular with regard to the former, it is a particular advantage that most farmers in the community own their farms and thus have less to pay in rent. The Mývatn community thus enjoys much freedom.

Translation A.E.J. Ogilvie

The passage is taken from a farm evaluation and description, in Icelandic called a *jarðamat*, produced in the year 1849. The purpose of the evaluation was to establish the value of farms to ascertain the taxes due. These were valued according to the productive capacity of the individual farms in terms of being able to sustain a flock of sheep or cattle. Additional farm assets such as the availability of birds' eggs or access to fishing would also count toward the value of the farm. However, the *jarðamat* worked in the favor of the farming community in that it was a long-standing mechanism to limit the somewhat arbitrary demands of landowners on the tenants. With regard to the comment that the nearest trading station is 6–8 miles away, this refers to the Danish mile which at the time was equal to approximately 7.5 km. Hence the distance referred to is 45–60 km away. It is undoubtedly the coastal town of Húsavík that is referred to here, in fact some 60 km from Mývatn.

Regarding land ownership, the passage states that "... most farmers in the community own their farms and thus have less to pay in rent. The Mývatn community thus enjoys much freedom." In the past, property in Iceland was

FIGURE 4.1 Shapes and contrasts of the Mývatn district. Skútustaðagígar, a cluster of pseudo craters on the southern edge of Lake Mývatn. *Photo taken by Örn Friðriksson.*

frequently owned by large landowners including the Church and the Danish King. In Mývatn, some farms were often partly owner occupied, with the farmers owning as much as half of the property. However, according to research on land ownership in the region, farmers started to buy their farms in the 19th century, and by around 1900 many farms had become farmer owned (Teitsson, 1973, pp. 51–55). However, prior to this, landowners in Mývatn had often been absent and, largely as a result of the relative isolation of the community, individual farms functioned as independent units for centuries, without the interference of large landowners, possibly even from settlement times. The farmers thus did enjoy a certain freedom, as suggested in the description, and consequently also held responsibility for their land far beyond their official ownership status. Parish documents show continuity in that the same families lived in the area for generations.

Despite its relatively isolated and mountainous location, archaeological remains have shown that the area was settled early, around the same time as the settlement period began in the rest of Iceland around AD 870 (Vésteinsson and McGovern, 2012). The region is the only high-altitude community in Iceland that has persisted continually from settlement over 1100 years ago to modern times. In the 19th century, 22 farms existed in the district. At the present time there are 23; however, although many farms have the same names and locations, land boundaries have changed somewhat. The church-farm at Skutustaðir is likely to have had the same location from early times (see Fig 4.2). What made the Mývatn region different from the other early settled mountainous regions of Iceland was undoubtedly its water, and the management of its water-related resources.

At 65.6039°N, 16.9961°W, Mývatn can be classified as a sub-Arctic/Alpine landscape. Lake Mývatn is the community in Iceland that is highest above sea

FIGURE 4.2 The vicarage at Skútustaðir with Mount Búrfell in the background. *Photo taken by Örn Friðriksson.*

level (a.s.l.), with farms ranging from 260 to 345 m a.s.l. Because of its outstanding natural beauty and interesting geological features, the area is regarded as one of Iceland's most valued natural treasures, and is one of Iceland's most popular tourist destinations. Mývatn and its tributary river, the Laxá (meaning "salmon river") were protected by law in 1974, and in 1978 placed on the RAMSAR list of wetlands of international importance (http://www.ramsar.org/). The rationale for nature conservation in the Lake Mývatn region includes rare and red-listed and endangered species of flora and fauna, and an unusually large biodiversity considering its geographical position, as well as unique geological phenomena.

At the present time, the Mývatn ecosystem is seriously threatened, its food webs disrupted, and its conservation value in danger of decline. After over a millennium of apparent sustainable use and management of water and

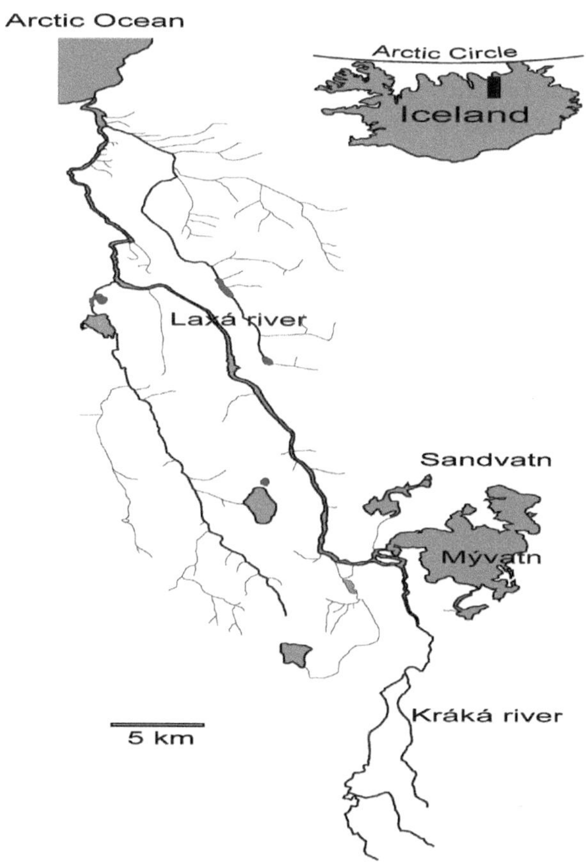

FIGURE 4.3 Location map of the Mývatn area. *Courtesy of Árni Einarsson.*

wetland-related resources, a variety of developments have disrupted its former balance. These include mining operations and demand for its hydro and geothermal power resources. In recent decades, both the economic and social life of the area has changed radically. This has been accompanied by a very rapid increase in tourism. Before exploring the connections between water and sustainability in the region in the past, these topics are placed in the context of available data sources, as well as discussions on the physical environment and related features, the climate, history, economy, and ecology of Iceland in general, and for the Mývatn area in particular. Current issues and concerns are also discussed in this chapter. For a location map see Fig. 4.3.

DATA SOURCES

Documentary Historical Sources

Iceland is well known for its rich literary tradition that includes a wealth of written works encompassing a wide variety of genres (Ogilvie, 2005, 2010). Lack of space allows only minimal discussion of the different types of sources listed in the following, except to emphasize that, as a fortunate result of the Icelandic penchant for meticulous environmental and social observation, they constitute a vast treasure chest of information on the past. For an example see Fig. 4.4. In addition to the local traditions of writing, in the late-17th century the authorities in Denmark began to commission a number of studies of the

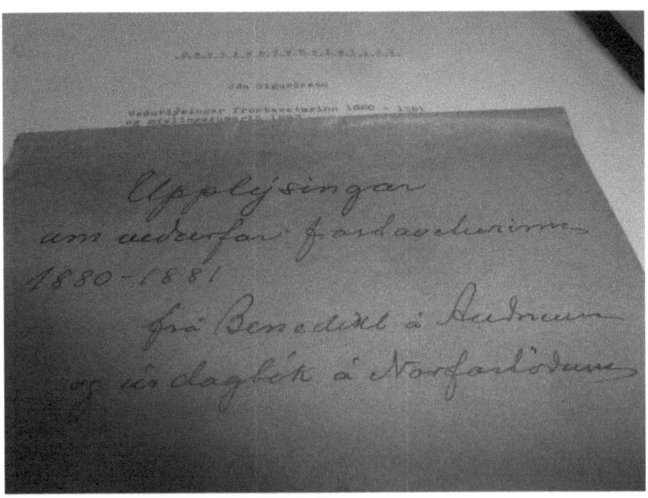

FIGURE 4.4 Current research by the authors of this chapter relies heavily on the extensive documentary historical records available for Iceland. These manuscripts in the archives in Húsavík refer to the severe winter of 1880–81. The 1880s were very severe on the whole with much sea ice. *Photo taken by Astrid Ogilvie.*

situation in Iceland, partly in the wake of difficult climatic conditions and famines. The most comprehensive and valuable of the resulting documents is the *Jarðabók*, a register of land values and human and livestock populations for farms in the whole of Iceland, compiled by Árni Magnússon and Páll Vídalín over the period 1702–14 (Magnússon, 1913–1943). The information contained in *Jarðabók* has been fully digitized. Other primary sources that have been published include numerous censuses, the later Icelandic annals (*Annálar* 1400–1800) covering the period from around 1600 to 1800, and the *Sýslu-og sóknarlýsningar* (1979) which were surveys of the countryside from the first half of the 19th century. Other extremely valuable sources, such as the annual reports to the Danish authorities of the District Sheriffs and Governors, are still in manuscript form. These are currently being analyzed by Astrid Ogilvie (2007, 2008).

Other unpublished records include personal letters written in the Mývatn region and surrounding areas that belong to the family of lead author Ragnhildur Sigurðardóttir who is from Mývatnssveit, and whose grandfather and great-great-grandfather have left an extensive legacy of personal documents from the late-19th to the mid-20th century. A number of published scholarly studies are also valuable sources for the period. Examples are: the seminal two-volume work on Icelandic agriculture by co-author Árni Daníel Júlíusson (2013a,b); a comprehensive study of Suður-Thingeyjarsýsla focusing on social and cultural activity during the second half of the 19th century, with an emphasis on political life and trade, by Gunnar Karlsson (1977); and a study of settlement and farm ownership during 1703–1930 by Björn Teitsson (1973) that describes a period of contraction in the 18th century, expansion in the 19th century, and the beginning of consolidation in the 20th century. Partly as a result of Iceland's literary heritage, many farmers were self-taught polymaths, and many of the historical documents available were composed by learned farmers. Some documents are still in private ownership; however, most are located in the National Archives and the National Library in Reykjavík, as well as in local libraries and museums. Also vital for historical research on the Mývatn area are the Icelandic manuscripts and documents stored in the Danish Royal Library and various archives in Copenhagen. These are currently being researched by the authors of this chapter.

Archaeological Information

Data drawn from the archaeological record illuminate the documentary evidence, and, as a source of both social and ecological data, provide further insights into the characteristics of sustainable or exploitative economies. In particular, current research in the Mývatn area benefits from the collaborative international work that has been ongoing in the region for over three decades (Ogilvie and McGovern, 2000; Friðriksson et al., 2004; McGovern et al., 2006, 2007; Lucas, 2009; Hicks, 2014; Hicks et al., 2016). This research has largely been drawn from the discipline of zooarchaeology. This involves the study and analysis of faunal remains that include bones, shells, hair, and DNA. For research in the Mývatn area, analyses of middens, in particular, reveal large quantities of animal bones. These can be used to understand shifts in production centering on livestock, and to illuminate herding strategies, herd demography, and the animal products that were the primary focus of the farmers' seasonal rounds. Furthermore, evidence regarding the habits of consumption helps to show the trading partners interacting with Mývatn. The fish and seal bones that are found reflect an emphasis on marine resources, suggesting trade with the nearby coastal regions. Of particular interest, however, are the numerous deposits of shells from birds' eggs suggesting a sustainable harvesting of the eggs of Mývatn waterbirds (Hicks et al., 2016).

A particularly useful archaeological tool is tephrochronology. This dating technique was developed by the Icelandic geologist Sigurður Thórarinsson (1944, 1970). The method involves using the ash layers from volcanic eruptions and Thórarinsson chose the name from the Greek word "tephra" meaning ash. The term *tephra* is used to describe all of the solid matter produced from a volcano during an eruption (Thórarinsson, 1944). As noted earlier, Iceland is a volcanic island and eruptions occur with relative frequency. Tephrochronology is thus particularly useful in an Icelandic context, but has been developed in other areas of volcanic activity. Tephra is a very fine material and can be blown across large areas. Research has identified volcanic ash of Icelandic origin in many parts of the world (Demarée and Ogilvie, 2001). The technique involves using tephra from a single eruption to create a chronological framework. The basis on which this rests is that any volcanic event will produce ash that has a unique geochemical "fingerprint" that will allow the deposit to be identified in more than one place. A great advantage of the technique is that it is usually fairly easy to see the tephra layers in soils and sediments (see, e.g., Fig. 4.5). Tephrochronology thus provides a precise and accurate dating tool for archaeological and paleoenvironmental events (Dugmore et al., 2004; Lowe, 2011; Dugmore and Vésteinsson, 2012; Streeter and Dugmore, 2013; Batt et al., 2015). It can also be used to corroborate and illuminate documentary historical records (Hicks et al., 2016).

PHYSICAL ENVIRONMENT AND NATURAL SYSTEMS

Climate

The island of Iceland is located in a climatologically sensitive area close to major and contrasting features of the Northern Hemisphere's atmospheric and oceanic circulations – at the intersection of cold Polar air and warmer Atlantic air, and the relatively warm Irminger current and the colder East Iceland current. Because of this, Iceland is very sensitive to minor fluctuations in the strength of these different air masses and ocean currents. This is one of the

FIGURE 4.5 An archaeological dig at the site of Skútustaðir shows volcanic ash layers as used in tephrochronological dating. *Photo taken by Astrid Ogilvie.*

main causes of the variability of the climate of Iceland on all time scales. However, as a consequence of Iceland's maritime position, and the relatively warm waters of the North Atlantic, the climate of Iceland is fairly mild and the annual temperature range is small (Eythórsson and Sigtryggsson, 1971). The seasonal boundary of the Arctic drift ice also lies close to Iceland, and is an important feature of its climate, especially the north of Iceland (Ogilvie and Jónsson, 2001). The features of the general atmospheric circulation that influence the region of Iceland include the Iceland Low, an area of frequent storms and intensification of storm systems, and the upper air Baffin Island trough.

The climate of the Mývatn area is generally fairly dry on the whole, compared to other parts of Iceland, with annual precipitation of around 435 mm. The winter season is relatively cold, with lake ice typically persisting from early November to mid-May and a continuous snow cover for 3–4 months. Ice on the Grænavatn lake is shown in Fig. 4.6. The short summer is relatively warm compared to coastal areas in the same part of the country. A factor in the unique productivity of the lake is its temperature and insolation (Björnsson and Jónsson, 2003). As it lies in the rain shadow of the Vatnajökull glacier, the Mývatn region receives more incoming solar radiation than areas outside of the shadow. Combined with the shallowness of the lake and the heat magnifying effect of solar radiation on the biological turbidity of the lake, the water temperature can reach up to 15°C in summer. This is in great contrast to the 1–2°C of the main water body during the winter. The lake is usually covered with ice 140–250 days of the year, with the exception of the spring areas on the eastern bank of the lake, which provide valuable ice-free habitat for birds during the winter months (Ólafsson, 1991).

FIGURE 4.6 Ice on Lake Grænavatn, adjacent to Lake Mývatn. The farm of Grænavatn in the background. *Photo taken by Örn Friðriksson.*

Climate Variability in the Past

Iceland's vulnerability to climate impacts in the past, and potentially in the future, is due, in large measure, to the variability of the climate (Ogilvie, 1991, 1992, 2001; Ogilvie and McGovern, 2000; Ogilvie and Jónsson, 2001). Analyses of historical climate data for Iceland as a whole show highly variable temperatures and sea-ice incidence over the past ≅1000 years (Ogilvie, 1984a; see also, Ogilvie and Jónsson, 2001). Famines compounded by difficult climatic conditions occurred frequently, but notably in the early and late 1600s, 1740s, and 1750s. The 1780s are likely to have been the most difficult decade of the 18th century, but this was largely due to volcanic activity, in particular the *Lakagígar* eruption of 1783–84 (Bjarnar, 1965; Ogilvie, 1986; Steingrímsson, 1998; Demarée and Ogilvie, 2001). The last great subsistence famine occurred in the 1880s (Ponzi, 1995). During these difficult periods, the Mývatn region was no less affected than other regions. Although it was situated some distance from the coast, sea-ice years also brought colder conditions there.

Physical Environment, Geology, Hydrology

Much of Iceland is uninhabitable and barren, with sandy deserts, mountains, and glaciers (glaciers currently comprise approximately 11% of the area but are fast diminishing) and settlement has traditionally been focused on the coasts, with a notable exception of the Mývatn area, and other inland valleys with suitable conditions for farming. In geological terms, Iceland is young, and it is also one of the world's most active volcanic regions. See, e.g., Fig. 4.7. Lake Mývatn is situated on the western side of the mid-Atlantic ridge, close to its center, and has been affected by frequent eruptions and earthquakes, impacting both human and natural systems. The lake that dominates the landscape is one of the largest in Iceland and is 37 km^2 in size. The lake takes the form of a large shallow basin; however, it is geographically divided into two main water bodies: (1) Ytri-Flói, which consists of about 22% of the total area of the lake, situated to the northeast and (2) Syðri-Flói, which constitutes the southern and western parts of the lake. Two adjacent lakes also occur in the system: (1) Sandvatn ytra, which lies to the north and west and (2) Grænavatn to the south and east. Because of the fractured bedrock of the plate boundary, the lakes receive most of their water from abundant springs, the largest of which lie to the east of Lake Mývatn. Grænavatn is also almost completely spring fed, but between Grænavatn and Mývatn lies the only river, Grænilækur, which has an outflow into the lake. The Mývatn region is by far the most voluminous spring area in northern Iceland. In total, approximately 35 m^3/s feed the lake from its springs, resulting in a uniform inflow of water all year round (Ólafsson, 1991). The surface of the lake lies at 277 m a.s.l., which, in Iceland's subarctic location, is considered to be a mountainous area. The mean residence time of water in the lake is very short, only 27 days (Ólafsson, 1991).

The interplay between the hydrology and geology of the Lake Mývatn area has formed a unique landscape characterized by an abundance of pseudocraters

FIGURE 4.7 Hveraröndm, a geothermal area on the eastern side of Mount Námafjall in the Mývatn region. The hot springs in the area are known for their variety, including sulfurous mud springs (solfataras) and steam springs (fumaroles). The geothermal area gets its energy from the Krafla volcanic system, which feeds the lake with its warm spring water. The proposed Bjarnaflag power station is located on the western side of Mount Námafjall, about 3.5 km from the lake. *Photo taken by Astrid Ogilvie.*

and curious lava formations. Pseudocraters are rare geological phenomena, structured in a similar way to tephra volcanic cones, but formed as a result of steam explosions of shallow lakes or water-rich sediments under hot flowing lava (Thórarinsson, 1951). These pseudocraters, as well as other lava formations, form the backdrop to the landscape, both within and on the edges of the lake. The pseudocraters at Lake Mývatn are among the most notable geological features in Iceland, and are far more striking than other similar formations. Their origin lies primarily in two major volcanic eruptions that occurred c. 2000 years ago (Sæmundsson, 1991). A significant volcanic eruption also occurred in more recent times, in 1729, which also contributed to the formation of the current landscape. It also caused the permanent desertion of two farms in the region, as the farmhouses and arable land became covered with new lava. The most recent volcanic period, from 1975–1984, resulted in increased scientific understanding of plate tectonics on spreading ridges (Einarsson, 1991). Over this period, a total of nine eruptions and number of subterranean lava flows occurred, as well as the formation and spreading of fissures and faults. See, e.g., Fig. 4.8.

Mývatn Ecology

The Lake Mývatn ecological system is among the most complex of all the natural phenomena in Iceland. The name *Mývatn* literally means "Midge

FIGURE 4.8 Volcanic eruption in July 1978 at Gjástykki in connection with a series of eruptions, the Krafla Fires (*Kröflueldar*) that occurred in the Mývatn district from 1975 to 1984. *Photo taken by Ragnhildur Sigurðardóttir.*

Lake" and refers to the tremendous quantities of around 50 species of flies or midges (e.g., *Chiromidae* and *Simuliidae*), that breed in the lake and outflowing river Laxá (Lindegaard, 1979; Lindegaard and Jónasson, 1979). The midge population is of vital importance for the local ecosystem. However, see e.g., Fig. 4.9. Not only does it provide food for the waterbirds that flock to the area, as well as the fish in the lake, but their population fluctuates dramatically, and they are sometimes to be found in such great numbers that when they die their remains blanket the surrounding landscape with nutrients that are especially high in nitrogen and phosphorus, greatly increasing productivity (see e.g., Ives et al., 2008; Einarsson, 2010).

The chemistry and temperature of the spring water varies depending on the location of the springs, from cold freshwater with a constant temperature of 5°C in the southern part, to warm springs of 15–25°C with water of geothermal origin in the northern part. The spring water is very rich in phosphates, which partially explains the natural eutrophic nature of the lake (Ólafsson, 1979, 1991). The springs rich in nutrients create an excellent naturally eutrophic system. Another important factor in the ecology of the lake is its shallowness, with photosynthetic capacity extending throughout the lake bottom, with the deepest natural point being 4 m.

The foundation of the food chain in the lake is formed by diatoms. These are unicellular phytoplankton organisms, covered with a cell wall made of silica. The abundance of silica in the springs feeding the lake, and the richness of other nutrients, such as phosphorus, make the lake particularly suited for a high production of these organisms, leading to thick sediments of silica shells covering the lake bottom and giving rise to the mining of these layers during the years 1967–2003. High levels of phosphorus in the water also periodically

FIGURE 4.9 Archaeologist Tom McGovern takes schoolchildren on a field trip to learn about the archaeology of the region. The midges are in evidence! *Photo taken by Astrid Ogilvie.*

FIGURE 4.10 Lake balls in Lake Mývatn. Lake balls (*kúluskítur* in Icelandic) are a very rare growth form of the green algae *Aegagropila linnaei*. The species itself is common, but colonies of large lake balls (around 12 cm in diameter or more) are extremely rare and only occur in Lake Mývatn and two other lakes on Earth. *Photo taken by Isamu Wakana and provided by Árni Einarsson.*

results in a bloom of blue-green algae, *Anabena flos-aquae*, which have the capacity of mineralizing gaseous nitrogen from the atmosphere. These algal blooms can have drastic impacts on other parts of the ecosystem, as they affect the amount of incoming solar radiation to the lake bottom that limits photosynthesis from benthic flora (Einarsson et al., 2004).

Perhaps the most notable feature of the lake flora is a ball-shaped live form of green algae with a velvety texture (*A. linnaei*) called *kúluskítur* in Icelandic, *marimo*, from the Japanese, meaning "water plant ball," and "Lake ball," "Mossimo," or "Moss balls" in English. See e.g., Fig. 4.10. The species itself is common, but colonies of large lake balls (around 12 cm in diameter or more) are extremely rare and have been found only in Lake Akan in Japan, a lake in the Ukraine, and in Lake Mývatn (Einarsson, 2014). In Japan, the lake balls have been listed as a special national treasure. They used to grow over large sections of Lake Mývatn; however, the formerly extensive colony has been contracting over the last 30 or so years and there have appeared to be none left (Einarsson et al., 2004; Einarsson, 2014). However, after storms in the spring of 2016, small lake balls were observed washed to shore indicating that they have not yet become extinct from the lake (http://www.visir.is/kuluskitur-finnst-a-ny-i-myvatni/article/2016160609681).

HUMAN SYSTEMS: ECONOMY AND SOCIETY

Natural Resources

The original settlers to Iceland in the late-9th century brought with them a way of life that focused on a farming economy based on animal husbandry. Here it is of note that the only native land mammal in Iceland is the Arctic fox, and the livestock brought in by the settlers were the first grazing mammals in Iceland. Cattle, sheep, and horses were the main domestic animals. In early times, pigs and goats were also kept, but these soon became largely unviable, partly due to the unsuitability of the climate, but also because their grazing habits had a negative effect on the growth of young plants, in particular trees and shrubs. It appears that barley was grown in early settlement times, but this activity, although continued sporadically, possibly as late as the 15th century, was not a significant crop (Ogilvie, 1991; Júlíusson, 2013a). Thus, from settlement times onwards, the main crop was the grass that was harvested as hay and given to the livestock during the winter and early spring. The importance of the hay crop cannot be overemphasized, and the single most important activity for the farmer was to gather enough hay to keep the livestock fed over the winter. If the harvest failed, and the livestock starved, then the human population was also subject to major stressors leading to malnutrition, social dislocation, and ultimately death. Iceland's economy has thus been built primarily upon the extraction of natural resources (mainly fishing and farming) requiring sustainable management of its fragile ecosystems to avoid degradation of these ecosystem services.

Farming in the Mývatn region is both historically and currently predominantly based on sheep husbandry, with a number of farms practicing the

FIGURE 4.11 Sheep round up at the farm Baldursheim (*Baldursheimsrétt*) in the 1980s. Sheep farming has been the main agricultural practice in the Mývatn region since settlement. Each September sheep are herded from the outlying pastures (*afréttir*) into pens (*réttir*) where each farm or farmer has a separate section into which his or her sheep will be put into after ownership of each sheep has been identified. Two main pens occur in the Lake Mývatn region, one at Baldursheimur farm southeast of the lake, and another at Reykjahlíð northeast of the lake. Both of these are located close to the edge of the outlying pastures, in the highlands south and east of lake. *Photo taken by Ragnhildur Sigurdardottir.*

traditional mixed farming of dairy and sheep. The sheep provide wool and meat, and in the past were also important for the production of tallow and milk. See e.g., Fig. 4.11 for an illustration of the autumn round up. After a peak in sheep production in Iceland around 1980, followed by a state-wide production quota established to limit the nationwide overproduction of sheep products, the number of sheep declined rapidly in the Mývatn region during the 1980s, from over 8000 in 1981 to 4400 in 2014 (Statistics Iceland: http://www.statice.is/). In the past, the inhabitants of the region traveled to the closest coastal areas (around Húsavík) to undertake fishing for cod, sharks, and seals.

Governance and Land Tenure

In addition to the constraints of the natural world, governance by state and church and the system of land tenure in use in the Mývatn area were clearly of importance for farm management and the economy (Teitsson, 1973). Prior to c.1860, Icelandic society was almost completely rural in character, and Mývatn was no exception. No towns or villages existed then, only isolated farmsteads and a few fishing hamlets on the coasts. Settlement was concentrated in the coast and lowland areas and in the valleys that cut into the higher land of the interior. The Icelandic land registers reflect a society that consisted primarily of freehold farmers and tenant farmers who were extremely poor. The vast majority of farms were run by tenants, leasing their farms from the Danish king, or the church. The census taken in 1703 records that a considerable proportion of the population (about 1 in 7) consisted of landless laborers (Magnússon, 1913–1943, pp. 18–23). In the 1849 farm registry (*jarðamat*) quoted previously, ownership of land had changed hands for most farms in the Lake Myvatn region, and was in the ownership of the individual farmer.

PAST PERSPECTIVES

Resources and Sustainability

> *Öllum þessum Sveitarinnar helstu gæðum er þannig varið, að þau útheimta sérlega atorku sem því aðeins geta komið að liði, að sveitarmenn hafi bæði nægilega krafta og gott samheldi. [The utilization of the district's resources requires the population of the area to be exceptionally hardy. However, this would be to little avail if the inhabitants did not have strength enough and the ability to cooperate well.]*

Sustainability, or sustainable development, as defined by the Brundtland Commission (1987) is "... development that meets the needs of the present without compromising the ability of future generations to meet their own needs" (http://www.un-documents.net/ocf-02.htm). At the 2005 World Summit

FIGURE 4.12 Inset of areas of erosion, fertile fields, and Lake Mývatn in the background. *Photo taken by Astrid Ogilvie.*

on Social Development, it was noted that this requires the reconciliation of environmental, social, and economic demands (United Nations General Assembly, 2005; https://daccess-ods.un.org/TMP/2200996.57773972.html). Although such recommendations are firmly rooted in modern times, they fit well with many of the practices of the inhabitants of the Mývatn area in the past. Part of the success of the Mývatn region to sustain a continuous human population since the first settlement of Iceland is undoubtedly due to the availability of resources such as the eggs of waterbirds and Arctic char from the lake, making food resources more variable than in most other higher elevation areas habituated by the early settlers. Wetlands in the region helped to provide a reliable hay harvest; however, the dryer ecosystems in the region did suffer from erosion and soil degradation. See e.g., Fig. 4.12.

Clearly, living conditions were generally harsh in the past in Iceland, and the Mývatn area was no exception. To ensure long-term survival in the community, farmers needed to be invested in using their most vital natural resources as sustainably as possible, and most of these were directly connected to water. The complex food web and biogeochemistry of the lake created unique wildlife resources on which the people depended, and adjacent wetlands produced the most important sources of hay for the animal husbandry practiced in the area. Drawing on one of the valuable historical sources noted earlier, the seasonal round may be paraphrased thus:

> *During Spring, the most important activities relating to subsistence involve manuring the homefields, acquiring coal and firewood, overseeing the Eider duck nesting (for the harvesting of down) building and maintaining houses, fishing,*

and harvesting valuable herbs in the mountains. In July and August, and partly in September, the hay is harvested, at least until snow and frost prevents this. The difficulty is that these activities have to take place over 3 months as the Spring and Autumn here are usually extremely severe. In Winter, people work trade goods out of wool and make tackle for fishing and sealing.

Translated and paraphrased by Astrid Ogilvie from a Sheriff's letter for 1779, see references.

The *jarðamat* account (1849) echoes these activities, listing the most valuable resources of the district, in particular grass, fish and birds' eggs, and also the woodlands and peatlands used primarily for firewood, and Iceland moss, *Cetraria islandica* (presumably the "valuable herbs" described earlier, despite being a lichen and not a herb per se) highly valued for its nutritional and medicinal properties. Both the woods and the moss are noted as diminishing in 1849, in contrast to the lake and wetland resources that were described as abundant, albeit often challenging to harvest. The varying resources are discussed in the following section, with a relevant quotation from the *Jarðamat* at the start of each section.

Grass and Hay

... einnig má telja afréttarlönd er sveitinni tilheyra. Yfirhöfuð eru landkostir töluverðir, en þó mjög misjafnir og heyskapur eins. [Outlying pastures also belong to the district. On the whole, the habitable land is reasonably fertile, but it is variable, and the hay production also.]

The context here is that from early settlement times on Iceland, the success or failure of the all-important grass crop, coupled with winter rangeland grazing, was the one aspect of the economy on which all else rested. With its North Atlantic location, marginal for agriculture, grass was the only viable crop, and, as noted earlier, the economy focused primarily on animal husbandry until comparatively early times. The successful harvesting of hay was thus the farmers' most important annual task. If there was not enough hay in the winter to feed the livestock they could die, and this could lead to famine and death among the human population. This unfortunate train of events occurred many times in Iceland's history, and not least in the Mývatn district (Friðriksson, 1969; Bergthórsson et al., 1988; Júlíusson, 2001; Ogilvie and McGovern, 2004).

The grass crop was not cultivated but grew naturally in three distinct areas, each of which was important for Icelandic farming. The three different types of grass fields were (and are): the so-called *tún*, or homefield, close to the farm; the *engi*, or meadow, slightly further afield; and the outlying pastures called *úthagar* or *afrétti*. In some ways the *tún* may be said to have been the most valuable; its importance is reflected in that it is the root of the English word "town." As time went on it became normal practice to manure the

FIGURE 4.13 Dairy cows at Lake Grænavatn. *Photo taken by Astrid Ogilvie.*

homefield, and the higher quality hay produced on it was given primarily to the cattle who required far more hay than the more low-maintenance sheep. The sheep usually roamed further afield and were rounded up in the autumn for slaughter. A further aspect of the farming economy was the practice of transhumance, with cattle and sheep being taken to richer summer pastures or shielings; *sel* in Icelandic. Another economically important grass component is the *Framengjar*, an area of wet meadows and ponds adjacent to the southern edge of Mývatn. See e.g., Fig. 4.13.

The way the grass grows in Iceland makes it an excellent sheep-rearing country. In the early part of the summer, new growth is available on low-lying ground, whereas on higher ground growth occurs later in the season, thus providing a continuous supply of grass (Friðriksson, 1973). In the past, haymaking was often a difficult and laborious task. The reaping of grass for hay was done by hand in the autumn, and was more or less a continuous process, often from early July to early September, depending on the state of the ground and the weather. First the grass on the homefields was mown, then the outlying pastures. After it was cut the hay would be left on the ground to dry. It was then usually turned regularly using rakes to accelerate the process. See Fig. 4.14 for an illustration. Later it would be collected for winter storage (Ogilvie, 1984b). Tools for haymaking remained virtually unchanged from early times to the 19th century. The better quality hay from the homefield would generally be reserved for cattle. Until around the mid-20th century the uncultivated vegetation provided horses and sheep with almost all their fodder, and also supplied the cattle in summer.

FIGURE 4.14 Raking hay at Baldursheimur farm in 1953. The photograph shows Hólmfriður Hemmert, grandmother of Ragnhildur Sigurðardóttir and a young man named Sveinn. *Framengjar* (outlying pasture) wetlands in the background. *Photo taken by Sigurður þórolfsson, father of Ragnhildur Sigurðardóttir.*

Waterbirds of Lake Mývatn

Í vatni því er í Sveitinni liggur og kallað er Mývatn eru margar eyjar og hólmar og í sumum þeirra allgóður heyskapur og nokkurt andarvarp. [A lake in the community is called Mývatn where there are many islands and islets. On some of them excellent hay is produced, and ducks also nest there.]

The variety and density of duck and waterbird species is one of the most significant elements of the ecology of Lake Mývatn and is perhaps the feature of the ecosystem that is best known internationally. Possibly the earliest report on the birdlife of Mývatn, listing different species, was made in 1747 by Jón Benediktsson, an Icelandic Sheriff (Benediktsson, 1957). International recognition of Lake Mývatn as a unique breeding habitat for ducks may be said to have resulted from a traveler, Faber, who visited the area in 1822 and subsequently published an account (Faber, 1822). Throughout the 19th century, numerous travel narratives describe the abundant duck fauna at Mývatn.

All breeding duck species in Iceland, with the exception of a last settler, the Common Shelduck (*Tadorna tadorna*), breed in the Mývatn region. Between 10 and 15,000 pairs of waterbirds from 20 species nest by the lake, including 16 species of ducks. In total 28 duck species have been identified on the lake, some of them irregular visitors, however (Garðarsson, 1991). The composition of duck species is unique in its mix of Eurasian and North American as well as arctic and boreal species (Jónasson, 1979). Most of the ducks are migratory, arriving in late April to early May. Barrow's Goldeneye (*Bucephala islandica*) is the most characteristic bird on lake Mývatn and breeds only in this area. The Tufted Duck (*Aythya fuligula*) is the most common, but Scaup (*Aythya marila*), Wigeon (*Anas penelope*), Mallard (*Anas platyrhynchos*), Teal (*Anas crecca*), Red-breasted Merganser (*Mergus serrator*) and the Red-necked Phalarope (*Phalaropus lobatus*) are also prevalent. Whooper Swans (*Cygnus cygnus*) and Greylag Geese (*Anser anser*) are often seen. Ptarmigan (*Lagopus mutus*) are particularly common. Birds of prey also occur in the area and include Gyrfalcons (*Falco rusticolus*), Short-eared Owls (*Asio flammeus*) and Merlins (*Falco columbarius*).

The earliest account of egg collecting as an important natural resource in the lake Mývatn area is in the *Jarðabók* from 1712 (Magnússon, 1913–1943). In fact, archaeological evidence reveals that collecting eggs has been an ongoing activity since the area was first settled over 1100 years ago. Hunting of ducks, however, has hardly been practiced at all (McGovern et al., 2006). Analyses of eggshells from middens at Skútustaðir by co-author Megan Hicks show that all duck species whose eggs have been collected have persisted throughout these centuries of resource management, with the addition of a few new species as they have settled the area (Hicks et al., 2016). Thienemann (1827) recounted that the people of the Mývatn community regulated their egg collection by never emptying any nests, and Shepherd (1867) described that tradition decreed that at least 4–5 eggs were left at each nest. This practice, which is likely to have had a much longer history, prevented overharvesting of eggs during years when fluctuating duck populations were low. In this regard, the breeding success of many of the duck species has been shown to follow the productivity of chironomid larvae on the lake bottom (Einarsson et al., 2006). Duck eggs are still collected for human consumption in the Mývatn area, but the importance of this resource has declined in modern times. Today an estimated 10,000 eggs are collected annually (Umhverfisstofnun, 2011), which contrasts with the 41,000 eggs collected in 1941 (Guðmundsson, 1979).

Fish Species of Lake Mývatn

> *Silungsveiði er og í vatninu á flestum tímum ársins. [Year-round Arctic char fishing occurs in the lake.]*

The high primary productivity of the lake and the tremendous quantities of midge larvae on the lake bottom have made the lake an exceptionally favorable habitat for Arctic char, *Salvenius alpinus* (Jónasson, 1979). In addition to the char, brown trout (*Salmo trutta*) and stickleback (*Gasterosteus aculeatus*) populate the lake, with the char being the most important species throughout documented history. The outflowing river Laxá ("salmon river") has, as indicated by its name, a fourth species, Atlantic salmon (*Salmo salar*). Lake Mývatn used to be among the most productive fishing lakes in Iceland, with abundant stocks of Arctic char forming a major resource in the region, creating a fishing, and fishprocessing culture characteristic to the area.

The earliest account of fishing being of particular note at Lake Mývatn is from the saga of the saintly Bishop Guðmundur Arason (1161–1237), commonly referred to as *Guðmundur góði* or "Guðmundur the Good." His story is rich in apparent miracles that were said to be performed for the sake of the people in Iceland. One such miracle was to throw a net into Lake Mývatn, and to create an abundant fishing lake that before had only had poor fishing (*Jarteinabók Guðmundar biskups*, 1948, pp. 437–478).

Archaeological research in the area shows that catching Arctic char for human consumption has been an ongoing activity since settlement over 1100 years ago (McGovern et al., 2007; Hicks, 2016). Historical accounts from the 18th and 19th centuries show repeated fluctuations in fish stocks in the lake. In a letter written by a local farmer, Stefán Stefánsson, to the fisheries scientist Bjarni Sæmundsson (published in a paper by the latter; see Sæmundsson, 1923), it is stated that a *Veiðifélag* or a fishing monitoring club was founded in 1905 regarding the fishing in Mývatn. The founding of this club also initiated a long-standing monitoring of fish catches in the lake, as well as the start of experiments with fish breeding to improve the productivity of the lake. Two earlier attempts to regulate the fishing in the lake had been made in the 19th century, in 1889 and 1895 (Sæmundsson, 1923). In the letter from Stefán Stefánsson, published as a part of Sæmundsson's paper, Stefán described booms and busts of fishing in the lake during the 19th century, from the time he could first remember (he was born in 1854) but also from oral accounts by elders in his community. He also provided poems on the fishing as evidence of either a lack or abundance of fish based on the tone of the verses. Stefán also mentioned that the year 1849 was particularly bad for fishing in the lake. It is interesting to compare this with the *Jarðamat* from 1849, the account that has served to illuminate this chapter. Table 4.1 lists farms in the community, and includes an assessment of fishing in the lake as a natural resource for each farm. For some of the traditional fishing farms, such as Grímsstaðir, Álftagerði, and Geiteyjarströnd, fishing is not mentioned as a resource. Other farms, which should have been important fishing farms, such as Ytri Neslönd, Syðri Neslönd, Reykjahlíð, and even Vogar, were accorded small and insignificant fishing resources. Only five farms stated that fishing was an economic option in their resource management.

By reading the *jarðamat* from 1849 alone, one would not be inclined to describe Lake Mývatn as a fishing lake with tremendous productivity. However, as noted earlier, in his letter to Bjarni Sæmundsson, Stefán Stefánsson wrote that 1849 was a particularly bad year for fishing. He even recounted a story that he remembered regarding 1849 to illustrate this. He wrote that when he was young he heard a woman tell his mother that when she was pregnant with her first child, she wanted nothing as much as some trout, but none was to be had. He could be certain that it was the year 1849 because the child that the

TABLE 4.1 Fishing as a Resource by Farm According to Jarðamat (1849)

Hofsstaðir[b]	Fishing not mentioned
Geirastaðir[a]	*Veiðiskapur bæði í Mývatni og Laxá stundum til töluverðs hagnaðar* [fishing both in Mývatn and Laxá sometimes in a reasonable quantity]
Vindbelgur[a]	*Silungsveiði til góðs hagræðis einkum á vetrum* [trout fishing good, especially during the winter]
Ytri Neslönd[a]	*Silungsveiði lítil* [trout fishing poor]
Syðri Neslönd[a]	*Ekki teljandi silungsveiði* [trout fishing not worth mentioning]
Grímsstaðir[a]	Fishing not mentioned
Reykjahlíð[a]	*Silungsveiði er síður teljandi kostur vegna óstöðugleika* [trout fishing inconstant/variable]
Vogar[a]	*Silungsveiði er einnig nokkur en kostnaðarsöm vegna sérlegs slits á netjum og bittu* [trout fishing reasonable but expensive because of damage to the nets and hooks]
Geiteyjarströnd[a]	Fishing not mentioned
Kálfaströnd[a]	*Silungsveiði er þar talsverð sumar og vetur en mjög kostnaðarsöm* [trout fishing considerable in summer and winter but costly]
Garður/Brjánsnes[a]	*Silungsveiði á vetrum til hagræðis* [trout fishing in the winter reasonable]
Grænavatn	Fishing not mentioned
Skútustaðir[a]	*Við Mikley er dráttarveiði á sumrum og einnig er netjaveiði framan af vetri til hagræðis* [at Mikley (the largest island in the lake) there has been reasonable fishing with nets (involving 2 people dragging the nets together) and the usual fishing with nets in early winter]
Álftagerði[a]	Fishing not mentioned
Haganes[a]	*Silungsveiði er þar til hagræðis bæði í vatninu og ánni* [trout fishing reasonable both in the lake and river]
Arnarvatn[b]	Fishing not mentioned
Sveinsströnd/Litlaströnd	Fishing not mentioned

Continued

TABLE 4.1 Fishing as a Resource by Farm According to Jarðamat (1849)—cont'd

Baldursheimur	Fishing not mentioned
Árbakki	Fishing not mentioned
Gautlönd	Fishing not mentioned
Helluvað[b]	Fishing not mentioned
Hörgsland	Fishing not mentioned

[a]Farms that all have direct access to the lake and are traditional fishing farms.
[b]Farms that have direct access to the river Laxá. The other farms have nondirect access to the major fish-producing water bodies, but most of them have access to smaller fishing lakes in the region or tributaries, such as river Kráká.

woman was carrying was still alive at the time when he was writing, and she was now aged 72.

Stefán noted other fluctuations in the fish stocks during the 19th century, and described an increase in fishing of Arctic char in 1864–74, which peaked in 1878–79. He estimated that between 100 and 200,000 fish were caught per year during those years, or even more. A few years earlier, in 1859, a great loss of livestock occurred in Mývatn and surrounding communities and again due to the severity associated with sea ice off the coasts in 1869. Stefánsson mentioned that during those harsh years for livestock farming, the lake resources prevented people from the entire county from starving to death. He also stated that an unusually warm summer in 1880 killed huge quantities of char in the lake.

Other disturbances affecting the fishing stocks of the lake include those reported by Eggert Ólafsson and Bjarni Pálsson who traveled through Iceland in the years 1752–57, having been appointed by the King of Denmark and the Danish Scientific Academy (Ólafsson, 1772). They traveled to Lake Mývatn in 1752 and reported meeting farmers who remembered fishing in the lake prior to the 1729 eruption, when lava entered the northern side of Ytri Flói of the lake. According to the farmers, the fishing in the lake was severely affected by the eruption, due to the heating effect it had on the water body of the lake. By 1752, the fishing had not yet recovered to preeruption standards.

Both before and after the establishment of the fishing management association of Lake Mývatn, comprehensive monitoring had been conducted on annual fish catches. From 1900 to 2014, the average catch was 27,798 fish/year (Guðbergsson, 2015). The fishing had, however, fluctuated considerably throughout the last century, peaking in the 1920s, when it exceeded 100,000 fish/year, to about 31,000 fish/year from 1930 to 1969. The char catch has reflected the availability of midge larvae in the lake, and has, in tandem with the larvae, decreased rapidly since c.1970. From that time, up to 2014, the catch has only been 13,288 fish on average per year. A 10-year average between 2004 and 2014 has only seen 3623 fish/year, with an ultimate low catch of 2401 fish in 2014, including only 645 Arctic char and 175 brown trout. The stock has crashed seriously, and fishing of char has been heavily regulated since 2011 (Guðbergsson, 2015).

Woods, Iceland Moss, and Edible Plants

Hér og hvar eru dálitlar skógarleifar og grasatekja. [Here and there are small wooded places and areas where Iceland moss may be found.]

Research suggests that when Iceland was first settled in the late-9th century woodlands were far more extensive than they later became (Loftsson, 1993). An early historiographical source, *Íslendingabók* ("The Book of Settlements") states "At that time Iceland was covered with woods from the mountains to the seashore" (*Íslendingabók*, 2006). The traditional view has been to follow this statement and assume that Iceland was quickly deforested by the early settlers and their grazing animals with the native birch and willow being cleared for pasture land, charcoal production, and negatively impacted by the livestock. Although it is likely that approximately one-third of the Iceland was wooded when the settlers first went there, and the forested area is now approximately 0.3%, current research suggests that the history of woodland exploitation is more complex than previously thought. Thus, for example, a study by Church et al. (2007) identified two distinct phases of charcoal production and woodland exploitation (in the vicinity of þorskmörk in southern Iceland) during the Viking Age and early medieval era, the first within the first two centuries of settlement (AD 870–1050) and the second phase more than a century later (AD 1185–1295). Evidence also exists that erosion in the 18th century had a negative impact on woodlands (Ogilvie and McGovern, 2000).

The main woodland species in Mývatn, as elsewhere in Iceland, consists of willow species such as woolly willow (*Salix lanata*) and tea-leaved willow (*Salix phylicifolia*) as well as dwarf birch (*Betula nana*). According to the land registry, *Jarðabók* (Magnússon, 1913–1943) certain farms, for example, Helluvað, Gautlönd, Arnarvatn, Grænavatn, Reykjahlið, Fagranes, and Grímsstaðir had enough wood and "wood hay" (meaning willow leaves and branches) for fuel. A more valuable form of birch is the taller *B. pubescens* (see e.g., Fig. 4.15) that would have more diverse uses as it provided fuel, charcoal, and timber, and a prime habitat for rock ptarmigans (*Lagopus muta*), a wildfowl that was widely harvested in the Mývatn region since the settlement of the area (McGovern et al., 2006, 2007). According to the *Jarðamat* (1849), the main use of the birch woodlands was to make charcoal for home and for blacksmithing for the purpose of sharpening the scythes (for reaping the hay). These were heated to sharpen and harden them before using them in the

FIGURE 4.15 Birch woodlands (*Betula pubescens*) in the foreground of Láke Mývatn. *Photo taken by Astrid Ogilvie.*

hayfields. In 1849, all farms having birch woodlands claimed that they only had small remnants left, and that they were only sufficient for the above-mentioned charcoal-making purposes for their individual households (*Jarðamat*, 1849). The farms Geirastaðir, Vindbelgur, Grímsstaðir, Reykjahlíð, Vogar, Geiteyjarströnd, Kálfaströnd, and Grænavatn all mention having these resources, whereas the farm of Skútustaðir claimed that they had the right to use forests on the land of other farms, while mentioning also that those resources were already depleted. It is noteworthy that, in spite of ongoing soil erosion in the Mývatn area, and the very active erosion processes mentioned in *Jarðamat* (1849), the same farms that were said to have forest resources in 1849 still have birch woodlands today (Sigurðardóttir, 1992).

The vegetation in the Mývatn area includes several species collected for human consumption: crowberries (*Empetrum nigrum*), bilberries (*Vaccinium myrtillus*), and bog bilberries (*Vaccinium uliginosum*). These two latter are called *aðalbláber* and *bláber*, respectively, in Icelandic. In English both are commonly referred to as "blueberries." Angelica (*Angelica archangelica*) was traditionally harvested for food in the Mývatn area, especially the roots. All of those species have been found to have medicinal purposes and are used today as commercially available supplements. For typical vegetation see e.g., Fig. 4.16.

Of particular interest is the Iceland moss (*C. islandica*) that is noted in the *Jarðamat* (1849). This is a lichen species, occurring in the mountainous heathlands of Iceland, and which was an important source of food. This moss was ingested as tea, replaced grains in some traditional food recipes, or was cooked in milk to produce a soup. Traditional medicinal uses included cures for stomach problems and respiratory illnesses. Iceland moss has been found to

FIGURE 4.16 Typical dry heathland vegetation of the Mývatn area, with willow shrubs (*Salix phylicifolia* and *Salix lanata*) dwarf birch (*Betula nana*) and bilberries (*Vaccinium uliginosum*). *Photo taken by Astrid Ogilvie.*

activate the immune system (Olafsdóttir et al., 1999) and has been used to treat a variety of ailments from tumors (Hartwell, 1971) to sore throats and coughs, tuberculosis, bronchial asthma, and gastric disturbances (Kartnig, 1987) and is currently processed as a both a herbal medicine and as the active ingredient in throat lozenges. The consumption of Iceland moss has thus no doubt played a role in the health and survival of the population in Iceland in the places that had access to lichen collection areas such as the Mývatn district.

Jarðamat (1849) stated that the Iceland moss was regarded as one of the more important resources, but that it was diminishing because of overharvesting. Farmers who had access to this stated that they only had enough for home consumption and complained about people from lower elevation communities coming on their lands to harvest without their permission.

CURRENT CONCERNS

Energy Resources and Related Issues

It has long been evident that harnessing of energy through both hydro and geothermal power should be feasible due to the volcanic activity and hydrological conditions of the region. However, although the heavy agricultural focus of the region prevailed, it was not viable to develop these energy resources significantly. Nevertheless, well before the restrictions on sheep husbandry took effect during the 1980s, the Icelandic government showed interest in harnessing and mining the natural resources of the Mývatn region. A relatively small hydropower station was built in three stages in 1939, 1953, and 1973 in the outflowing Laxá river, producing in total 27.5 MW of energy. Explorations for geothermal energy at the Krafla volcanic system started in 1974. The Krafla power station started operating in several steps from 1978 to 1999, now producing altogether 60 MW of energy.

In 1966, in collaboration with an American corporation, and several municipalities in northern Iceland, the Icelandic government established a company to mine the diatomite from the lake. The mining and processing of diatomite operated in the Lake Mývatn area from 1967 continuously until 2003. In connection with the mining operations, a small 3 MW geothermal power station was established at Bjarnaflag to provide energy for the factory. This mining operation made use of approximately 2–4 m thick sediments of the skeletal remains of unicellular organisms, or diatoms, living on the lake bottom, that have been built up over the past c.4000 years since the lake was originally formed. These remains form the substance known as *diatomite* that has a fine particle size and a low specific gravity. It is extremely useful as a filter media, an absorbent, and as a filler for rubber, paint, and plastics, as well as an insecticide and pesticide. When crushed into a powder, it is usually called "diatomaceous earth." This presence of diatomite in Lake Mývatn provided the opportunity for a mining industry to be developed.

These industrial developments, in particular the mining operation, and the 1973 hydropower development in the Laxá river and the Krafla power station, created massive political debate, both within the Mývatn community, and in Iceland as a whole. Fig. 4.17 shows the Krafla power station. It is likely that these debates influenced the environmental movement in Iceland that led to the establishment of the nature conservation area in the region in 1974. The debate focused around a massive hydropower plan suggested by the government

FIGURE 4.17 The Krafla power station. Lake Mývatn in the background. *Photo taken by Ragnhildur Sigurðardóttir.*

involving the movement of large glacial water bodies into the Mývatn area, narrowly bypassing the lake, and creating a reservoir with a higher water table in the lake and then filling up of the adjacent valley. These plans created strong resistance among Mývatn residents who decided to take matters in their own hands. Gathering together one summer night, unknown residents used dynamite to destroy the dam that would have increased the water table of the lake. This act resulted in a very strong communal bond among the people of the Mývatn region and a much reduced hydropower station with minimal effect on the lake itself.

The situation as regards the diatomite-mining operation had, however, a very disruptive effect on the Mývatn community, creating a social imbalance, disputes, and unrest that escalated with time and did not begin to heal until the mining operation closed in 2003. The main protagonists were the traditional farming community on the one hand, and the inhabitants of the newly formed village by the diatomite plant on the other. In this community of about 400 inhabitants, a separate sports club and a school for children were established and operated separately for several years. Surprisingly, therefore, the closing of the mining operation in 2003 had only minimal effects on the economy of the community (Arinbjarnarson, 2007).

Drastic fluctuations within the food chain have been observed since c.1970 resulting in serious consequences for the top predators, such as birds and fish (Einarsson et al., 2004). Numbers of the previously abundant Arctic char have crashed, following a decline in benthic midge larvae (Einarsson et al., 2004; Guðbergsson, 2015). The food sources of the brown trout, a relatively minor component in the lake fauna, include freshwater invertebrates and sticklebacks and the trout have therefore been less affected by the crash of the midge larvae in the lake (Guðbergsson, 2015). However, this could change as the stickleback population is reported to have crashed in spring 2016 (http://www.visir.is/hornsili-i-myvatni-i-sogulegri-laegd/article/2016160329132).

The fluctuations in the lake ecology have long been suggested to be the result of the diatomite mining in the lake (Garðarsson, 1991; Einarsson and Hafliðason, 1988; Einarsson et al., 2002). This has added to the tensions within the Mývatn community and between environmentally conscious Icelanders and the government. The burden of proof has been placed on ecologists researching the lake. However, to ascertain causes and effects proved to be difficult because of the complexity of the lake system, and was not fully realized until after the mining operations had already stopped (Ives et al., 2008).

By the year 2000, approximately 40% of the diatomaceous mud at the bottom of the northern part of the lake had been dredged, changing the depth from 1 m to 3–4 m, and thus losing an essential bioactive layer of mud from the surrounding area into the deep manmade troughs (Einarsson et al., 2004). In addition to affecting the benthic communities in the lake, this consequently affected other groups of species in the food chain. Furthermore, it is thought to have contributed significantly toward eutrophication over the natural levels in that the turbidity from the pumping of 2–3 m of the sediments may have affected primary productivity on the lake bottom, and released nutrients that had previously been sequestered within the deep sediments (Umhverfisstofnun, 2011).

It has been suggested that the most likely single cause for the collapse of the Lake Mývatn algal mat and the lake-ball colonies is due to increased intensity or duration of cyanobacterial blooms. This, in turn, may be attributed to the mining operation in the lake which may have altered the characteristics of these blooms and hence made the lake uninhabitable for the algae (Einarsson, 2014). At the moment of writing (June 2016), however, as noted earlier, some regenerating lake balls have been found.

Even though the mining operations have ceased, questions regarding its environmental footprint on the lake have still not fully been resolved, nor have any mitigating measures been taken. Other potential sources of pollution from industrial operations, such as the geothermal power stations have also not been fully assessed (Sigurðardóttir, 2010). These power stations at Lake Mývatn harvest energy through volcanic activity, whereby groundwater is heated by the ever-present molten lava situated 3–7 km under the surface. The superheated water and steam carry a significant load of chemicals to the surface (Einarsson, 1991; Ármannsson, 1993; Kristmannsdóttir and Ármannsson, 2004). In connection with the mining operation, these chemicals were deposited (dumped) in manmade ponds on the top of the fractured bedrock area on the eastern side of the lake. After the mining operation ceased in 2003, the water from the boreholes have been repurposed to feed a bathing pool for the tourism industry. Groundwater from both of these geothermal fields feed the warm springs in the lake (Ólafsson, 1991; Einarsson et al., 2004).

Despite the dire straits of the lake, and the potential polluting effects of the geothermal liquids, as of the present date 2016, the Icelandic government intends to increase the size of the two geothermal power stations as part of their plan for new energy developments in Iceland. According to this plan, the Bjarnaflag power station will increase from 3 to 80 MW, and the Krafla power station will also increase by adding 150 MW to the current 60 MW it produces. See e.g., Fig. 4.18. Multiplying the energy harnessing adjacent to the lake will, in turn, multiply the deposition of densely chemical liquids from the boreholes on the surface. These include both bioactive elements and heavy metals, and may subsequently be transported via groundwater or by air into the already stressed lake ecosystem.

Trends in Tourism

After having been virtually isolated for a large part of the year for over 1000 years, the Mývatn region is now well connected to the road system in Iceland. Despite its isolation in the past, its uniqueness and beauty have

FIGURE 4.18 Geothermal area in the Mývatn region with Bjarnaflag power station in the distance. The pond in the foreground consists of mineral-rich water pumped from the ground and left to settle and evaporate on the surface, potentially feeding shallow groundwater, and subsequently Lake Mývatn with mineral-rich water. This may alter the chemical cycling and the ecological processes of the lake. *Photo taken by Astrid Ogilvie.*

attracted visitors from the 18th century onwards. However, in recent years, tourism in the Lake Mývatn region has expanded dramatically, following a national trend (all tourism statistics are from http://www.ferdamalastofa.is). In 2014, the Mývatn region was the ninth-most popular summer destination for foreign tourists, and 35% of all tourists visited the area. An illustration of the phenomenon may be seen in Fig 4.19. Also, 10.1% of winter guests visited the area. As regards Icelandic tourists, Myvatn was the eighth-most popular destination and more than 17% of Icelanders who traveled domestically visited the Mývatn region. During the three summer months of 2014, the area received some 140,000 tourists. Based on statistics of road traffic, this represents a doubling of numbers since the summer of 2001. A total of 408,604 foreign tourists traveled to Iceland in 2014, which gives an estimate of 139,000 foreign tourists visiting Mývatn in that year. In 2014, the number of inhabitants in the Skútustaðir municipality was 378. In other words, there were 368 visitors for every inhabitant. The mass tourism has been suggested to be the major factor affecting the overeutrophication of the lake, as septic systems in the area lack the capacity to process the waste created by the huge increase of tourism in the area that doubled from 2001 to 2014. The polluting effects of the tourism industry have been contested (Jónsson, 2016); nevertheless, repairing the septic systems is currently being discussed as the major mitigating measure to try to reverse the disturbing situation regarding the lake (http://www.mbl.is/frettir/innlent/2016/06/21/minnka_tharf_alagid_a_myvatn/).

FIGURE 4.19 Tour buses converging on the area. Lake Mývatn is one of the most popular tourist destinations of Iceland. *Photo taken by Astrid Ogilvie.*

WATER AND SUSTAINABILITY: POSTSCRIPT

The 1849 *Jarðamat* farm evaluation report assessed the number of livestock that each farm could sustainably carry through both favorable and unfavorable climatic conditions, taking into account the vulnerability of the farming system to the variable climate. The report also estimated the amount of hay the farmers could harvest from their fields and wetlands and the threats the environment placed on each farming unit. Water was clearly of supreme importance to the mid-19th-century farmer, as preserving wetlands from erosion to maintain access to wetland hay proved to be the most challenging task some of the farms endured each year. The lake and its fauna also represented vital resources, with char fishing and collecting duck eggs stated to be essential for the viability of the community. As noted earlier, archaeological research on middens from Lake Mývatn farms has shown that harvesting these resources has been continuously ongoing from the start of the settlement in the area from the late-9th century. These resources thus appear to have been sustainably harvested for 1100 years, until the instability of the ecosystem and the degradation of the lake set in around 1970. It is possible to see the environmental history of Lake Mývatn and the community of the region as a microcosm of social change, both in Iceland and internationally, from an agricultural society to a modern one focusing on industry and tourism. However, the transition at Lake Mývatn is compounded by the intensity of all the disturbances placed on the system. In all cases, the mining and energy operations were promoted in an effort to create jobs and increase economic growth. However, it is possible that there will be long-term negative effects on the economic sustainability of the area. The long-term consequences on the sustainability of the lake and its natural systems are unclear, but the seriousness of the current ecological crisis makes finding mitigating actions to be of huge importance.

The fate of Lake Mývatn and its community shows how important it is to look at water and social systems in vulnerable mountainous regions holistically and in a cross-disciplinary way, including understanding the need to preserve both nature and its human inhabitants. Such an approach will help to understand the vulnerability of such systems and the need to look beyond short-term gain and favor long-term sustainability. Lessons may also be learned from the past as regards how an ecological system can be pushed beyond a productivity threshold to land degradation. Such studies allow for a close analysis of the survival of human communities and the resilience of natural ecosystems in vulnerable systems. In the case of Lake Mývatn all measures are important, but reversing the trend of ecological disruptions since about 1970s may prove to be difficult.

However, signs exist that the problems that beset the lake are being taken very seriously. The Lake Mývatn ecosystem and the Mývatn nature reserve have been a source of serious concern in the past few years. The Environmental Agency of Iceland published its first report on the status of nature reserves in Iceland in 2010, placing the Mývatn reserve on an orange list of reserves in danger (UST, 2010). In its next report published in 2013, the Mývatn reserve was upgraded to the red list of reserves in serious danger, where it still resides (UST, 2013). The rationale for the listing is based on the status of the lake balls in the lake, anthropogenic eutrophication, the crash of the stocks of Arctic char, and the pressure of increased tourism on septic systems and trampling on sensitive vegetation and geological formations. However, as of May 2016, the Ministry of the Environment in Iceland has formed a taskforce to mitigate the ongoing environmental crisis in the area. This is a hopeful sign for the future preservation of the national wetland mountain treasure centered on Lake Mývatn.

ACKNOWLEDGMENTS

The authors wish to acknowledge: National Science Foundation (USA) Award 1446308 "Investigations of the Long Term Sustainability of Human Ecodynamic Systems in Northern Iceland" and RANNÍS (Research Council of Iceland) Award 163133-051 "The Mývatn District of Iceland: Sustainability, Environment and Change ca. AD 1700 to 1950." They also wish to thank their many "Mývatn" colleagues for collaboration and friendship, in particular Professors T.H. McGovern, I.A. Simpson, Árni Einarsson, and Christian Keller. Thanks are also due to the residents of the Mývatn district for their cooperation and help.

REFERENCES

Annálar 1400–1800 1-VII. 1922–1998, Hið íslenzka bókmenntafélag, Reykjavík.

Arinbjarnarson, H., 2007. Samfélagsleg áhrif kísiliðjunnar í Mývatnssveit (B.A. thesis). University of Akureyri, Akureyri. http://hdl.handle.net/1946/737.

Ármannsson, H., 1993. Jarðhitakerfið í Námafjalli. Efnafræðileg úttekt. Orkustofnun. OS-93053/ JHD-29 B. 30 pp.

Batt, C.M., Schmid, M.M.E., Vésteinsson, O., 2015. Constructing chronologies in Viking Age Iceland: increasing dating resolution using Bayesian approaches. Journal of Archaeological Science 62, 161–174.

Benediktsson, J., 1957 (Originally published 1747). Description over Nordersijssels udi Island Situation, Bjærge, Klipper, Fjælde, Dale og adskillige Producter og Beskaffenhed til Land og Vand. In: Sýslulýsingar 1744–1749, Sögurit 28, pp. 218–251 (Reykjavík).

Bergthórsson, P., Björnsson, H., Dýrmundsson, Ó., Guðmundsson, B., Helgadóttir, A., Jónmundsson, J.V., 1988. The effects of climatic variations on agriculture in Iceland. In: Parry, M.L., Carter, T.R., Konijn, N.T. (Eds.), The Impact of Climatic Variations on Agriculture, Assessments in Cool Temperate and Cold Regions, vol. 1. Kluwer Academic Publishers, Dordrecht, Boston, London.

Bjarnar, V., 1965. The Laki eruption and the famine of the mist. In: Bayenschmidt, C., Friis, E.J. (Eds.), Scandinavian Studies, The American-Scandinavian Foundation. University of Washington Press, pp. 410–421.

Björnsson, H., Jónsson, T., 2003. Climate and climatic variability at Lake Myvatn. Aquatic Ecology 38, 129–144.

Church, M.J., Dugmore, A.J., Mairs, K.A., Millard, A.R., Cook, G.T., Sveinbjarnadóttir, G., Ascough, P.A., Roucoux, K.H., 2007. Charcoal production during the Norse and early medieval periods in Eyjafjallahreppur, Southern Iceland. Radiocarbon 49 (2), 659–672.

Demarée, G.R., Ogilvie, A.E.J., 2001. Bon baisers d'Islande: climatological, environmental and human dimensions impacts in Europe of the *Lakagígar* eruption (1783–1784) in Iceland. In: Jones, P.D., Ogilvie, A.E.J., Davies, T.D., Briffa, K.R. (Eds.), History and Climate: Memories of the Future? Kluwer Academic/Plenum Publishers, New York, Boston, Dordrecht, London, Moscow, pp. 219–246.

Dugmore, A.J., Larsen, L., Newton, A.J., 2004. Tephrochronology and its application to Late Quaternary environmental reconstruction, with special reference to the North Atlantic islands. In: Buck, C.E., Millard, A.R. (Eds.), Tools for Constructing Chronologies. Springer–Verlag, London, pp. 173–188.

Dugmore, A.J., Vésteinsson, O., 2012. Black sun, high flame and flood; volcanic hazards in Iceland. In: Cooper, J., Sheets, P. (Eds.), Surviving Sudden Environmental Change: Answers from Archaeology. University Press of Colorado, Boulder, pp. 67–90.

Einarsson, Á., 2010. The ecodynamics of Lake Mývatn. In: Andradóttir, H.Ó. (Ed.), 14th International Workshop on Physical Processes in Natural Waters, Reykjavík, Iceland, 28 June–1 July, 2010. Conference Proceedings, pp. 32–33.

Einarsson, Á., 2014. The Lake Balls of Mývatn. In Memoriam. http://www.ramy.is/wpcontent/uploads/2014/05/Lake_Ball_in_memoriam_final_draft_reduced_size.pdf.

Einarsson, Á., Hafliðason, H., 1988. Predictive paleolimnology: effects of sediment dredging in Lake Mývatn, Iceland. Verhandlungen des Internationalen Verein Limnologie 23, 860–869.

Einarsson, Á., Garðarsson, A., Gíslason, G.M., Ives, A.R., 2002. Consumer-resource interactions and cyclic population dynamics of *Tanytarsus gracilentus* (Diptera: Chironomidae). Journal of Animal Ecology 71, 832–845.

Einarsson, Á., Stefánsdóttir, G., Jóhannesson, H., Ólafsson, J.S., Gíslason, G.M., Wakana, I., Gudbergsson, G., Gardarsson, A., 2004. The ecology of Lake Myvatn and the River Laxá: variation in space and time. Aquatic Ecology 38, 317–348.

Einarsson, Á., Garðarsson, A., Gíslason, G.M., Guðbergsson, G., 2006. Populations of ducks and trout of the River Laxá, Iceland, in relation to variation in food resources. Hydrobiologia 567, 183–194.

Einarsson, P., 1991. Umbrotin við Kröflu 1975–89. In: Garðarsson, A., Einarsson, Á. (Eds.), Náttura Mývatns. Hið íslenska náttúrufræðifélag, Reykjavík, pp. 96–139.

Eythórsson, J., Sigtryggsson, H., 1971. The climate and weather of Iceland. The Zoology of Iceland 1 (3), 1–62.

Faber, F., 1822. Prodromus der Islandischen Orinthologie oder Geschichte der Vögel Islands. Copenhagen, 109 pp.

Friðriksson, A., Vésteinsson, O., McGovern, T.H., 2004. Recent investigations at Hofstaðir, northern Iceland. In: Housley, R.A., Coles, G. (Eds.), Atlantic Connections and Adaptations: Economies, Environments and Subsistence in Lands Bordering the North Atlantic. Oxbow Books, Oxford, pp. 191–202.

Friðriksson, S., 1969. The effects of sea ice on flora, fauna and agriculture. Jökull 19, 146–157.

Friðriksson, S., 1973. Crop production in Iceland, International Journal of Biometeorology 17 (4), 359–362.

Garðarsson, A., 1991. Fuglalíf við Mývatn og Laxá. In: Garðarsson, A., Einarsson, Á. (Eds.), Náttura Mývatns. Hið íslenska náttúrufræðifélag, Reykjavík, pp. 278–319.

Guðbergsson, G., 2015. Silungurinn í Mývatni Yfirlit yfir rannsóknir og veiðitölur 1986–2014. Veiðimálastofnun (Institute of Freshwater Fisheries), Reykjavik. VMST/15025 96 pp. Online at: http://veidimal.is/files/Skra_0072657.pdf.

Guðmundsson, F., 1979. The past status and exploitation of the Myvatn waterfowl populations. Oikos 32, 232–249.

Hartwell, J., 1971. Plants used against cancer. Lloydia 34, 386–438.

Hicks, M.T., 2014. Losing sleep counting sheep: early modern dynamics of hazardous husbandry in Mývatn, Iceland. In: Harrison, Maher (Eds.), Human Ecodynamics in the North Atlantic: A Collaborative Model of Humans and Nature through Space and Time. Lexington Books, Lanham.

Hicks, M.T., Einarsson, Á., Anamthawat-Jónsson, K., Edwald, A., þórsson, Æ.þ., McGovern, T.H., 2016. Community and conservation. Documenting millenial-scale sustainable resource use at Lake Mývatn, Iceland. In: Isendahl, C., Stump, D. (Eds.), The Oxford Handbook of Historical Ecology and Applied Archaeology. http://dx.doi.org/10.1093/oxfordhb/9780199672691.013.36.

Íslendingabók, K., 2006. The Book of the Icelanders. The Story of the Conversion. Trans. and introduction Siân Grønlie. In: Viking Society for Northern Research Text Series, vol. XVIII. University College London.

Ives, A.R., Einarsson, Á., Jansen, V.A.A., Garðarsson, A., 2008. High-amplitude fluctuations and alternative dynamical states of midges in Lake Myvatn. Nature 452, 84–87.

Jarðamat, 1849–1850. S.- Og N.-þingeyjarsýsla. National Archives of Iceland. Available online at: https://myndir.handrit.is/doShowDocument.jsp?lang=is&source=ICDB&sourceId=Jar%F0amat+1849-1850+S-+og+N-%DEingeyjars%FDsla&style=&navpanes=0&view=Fit.

Jarteinabók Guðmundar biskups, 1948. *Biskupa sögur.* Annað bindi. Hólabiskupar. Guðni Jónsson bjó til prentunar. Íslendingasagnaútgáfan – Haukadalsútgáfan, Reykjavík, pp. 437–478.

Jónasson, P.M., 1979. Ecology of eutrophic, subarctic Lake Mývatn and the River Laxá. Oikos 32, 1–308.

Jónsson, G.S., 2016. Mývatn ákoma og afrennsli. Environmental Ministry of Iceland, Reykjavik.

Júlíusson, Á.D., 2001. The environmental effects of Icelandic subsistence farming in the late middle ages and early modern period. In: Sigurðsson, I., Skaptason, J. (Eds.), Aspects of Arctic and Sub-Arctic History. University of Iceland Press, Reykjavík, pp. 279–288.

Júlíusson, Á.D., 2013a. þúsund ára bændasamfélag. Landbúnaðarsaga Íslands, vol. I. Skrudda, Reykjavík, pp. 800–1800.

Júlíusson, Á.D., 2013b. Bændur og nútími. Sveitasamfélagið á 19. og 20. öld. Landbúnaðarsaga Íslands, vol. II. Skrudda, Reykjavík, pp. 800–1800.

Karlsson, G., 1977. Frelsisbaráttu Suður-þingeyinga og Jón á Gautlöndum. Hið Íslenska bókmenntafélag, Reykjavík.

Kartnig, T., 1987. *Cetraria islandica* – Islandisches Moos. Zeitschrift für Phytotherapie 8, 127–130.

Kristmannsdóttir, H., Ármannsson, H., 2004. Groundwater in the Lake Mývatn area, North Iceland: chemistry, origin and interaction. Aquatic Ecology 38, 115–128.

Lindegaard, C., 1979. A survey of the macroinvertebrate fauna, with special reference to Chironomidae (Diptera) in the rivers Laxá and Kráká, northern Iceland. Oikos 32, 281–288.

Lindegaard, C., Jónasson, P.M., 1979. Abundance, population dynamics and production of zoobenthos in Lake Mývatn, Iceland. Oikos 32, 202–227.

Loftsson, J., 1993. Forest development in cold climates. In: Alden, J., et al. (Eds.), Forest Development in Cold Climates. Plenum Press, New York, pp. 453–461.

Lowe, D.J., 2011. Tephrochronology and its application: a review. Quaternary Geochronology 6 (2), 107–153.

Lucas, G. (Ed.), 2009. In: Hofstaðir: Excavations of a Viking Age Feasting Hall in North-Eastern Iceland. In: Lucas, G. (Ed.), Monograph Series 1. Institute of Archaeology, Reykjavik.

Magnússon, Á., 1913–1943. Jarðabók Árna Magnússonar og Páls Vídalín, I-V, Hinu Íslenska fræðafjelagi í Kaupmannahöfn (Copenhagen).

McGovern, T.H., Perdikaris, S., Einarsson, Á., Sidell, J., 2006. Coastal connections, local fishing, and sustainable egg harvesting, patterns of Viking age inland wild resource use in Mývatn District, Northern Iceland. Environmental Archaeology 11 (1), 102–128.

McGovern, T.H., Vésteinsson, O., Friðriksson, A., Church, M., Lawson, I., Simpson, I.A., Einarsson, Á., Dugmore, A., Cook, G., Perdikaris, S., Edwards, K.J., Thomson, A.M., Adderley, W.P., Newton, A., Lucas, G., Edvardsson, R., Aldred, O., Dunbar, E., 2007. Landscapes of settlement in northern Iceland: historical ecology of human impact and climate fluctuation on the millennial scale. Invited paper in special issue on the archaeology of global change. American Anthropologist 109 (1), 27–51.

Ogilvie, A.E.J., 1984a. The past climate and sea-ice record from Iceland. Part 1: data to A. D. 1780. Climatic Change 6, 131–152.

Ogilvie, A.E.J., 1984b. The impact of climate on grass growth and hay yield in Iceland: A. D. 1601 to 1780. In: Morner, N.-A., Karlen, W. (Eds.), Climate Changes on a Yearly to Millennial Basis. Reidel, Dordrecht, pp. 343–352.

Ogilvie, A.E.J., 1986. The climate of Iceland, 1701–1784. Jökull 36, 57–73.

Ogilvie, A.E.J., 1991. Climatic changes in Iceland A. D. c. 865 to 1598. In: Bigelow, G.F. (Ed.), The Norse of the North Atlantic, Acta Archaeologica, vol. 61–1900. Munksgaard, Copenhagen, pp. 233–251.

Ogilvie, A.E.J., 1992. Documentary evidence for changes in the climate of Iceland, A. D. 1500 to 1800. In: Bradley, R.S., Jones, P.D. (Eds.), Climate Since A. D. 1500. Routledge, London and New York, pp. 92–117.

Ogilvie, A.E.J., 2001. Climate and farming in northern Iceland, ca. 1700–1850. In: Sigurðsson, I., Skaptason, J. (Eds.), Aspects of Arctic and Sub-Arctic History. University of Iceland Press, Reykjavík, pp. 289–299.

Ogilvie, A.E.J., 2005. Local knowledge and travellers' tales: a selection of climatic observations in Iceland. In: Caseldine, C., Russell, A., Harðardóttir, J., Knudsen, O. (Eds.), Iceland – Modern Processes and Past Environments. In: Rose, J. (Ed.), Developments in Quaternary Science, vol. 5. Elsevier, Amsterdam-Boston-Heidelberg-London, pp. 257–287.

Ogilvie, A.E.J., 2007. Bréf sýslumanna til stiftamtmanns og amtmanns: Environmental Images of Nineteenth-Century Iceland from Official Letters Written by District Sheriffs (Expanded web version of paper cited below). http://www.inor.is/index.php?m=N&id=ASTRIDGR1&sid=OGILVIE.

Ogilvie, A.E.J., 2008. *Bréf sýslumanna til stiftamtmanns og amtmanns*: environmental images of nineteenth-century Iceland from official letters written by district sheriffs. In: Wells, M. (Ed.), The Discovery of Nineteenth-Century Scandinavia. Norvik Press, Norwich, pp. 43–56.

Ogilvie, A.E.J., 2010. Historical climatology, *Climatic Change*, and implications for climate science in the 21st century. Climatic Change 100, 33–47.

Ogilvie, A.E.J., Jónsson, T., 2001. "Little Ice Age" research: a perspective from Iceland,. Climatic Change 48, 9–52.

Ogilvie, A.E.J., McGovern, T.H., 2000. Sagas and science: climate and human impacts in the North Atlantic. In: Fitzhugh, W.W., Ward, E.I. (Eds.), Vikings: The North Atlantic Saga. Smithsonian Institution Press, Washington, pp. 385–393.

Ogilvie, A.E.J., McGovern, T.H., 2004. Human ecology, local knowledge and interdisciplinary research in Mývatn, northern Iceland. In: 34th International Arctic Workshop, Program with Abstracts. INSTAAR, Boulder, Colorado, 10–13 March 2004, pp. 129–130.

Olafsdóttir, E.S., Ingolfsdóttir, K., Barsetr, Smestad Paulsen, H.B., Jurcic, K., Wagner, H., 1999. Immunologically active (1–>3)-(1–>4)-alpha-D-glucan from *Cetraria islandica*. Phytomedicine 6 (1), 33–39. © Gustav Fischer Verlag 1999. http://www.urbanfischer.de/journals/phytomedPhytomedicine.

Ólafsson, E., 1772. Vice-lavmand Eggert Olafsens og Land-Physici Biarne Povelsens Reise igennem Island, foranstalted af Videnskapernes Sælskab I Kiöbenhavn, og beskreven af forbemeldte Eggert Olafsen med dertil hörende 51 Kobberstökker og et nyt forfærdiget Kort over Island. Available online: https://ia600306.us.archive.org/19/items/ViceLavmandEgger000099709v1EggeReyk/ViceLavmandEgger000099709v1EggeReyk_orig.pdf.

Ólafsson, J., 1979. The chemistry of Lake Mývatn and River Laxá. Oikos 32, 82–112.

Ólafsson, J., 1991. Undirstöður lífríkis í Mývatni. In: Garðarsson, A., Einarsson, Á. (Eds.), Náttúra Mývatns. Hið íslenska náttúrufræðifélag, Reykjavík, pp. 141–165.

Ponzi, F., 1995. Ísland Fyrir Aldamót Harðindaárin 1882–1888. Iceland. The Dire Years: 1882–1888 From the Photographs and Diaries of Maitland James Burnett and Walter H. Trevelyan, Brennholt, Mossfellsbær, Iceland.

Shepherd, C.W., 1867. The North-West Peninsula of Iceland: Being the Journal of a Tour in Iceland in the Spring and Summer of 1862. Longmans, Green and Co., London.

Sheriffs, L., September 25, 1779. Is. J. 5 nr. 105. þingeyjarsýsla. þjóðskjalasafn Íslands (National Archives of Iceland), Reykjavik.

Sigurðardóttir, R., 1992. Soil Erosion and the Origin of Windblown Sand in Lake Myvatn Region NE Iceland (In Icelandic) (B.S. thesis). University of Iceland, Department of Geology.

Sigurðardóttir, R., 2010. Tradeoffs in Geothermal and Hydroelectric Power, Conservation and Ecological Resiliency in Iceland. In: Vogt, K.A., Patel-Weynand, T., Shelton, M., Vogt, D.J., Gordon, J.C. (Eds.), Sustainability Unpacked: Food, Energy and Water for Resilient Environments and Societies, Earthscan, London, UK.

Steingrímsson, J., 1998. Fires of the Earth. The Laki Eruption 1783–1784 by the Rev. Jón Steingrímsson. Introduction by Dr Guðmundur E. Sigvaldsson. English translation by Keneva Kunz. University of Iceland Press and the Nordic Volcanological Institute, Reykjavik.

Streeter, R., Dugmore, A.J., February 2013. Reconstructing late-Holocene environmental change in Iceland using high-resolution tephrochronology. The Holocene 23 (2), 197–207.

Sýslu-og sóknarlýsningar, 1979. Híns íslenska Bókmenntafélags 1839–1873. Sögufélag, Reykjavik.

Sæmundsson, B., 1923. Titill Fiskirannsóknir 1921–1922. Tímarit Andvari 48 (1), 67–112. http://timarit.is/view_page_init.jsp?gegnirId=000508579.

Sæmundsson, K., 1991. Jarðfræði Kröflukerfisins. In: Garðarsson, A., Einarsson, Á. (Eds.), Náttúra Mývatns. Hið íslenska náttúrufræðifélag, Reykjavík, pp. 24–95.

Teitsson, B., 1973. Eignarhald og ábuð á jörðum í suður-þingeyjarsýslu 1703–1930. Menningarsjóður, Reykjavík.

Thienemann, F.A.L., 1827. Reise im Norden Europas, vorzüglich in Island, in den Jahren 1820 bis 1821. Carl Heinrich Reclam, Leipzig.

Thórarinsson, S., 1944. Tefrokronoliska studier pa Island. (Tephrochronological studies in Iceland). Geografiska Annaler 26, 1–217.

Thórarinsson, S., 1951. Laxárgljúfur and Laxárhraun. A tephrochronological study. Geografiska Annaler Stockholm H 1–2, 1–89.

Thórarinsson, S., 1970. Tephrochronology and Medieval Iceland. In: Berger, R. (Ed.), Scientific Methods in Medieval Archaeology. University of California Press, Los Angeles, Berkely, London.

Umhverfisstofnun (The Environmental Agency of Iceland), 2011. Viðaukar við verndáætlun Mývatns og Laxár. Reykjavik. 96 p. https://www.ust.is/library/Skrar/Einstaklingar/Fridlyst-svaedi/Fridlysingar/verndar%C3%A1%C3%A6tlun%20M%C3%BDvatns%20og%20Lax%C3%A1r-www.pdf.

United Nations General Assembly, 2005. https://daccess-ods.un.org/TMP/2200996.57773972.html.

UST, 2010. Rauði listinn, svæði í hættu. Umhverfisstofnun. Reykjavík. http://ust.is/library/Skrar/Einstaklingar/Nattura/Skyrslur/astand_fridlystra_svaeda_skyrsla_umhverfisstofnunar_nov_2010.pdf.

UST, 2013. Rauði listinn, svæði í hættu. Umhverfisstofnun. Reykjavík. http://ust.is/library/Skrar/Einstaklingar/Nattura/Skyrslur/3021_Astand_fridlystra_svaeda_skyrsla_ust_jan2013_I_VEF.

Vésteinsson, O., McGovern, T.H., November 1, 2012. The peopling of Iceland. Norwegian Archaeological Review 45 (2), 206–218.

Chapter 5

Precipitation and Conifer Response in Semiarid Mountains: A Case From the 2012–15 Drought in the Great Basin, USA

S. Strachan[a]
University of Nevada, Reno, NV, United States
E-mail: scotty@dayhike.net

Chapter Outline

[a] Scotty Strachan grew up in the Great Basin and has performed climate- and conifer-related research activities in the region since 2002. His experiences with fieldwork operations and application of technology allow him to investigate mountain processes in unique ways. He co-manages the NevCAN stations presented in this chapter along with Gregory McCurdy and Bradley Lyles from the Desert Research Institute, both of whom can be blamed for Scotty's continued interest in automated science observatories. Scotty was partially supported by funding from the NSF-EPSCoR grant IIA-1301726, NSF grant 1230329, and the University of Nevada, Reno College of Science Dean's Office.

Mountain Ice and Water. http://dx.doi.org/10.1016/B978-0-444-63787-1.00005-6

INTRODUCTION

Precipitation in the Great Basin

The Great Basin hydroecological region of North America is a semiarid zone of interior drainage that extends from the crest of California's Sierra Nevada mountain range in the west to the peaks of the Wasatch Front in Utah (Fig. 5.1). The landscape of the Great Basin is dominated by mountains, but in a unique manner; hundreds of long, relatively narrow ranges of steep mountains are separated by broad, open valleys. Dozens of these mountain ranges exceed 3000 m in height, creating a stark contrast between the relatively dry desert below (typical base elevations of 1200–1800 m) and the subalpine settings above (Charlet, 1996; Grayson, 2011; Lanner, 1984). It is in these mountains where the annual water balances for local watersheds are determined, as distinct gradients of precipitation are present.

Frontal-type storm events originating over the Pacific Ocean deposit the bulk of rain or snow in the mountains through a combination of local orographic effects (Roe, 2005) and air mass evolution downstream of the Sierra Nevada rain shadow (Houghton, 1979). The presence of the Sierra Nevada mountain range and its influence on moisture blocking is the defining physioclimatic feature of the Great Basin in general, and is primarily responsible for its long-term semiarid nature. Proximity of the Sierra to the Pacific Ocean and abrupt elevation changes from sea level to over 3000 m ensure that a substantial portion of the moisture present in westerly flow is extracted orographically or by other mountain–atmosphere interactions such as blocking lateral airflow in the Sierra barrier jet (Daly et al., 1994; Dettinger et al., 2004; Lundquist et al., 2010; Parish, 1982). Although this modified westerly flow dominates the transport of moisture into the Great Basin, other synoptic mechanisms occur that contribute to overall precipitation totals in the region. Transiting "cutoff" low-pressure systems are disconnected from troughs in the northeast Pacific and move inland, most prominently in the springtime seasonal window (Nieto et al., 2005; Oakley and Redmond, 2014). Cyclonic circulation associated with these low-pressure systems over the continent can pull moisture from the eastern Pacific up into the Great Basin,

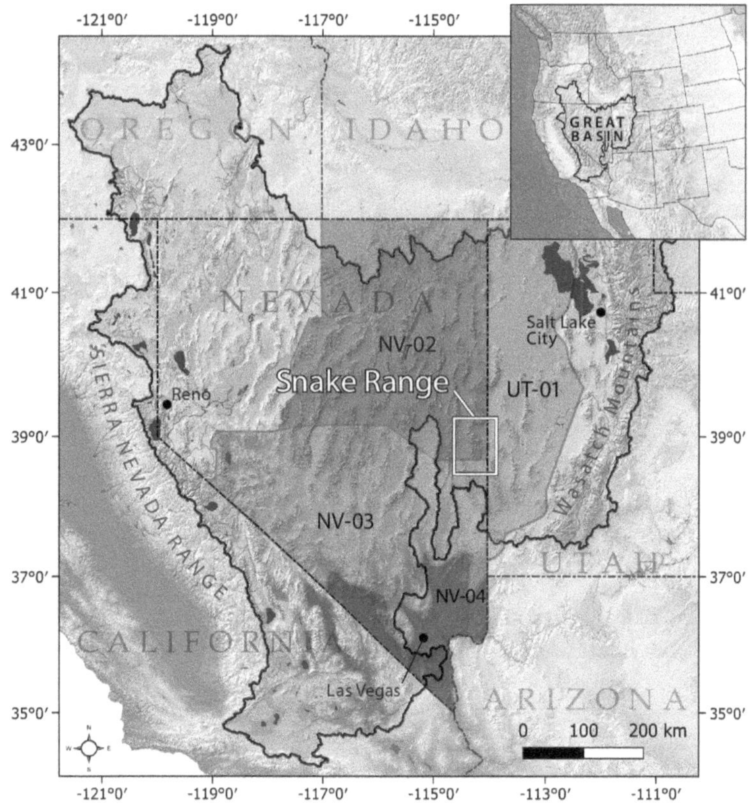

FIGURE 5.1 North America's Great Basin is a large region of interior hydrological drainage that is also associated with distinct ecologic and climatic zones. The observations highlighted in this chapter come from the Snake Range mountains, in the east-central portion at the intersection of three historical State Climate Divisions (NV-02, NV-03, and UT-01). NV-02 is considered to be an area dominated by winter- and springtime precipitation, while NV-03 and especially NV-04 are more influenced by North American Monsoon activity in the summertime.

bypassing blocking orography and creating "upslope" precipitation events on the leeward side of the mountain ranges (Reinking and Boatman, 1986). Another source of Great Basin moisture is the North American Monsoon (NAM), a summertime phenomenon of southerly moisture flow. Although the NAM is considered a dominant climate feature of the American Southwest, it also can strongly affect the Great Basin, particularly the south-central portion (Douglas et al., 1993; Means, 2013). Because NAM precipitation is mostly driven by daytime heating and convective activity, the effect of mountain ranges on precipitation amounts is not as dramatic as frontal–orographic interactions (Fig. 5.2). Impact of these various seasonal and circulatory mechanisms on mountain precipitation in the Great Basin remains generally uninvestigated because of extremely limited data (Houghton, 1979), resulting

FIGURE 5.2 Convective activity in the Great Basin is not necessarily limited to elevated slopes. Because the mountain ranges are generally narrow (15–20 km wide), convective cells can drop significant precipitation at any elevation. Frontal storms, on the other hand, usually impact upper elevations with greater precipitation volume as well as increased likelihood of snow due to orographic and elevation factors.

in high levels of uncertainty in climate models for the region (Brekke et al., 2013). Studies incorporating precipitation in the Great Basin are generally regional in scale and leverage interpolated models of precipitation (e.g., Dettinger et al., 1998; Wise, 2010; Wang et al., 2012), which in turn are mostly based on low-elevation observations and remain largely untested across elevation gradients. Nevertheless, these gradients, such as all mountain environments of the world, play a crucial role in local ecology and regional hydrology.

Valley environments in the Great Basin have a net water deficit; surplus in mountain ranges provides recharge, making snowpack a critical part of the annual water budget in this semiarid region (Welch et al., 2007). It is projected that mountains across the western US will retain less snowpack outside of the winter season in the future (Dettinger et al., 2004; Gleick and Chalecki, 1999; McCabe and Wolock, 1999), and significant changes to annual water budgets are possible (Knowles and Cayan, 2004), although not necessarily driven by temperature alone as annual-to-decadal precipitation amounts are certainly not static (Andrews, 2012; Hamlet et al., 2005). The impact of snow as a primary precipitation mode in North American mountains is dominated by the springtime runoff phenomenon, transporting water quickly to lower elevations; mountain-block infiltration and local aquifer recharge occur as a secondary effect. It is generally accepted that snowpack creates higher infiltration amounts as compared to rain (Maule et al., 1994), although this depends on the local maximum rates of soil infiltration versus typical rainfall or snowmelt rates. Interactions between precipitation and deeper groundwater recharge are also complex, and it is possible that future-projected changes in the snow and rain ratio favoring liquid precipitation do not necessarily equal greater shallow-aquifer recharge in snow-dominated regions (Huntington and Niswonger, 2012). Sparseness of data on precipitation, temperature, snowpack, soil, and groundwater trends in general hinders long-term assessment of these processes in many mountain regions (Stewart, 2009), and the Great Basin is certainly no exception given that it is likely the least-instrumented region in the United States (Mock, 1996), despite being one of the most mountainous (Fig. 5.3).

Relationships of Great Basin Vegetation Zones to Precipitation

Ecology of the Great Basin is ultimately tied to long-term precipitation inputs and their mechanisms. The hydrological area shown in Fig. 5.1 is actually separated into the colder, higher elevation Great Basin Desert ecoclimatic region in the north and the warmer, lower elevation Mojave Desert region in the south (Grayson, 2011). Both of these subregions depend on precipitation (especially snow) in mountain ranges to recharge soil moisture and eventually local aquifers, sustaining diverse ecological communities. The Great Basin Desert region is characterized by elevation-dependent, vegetative strata (Billings, 1951; Charlet, 2007), including the Sagebrush zone, the Pinyon-Juniper zone, the Montane zone, and the Subalpine zone (Fig. 5.4). These zones, and especially the upper-elevation forests, are populated with geographically isolated species (Charlet, 1996). The millennial-scale survivorship of individuals within these species (e.g., Great Basin bristlecone pine, *Pinus longaeva*; single-needle pinyon pine, *Pinus monophylla*; limber pine, *Pinus flexilis*; ponderosa pine, *Pinus ponderosa*; and various juniper species) indicates that in spite of notable swings in paleoclimatic conditions (that presumably altered seasonal normals of temperature and precipitation),

FIGURE 5.3 High elevations in the Great Basin run a net water surplus, and capture snow that ultimately finds its way to the valley floors, which run a net water deficit due to direct evaporation and evapotranspiration processes.

FIGURE 5.4 Vegetation in the Great Basin is generally stratified by elevation, although the zone boundaries vary highly with aspect, soils, and disturbance factors. This image shows the actual locations of the observatory sites used in this chapter's evaluation of precipitation in mountains.

enough regularity in moisture input persisted in the past to preserve these individuals and isolated populations in an ultimately water-limited context (Charlet, 2007; Lanner, 1984).

Details of past drought and pluvial periods at fine scales are largely unknown, although speculation can be made using paleoecological and paleoclimatic evidence (e.g., Wigand and Nowak, 1992; Nowak et al., 1994; Graham et al., 2007; Mensing et al., 2008; Strachan et al., 2012; Reinemann et al., 2014; DeRose et al., 2015). These sorts of paleorecords do not contain the ability to resolve actual precipitation amounts, phase (rain or snow), or even strict temperature regimes, although precipitation seasonality and general temperature ranges are inferred using present-day plant species' spatial-climatic distributions. In fact, the long-term relationship of wintertime snow and its seasonal persistence to primary Great Basin vegetative communities/zones is not precisely known, although Great Basin conifers typically associated with high-elevation, snow-dominated environments are sparsely located in locations with very little seasonal snowfall (Charlet, 1996) and a cool springtime growing season seems to be the single common feature across the region (Comstock and Ehleringer, 1992). These dichotomies create something of a problem for parameterization of predictive niche models as well as interpretation of plant-derived paleoecology records.

Between the precipitation and its ultimate fate lies the land surface, which in the Great Basin is comprised of a variety of soils that are typical of semiarid environments with complex geology: low to moderately developed organic horizons with textures that range from sand to clay (Miller et al., 2013; Sperry and Hacke, 2002). Shallow Great Basin soils can be subject to frequent hydrologic disturbances due to high interannual variability in the climate (Castelli et al., 2000). Arid and semiarid ecohydrology is viewed as being driven by precipitation pulse dynamics, or the sporadic arrival of water on the landscape, which are in turn strongly moderated by soils conditions (e.g., Rodriguez-Iturbe, 2000; Austin et al., 2004; Robinson et al., 2008). Reynolds et al. (2004) also point out that the characteristics of the storms which bring precipitation to the landscape are important, as the rainfall rate and total volume ultimately combine with antecedent soil conditions to form “effective” precipitation. Because seasonal snowfall is a standard feature of Great Basin mountains, it stands to reason that the form of the precipitation in this semiarid region (i.e., frozen, liquid, or mixed) must influence the seasonal soil moisture and therefore impact the near-surface ecohydrology as well as the amount of mountain-block recharge. As previously pointed out, however, the seasonal nature of snowfall, mountain temperature regimes, individual storm impacts to “effective” precipitation, and subsequent ecological responses are not well quantified through observation across Great Basin mountain gradients. Seasonally “effective” precipitation is clearly one of the most important aspects of semiarid ecohydrology, and therefore it is with this background in mind that we now turn to current climatic events and a suite of ecohydrological observations.

The 2012–15 North American Drought

The interval 2012–15 is noted for significant and persistent drought across large portions of North America, although not geographically or temporally contiguous (Bell et al., 2015; Overpeck, 2013). Generally, the causes of aridity during this period are associated with high temperatures exacerbating less than normal precipitation (Hoerling et al., 2014; Shukla et al., 2015; Swain, 2015), similar to the deep drought of 1934 (Cook et al., 2014). This balance of temperature and precipitation was not the same everywhere, but the worst effects of the drought occurred in California and the western Great Basin when antecedent dry and hot conditions from 2012 were followed up in 2013–14 by extremely low precipitation amounts. These anomalies were forced by wintertime atmospheric ridging that blocked storms from tracking into the American Southwest from the northern Pacific (Swain et al., 2014; Wang et al., 2014). As we will see from the following case studies, shifts in precipitation seasonality and effective moisture occurred in the central Great Basin during this time, and mountain tree species responded accordingly to alleviate the worst of the conditions.

OBSERVATIONS FROM A GREAT BASIN MOUNTAIN GRADIENT

Study Site and Data Descriptions

Four climate and ecohydrological monitoring stations were constructed in 2010–11 as part of the Nevada Climate-ecohydrological Assessment Network

(NevCAN; Mensing et al., 2013) within each major Great Basin vegetative zone in the Snake Range of eastern Nevada (Fig. 5.1) across a steep elevation gradient (Fig. 5.4). These sites were located in microsites that avoided local cold air pools, exposed ridges, and overly steep slopes (Fig. 5.5) to avoid biasing observations with highly localized topographic effects.

Management of the NevCAN stations and systems is a cooperative arrangement between three institutions of the Nevada System of Higher Education: the University of Nevada, Reno, the Desert Research Institute, and the University of Nevada, Las Vegas. Seasonal inspections and annual sensor audits are conducted onsite to improve quality assurance and reduce data gaps. Raw data are retrieved from datalogger systems through a high-speed terrestrial wireless, digital network to the Nevada Research Data Center (NRDC; Dascalu et al., 2014), where they are archived and distributed to the Western Regional Climate Center (WRCC) for independent quality control (QC) processing and storage. Meteorological sensors and dataloggers are mounted on 10 m aluminum towers in accordance with WRCC weather station deployment protocols based on World Meteorological Organizational

FIGURE 5.5 The 4 study sites in the Snake Range are part of a transect-type automated observatory that comprises a total of 12 stations in 2 mountain ranges. Because these 4 are separated by less than 10 km horizontal distance, but over 1500 m elevation, they allow study of processes such as storm precipitation characteristics across the mountain gradient without significant spatial uncertainty.

standards (WMO, 2008). Air temperature at 2 and 10 m heights is recorded from Type-T, thermocouple wire probes mounted in 6-plate Gill-type, passively aspirated radiation shields. Air temperature and relative humidity are monitored at 2 m using a Vaisala HMP50 combination probe mounted in similar shielding. Air pressure is monitored using a Vaisala PTB110 barometer. Wind speed and direction are monitored using an RM Young 05103 Wind Monitor. Incoming solar radiation is measured using an Apogee SP-110 pyranometer. The volumetric content of soil water is measured using Campbell Scientific CS-650, time-domain reflectometer probes at multiple depths and orientations. Soil temperature was measured using Type-T thermocouple wire at multiple depths. These sensors are controlled by Campbell Scientific CR3000 dataloggers using a nominal 3-s scan interval and recording statistics such as maximum, minimum, average, standard deviation, or sample values for each sensor at 1- and 10-min time frames. Vegetation sensors are controlled by Campbell Scientific CR1000 dataloggers running 30-s scan interval and recording Granier-type thermal dissipation probe (TDP) temperature differential values every 10 min. Bulk-catch precipitation gages consist of Geonor model T-200B (referred to hereafter as "Geonor" gages) weighing sensors and are mounted away from the meteorological towers on separate concrete bases. These gages incorporate passive Alter-type wind shielding to reduce undercatch. Liquid-catch gages of the tipping bucket type consist of Hydrological Services model TB4 (hereafter referred to as "tipping bucket" gages), and are mounted on the sides of the towers or on short extension arms without wind shielding. Hourly daylight images are automatically acquired using Canon point-tilt-zoom cameras controlled by software running at NRDC, requiring a functional network connection at the top of each hour. By combining observations from multiple sensor types and systems, a consistent picture of precipitation-related conditions on these sites can be assembled. Anomalies in one sensor data stream can be investigated in a supervised QC process using independent sensors as well as daylight imagery, reducing observation error and allowing for inclusion of real events (such as extreme precipitation intensity) that might otherwise be filtered out as noise in postprocessing procedures. All in all, these stations represent the state of the art in micrometeorological, automated observation and were designed with longevity in mind.

For the case studies in this chapter, 10-min precipitation data from each station were processed to identify the phase (frozen, liquid, or mixed) and to correct each gage for wind-driven undercatch. The phase-identification method used is a new, but fairly simple procedure (the Gage-Difference Method or GDM; Strachan et al., in preparation) where the relative catches of the mass-weighing Geonor sensor and the liquid-only tipping bucket are compared over short timescales and phase assigned in a multistage, data-filtering process that takes into account sensor noise and warm season catch differences. Final assignments are compared to air temperature parameters as an accuracy check and anomalies are inspected using daylight camera imagery. The method allows detection of phase under all temperature and relative humidity conditions, including warm season hailstorms. Very low-intensity rates of precipitation are excluded from this process due to sensor noise issues, and these types of events on these sites may account for 24% (high elevation) to 38% (low elevation) of the wintertime totals and 2% (high elevation) to 8% (low elevation) of the summertime totals based on a post hoc comparison with total seasonal mass collections. Being able to classify the majority of each precipitation event, however, allows for undercatch correction based on phase, which differs substantially between snow and rain (Goodison et al., 1998).

Summary Precipitation Characteristics

Mean Monthly Precipitation

Monthly precipitation normals for the State Climate Divisions in the study region indicate wet winters but large geographic divergences in spring and summertime moisture patterns (Figs. 5.1 and 5.6). The southern climate divisions show dry springtimes followed by wet summers (influence of the NAM). Climate divisions in the north, on the other hand, do not indicate a strong pulse of summertime moisture, but instead have historical peaks during the month of May. All of the climate divisions undergo high variability over the long term, as the interquartile ranges for each month demonstrate. The precipitation amounts represent the average of mostly low-elevation reporting stations over the long term, and although we would expect similar seasonal

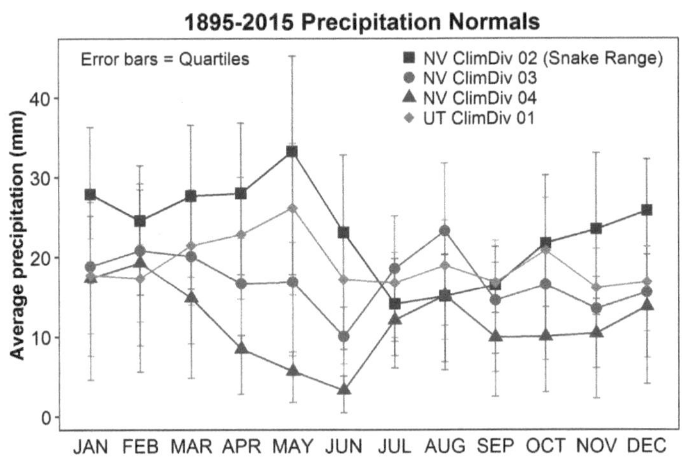

FIGURE 5.6 Precipitation climatologies for the central and southern Great Basin are derived from estimated precipitation of the climate divisional data from the full network (McRoberts and Nielsen-Gammon, 2011). The 25th and 75th percentiles are displayed as error bars, hinting at the tremendous interannual variability in hydrological input that characterizes the intermountain west.

patterns to emerge for the data from the Snake Range, absolute volume should differ with elevation.

Mean monthly, undercatch-corrected precipitation during the 2012–15 drought sequence is shown in Table 5.1, and it is clear that the differences in monthly totals between high and low elevations even over short distances are extreme.

From the table, it is apparent that during the drought years more precipitation fell during the summertime (July–September) than the wintertime, even when taking low-intensity events into account. The differences are particularly stark at the lower elevations, where wintertime inputs were quite low. These results do not match the long-term climatology for the NV-02 climate division (Fig. 5.6), and in fact more resemble the NV-03 division pattern.

Total Monthly Precipitation by Year

Even more useful in assessing the nature of precipitation in the Snake Range during the drought is the plot of total monthly catch for each site during each year (Fig. 5.7).

The annual patterns during the drought period seem to oscillate between the seasonal patterns of nearby climate divisions. The observed seasonal patterns contrast strongly with the climate division normals for NV-02, with a general failure of wintertime precipitation relative to other seasons. The years 2012, 2013, and 2014 most resemble climate division NV-03, whereas 2015 seems to match UT-01, with a strong springtime and weak winters.

Notable features include the May of 2015, a dry June across all years, and the appearance of summertime moisture mitigating dry conditions. July–September precipitation for 2012–14 appears to be relatively higher than the climatological normal would suggest, indicating stronger flow of summertime moisture into the region during otherwise drier years. Because these "drought" years are not "normal" where regional seasonal precipitation is concerned, assessment of annual, soil moisture curves, and response of shallow soil moisture to different precipitation episodes should provide insight into actual "effective" precipitation and the role of snow during these different seasons. Of particular interest in this regard is the monthly/seasonal fraction of frozen precipitation at each site, and associated observations of soil moisture seasonally available at each site. Monthly precipitation totals and their seasonal patterns do not necessarily tell us: (1) if that precipitation was frozen or liquid or (2) if that precipitation was "effective," that is, did it serve to recharge the local soil moisture adequately to mitigate dry conditions during the springtime growing season, as well as the summer and fall time frames.

Fractions of Frozen Precipitation Across the Gradient

Estimates of the percentage of total precipitation occurring in frozen form for each month in the years 2012–15 show high variability across the elevation

TABLE 5.1 Classified/Corrected Mean Monthly Precipitation (mm), 2012–15

Site	January	February	March	April	May	June	July	August	September	October	November	December
Sagebrush	6.3	9.7	4.6	10.2	20.1	2.8	21.8	41.8	29.6	17.6	10.8	22.1
P-J	12.2	17.7	14.7	23.4	32.1	5.5	29.6	39.2	38.7	27.5	13.0	31.7
Montane	25.7	27.8	29.9	43.5	46.1	7.4	37.4	44.8	44.7	35.2	26.4	46.5
Subalpine	40.2	41.8	49.8	69.3	71.6	12.2	53.4	56.7	54.7	46.9	46.7	75.3

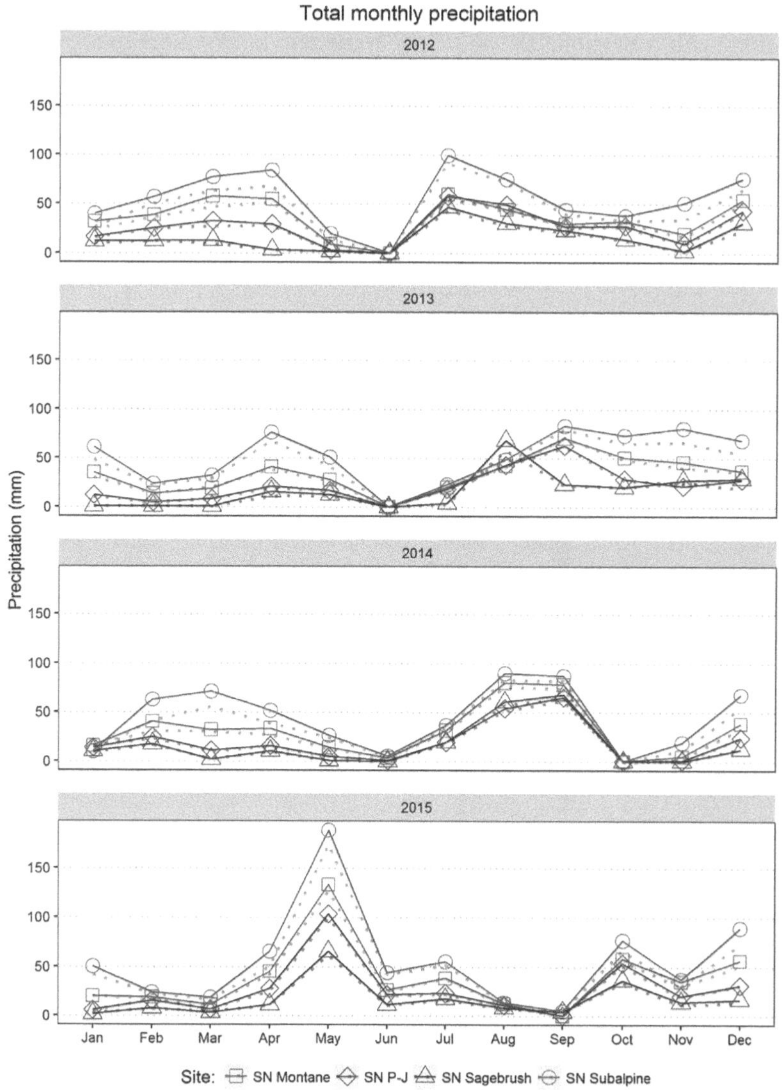

FIGURE 5.7 Total monthly precipitation at the Snake Range sites is shown for the corrected/classified precipitation data (*solid lines*), as well as the uncorrected, raw-gage catch (*dotted lines*). The annual patterns (except for 2015) diverge significantly from the divisional climatology, indicating that these were not "normal" years in terms of precipitation seasonality. Differences in precipitation total are evident as one moves up in elevation.

gradient (Fig. 5.8), and the general trend is definitely an increasing snow fraction with elevation, as would be expected in a midlatitude mountain environment.

In general, all sites receive frozen precipitation in fall, winter, and spring months, with the June–September window being nearly devoid of frozen events during the years of observation. Events classified as frozen during the

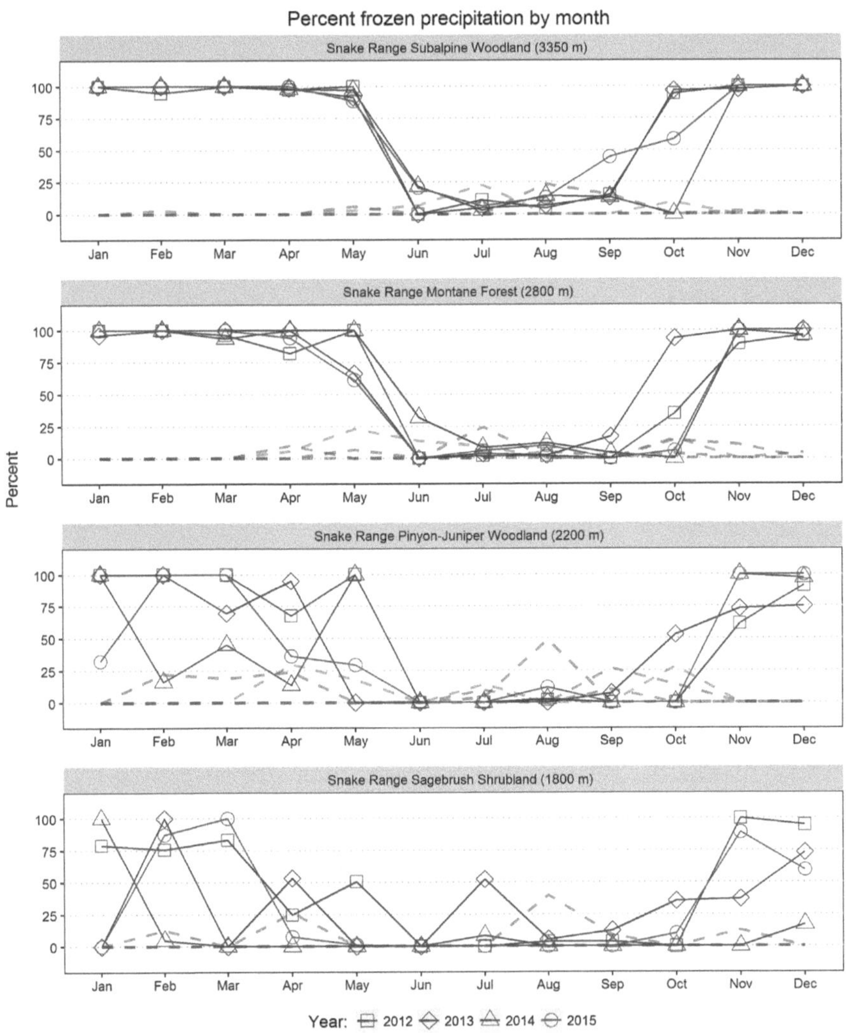

FIGURE 5.8 Percentage of total monthly precipitation classified as "frozen" for each site during the 2012–15 time frame. "Mixed" classification is shown as *dashed lines*. Low-elevation sites have a significantly lower proportion of frozen precipitation than the upper-elevation sites. Summertime "mixed" classifications are rain/hail events resulting from warm season, convective activity.

July–August time frame are either hail events or small-magnitude misidentifications by the GDM procedure due to wetting loss or hydrometeor bounce out of the tipping bucket orifice. Hail events classified as "mixed" are certainly visible (*dashed lines*, Fig. 5.8). Annual patterns for the two upper-elevation sites in the Snake Range are generally similar, with the exceptions of a 1-month lag in fall 2012 and a significant departure of the Subalpine site in September 2015 to October 2015. The Pinyon-Juniper site sometimes follows the seasonal pattern of the upper sites, and sometimes that of the lower Sagebrush site. The Sagebrush site is the only true valley location (Fig. 5.4), and therefore does not have orographic effects that modify precipitation processes on the mountain block. This site has substantially less precipitation in frozen form (19.7% overall) than the other sites (Table 5.2). Review of the raw-catch data prior to classification suggested that perhaps this site has a greater fraction of low-intensity precipitation that would be missed in the classified totals (up to 38% of the total) in December–March, which would most likely occur as snow. The lower elevation sites certainly receive less frozen precipitation during the transition seasons of April–June and October–November.

Of interest in particular, is the May–June transition, where a sharp decrease occurs in the frozen fraction, and the month of October, wherein sites can either undergo a major increase in the snow fraction (i.e., 2013, partially 2012), or not (i.e., 2014, partially 2015). This variable behavior in character of the transition season precipitation suggests a high level of control that synoptic conditions can exert on the snow line in the Great Basin on an interannual basis, and brings up questions of how local snowpack conditions, shallow soil moisture, and vegetation growth all respond to such variability in storms. In order to broadly evaluate interactions, we now turn to observations at the daily scale for all 4 years, and following that, individual case studies of interest at the hourly scale.

Daily Ecohydrological Patterns Across the Gradient

Observed Variables

Observations of rooting-depth (i.e., 10–20 cm) soil volumetric water content (VWC), continual wintertime snow cover, and conifer xylem activity (sap flow) provide a detailed picture of how precipitation and general atmospheric conditions impact near-surface ecohydrology under different seasons within the 2012–15 drought. A common measure of atmospheric demand for water, vapor pressure deficit (VPD; the difference between the saturated vapor pressure at a given temperature and the actual saturation amount) was calculated at the hourly time increment (Howell, 1995). Snowpack presence was determined by evaluating daily images of the ground surface above the soil–moisture probes, and delineating the wintertime window between snowpack establishment and melt-off (coverage longer than 7 days).

TABLE 5.2 Monthly Mean Frozen Precipitation Fraction (%), 2012–15

Site	January	February	March	April	May	June	July	August	September	October	November	December	Overall
Sagebrush	74.3	46.4	66.5	18.4	1.6	0.0	2.5	2.4	3.5	15.2	52.7	62.8	19.0
P-J	92.2	71.2	84.0	54.8	31.6	0.0	0.0	2.5	3.2	13.9	80.3	91.0	36.7
Montane	98.5	100.0	95.6	92.7	66.4	3.6	5.5	6.9	9.1	43.9	98.2	97.9	58.7
Subalpine	100.0	97.9	100.0	98.1	89.4	21.2	6.9	8.7	13.1	79.5	98.8	100.0	68.7

Short-term snow coverage that melted off within a week from early or late storms was not included in these date windows.

Rough estimates of soil VWC at field capacity (VWC_{FC}) and permanent wilting point (VWC_{PWP}) can be made for each site by alternatively examining the long-term mean VWC during wintertime snow cover or during the nongrowing season (maximum drainage) and the lowest VWC values in the driest portions of the growing season (maximum plant use), although actual VWC_{PWP} varies with potential evapotranspiration (Denmead and Shaw, 1962).

Sap flow data were obtained by applying the Granier TDP technique, which uses a combination of passive and heated electrical probes to measure movement of xylem sap in plant stems (Granier, 1987). Although sap flow measurements are often applied in a water balance context (e.g., Granier et al., 2000; Wilson et al., 2001; Williams et al., 2004), they are also useful indicators of plant activity in terms of transpiration and water stress when combined with basic atmospheric and soil measurements (Burgess et al., 2001; Pataki et al., 2000). Sap flow sensors were installed on trees at the Subalpine and Montane sites, with some replication across trees and species, but not enough to perform comprehensive whole plant water capacitance or estimate stand-level transpiration (Burgess and Dawson, 2008); rather, these installations were performed at the "pilot project" scale, to get a first-order assessment of tree growth activity across a variety of locations and in species that have not been studied in this fashion within the Great Basin. Ranges of values reported in sap flow literature vary substantially, and are usually evaluated over very short timescales (e.g., days), whereas this study presents years of measurements. Because of potential long-term changes in treetissue conductivity surrounding the TDP probes, and uncertainties related to intraspecies and stem radius replication (Nadezhdina et al., 2002), very low- or reverse-flow conditions (Burgess et al., 2001), and probe separation (Köstner et al., 1998), sap flow data are presented in either daily difference form ($\Delta T_{minimum(day)} - \Delta T_{hour}$; greater differences indicate higher maximum velocities), or in raw ΔT_{hour} values. For purposes of reviewing sensor and tree behavior in a relative sense over long time series, as well as instantaneous comparisons to other variables, maintaining these forms is sufficient.

Storm Event Identification

Because this chapter is expressly interested in highlighting impacts of precipitation to near-surface processes, multiple resolutions of time must be considered. The scope of this presentation is concerned with hourly to seasonal effects, and precipitation arrives through storm events that can last from minutes to days. Accounting for storm "pulses" as single contributing units is a prominent conceptual model in semiarid ecology (e.g., Weltzin et al., 2003; Chesson et al., 2004; Loik et al., 2004; Schwinning and Sala, 2004; Potts et al., 2006). Total storm precipitation amount, as well as the intensity at which the bulk of the storm precipitation arrives, are closely tied to the balance of runoff, infiltration, and surficial evaporation as a fundamental principle of surface hydrology (Horton, 1933). Thus, not all storms (or precipitation characteristics) are equal, hence we are looking for "tiered" ecosystem responses to different storm mechanisms that are subsequently filtered by local soils conditions (Reynolds et al., 2004).

Depending on the science question and scale of interest, "effective" precipitation is not strictly dictated by precipitation amount, but also must include information on infiltration as opposed to runoff (Yevjevich, 1967). The number of variables when directly calculating "effective,"or locally infiltrating, precipitation is substantial, including hillslope length, surface roughness, surface vegetation, local hydrological conductivity of soil, saturation/infiltration curves for the soil column, current saturation, rainfall rate, and so forth (Dunne et al., 1991). However, it is also common to infer effective precipitation using empirical data from measurements of the precipitation, runoff, and infiltration processes themselves (Beven, 2011). This presentation does neither in the modeling sense, but instead qualitatively assesses the impacts of each precipitation event to specific measured variables at each site. Following this line of investigation, "storm" events were classified as those time intervals where all days receiving precipitation were grouped into the same "storm" as long as the days were immediately adjacent. This grouping scheme also allowed for intrastorm gaps of one nonprecipitation day, because of the intensity selection bias in the GDM. In this fashion, each significant precipitation "event" is identified that should have discrete impacts on the local ecohydrology. Weighted, mean storm intensity was calculated for each grouped "storm" event by taking each hour with measured precipitation during the event, weighting that value using its percentage of the storm total, and summing the weighted values. The resulting weighted average provides a representative estimate of the hourly precipitation rate associated with the bulk of the storm's rain or snowfall.

Evolution of Daily Ecohydrology Across the Gradient

The ecoclimatic variables of the maximum and minimum, daily air temperature, VPD, shallow soil VWC, wintertime snow cover, precipitation phase and amount, and weighted mean storm intensity are shown in Figs. 5.9–5.12, across the elevation gradient for each year in the 2012–15 drought sequence. Notable gaps in the data are portions of soil VWC in 2012 and the substitution of air temperature at the 2 m height for 10 m in 2015 due to sensor malfunctions.

Following the daily data plots are specific "case study" examples of processes and interactions that were identified as being of ecohydrological relevance and until this study had not been directly quantified within the region. Hourly data in each case study show how diurnal cycles of temperature, VPD, and precipitation by phase are related to soil VWC and sap flow by species. These mechanisms can easily be placed in seasonal context by referring to the

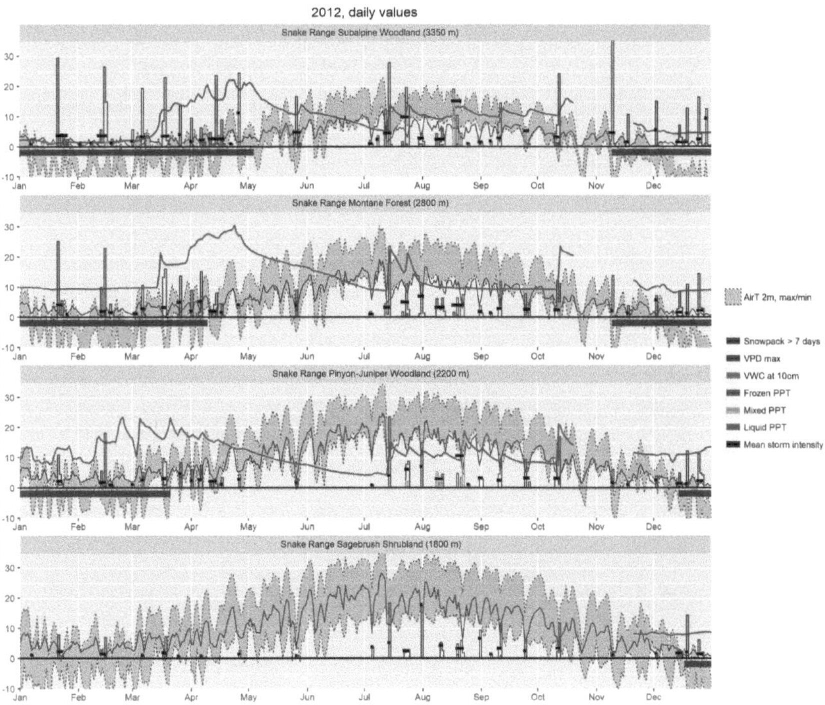

FIGURE 5.9 Daily ecohydrological observations across the elevation gradient of the Snake Range for the year 2012. Soil moisture data are absent at the Sagebrush site for most of the year, and short gaps exist at the other sites during November.

annual plots of daily data. The variables of sap flow and VPD were plotted as indices in the case studies, rather than in engineering units.

Sagebrush Site (1800 m)

For the Sagebrush site, soil moisture was not available during 2012 due to a sensor malfunction, and also a short data gap occurred in June 2014 due to battery failure. Notable features of this 4-year interval include, as indicated in the monthly summaries earlier, different precipitation seasonality each year. Persistent snow cover was not present in early 2012, and only three significant storms occurred prior to July. Summertime storms separated by about 1 week punctuated July 2012 to September 2012, presumably mitigating what otherwise was a very dry year. In 2013, snow cover was present in January, a remnant of a mid-December storm that persisted due to lower temperatures. Spring 2013 was only lightly punctuated by precipitation, but enough to prolong the drop of soil VWC. June and July of that year were extremely dry (as evidenced by the curve of shallow soil moisture) right up until an intense storm event in August deposited over 20 mm of liquid equivalent in 1 h. This

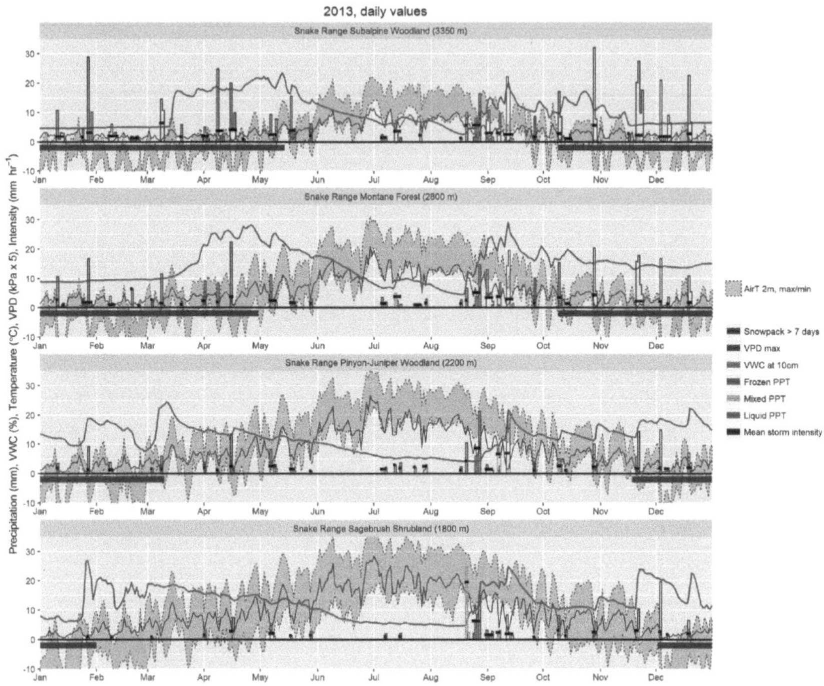

FIGURE 5.10 Daily ecohydrologic observations across the elevation gradient of the Snake Range for the year 2013.

was a mixed rain/hail event that was flagged as such by the GDM classification and confirmed with hourly imagery (Fig. 5.13). This event only partially affected shallow soil moisture, and it was not until a longer duration storm of moderate intensity a few days later that sufficient infiltration occurred to recharge soil moisture to springtime levels. Persistent smaller precipitation events throughout September, coupled with lower VPD during this period, maintained high soil VWC going into fall 2013.

Snow cover persisted into February 2014, in spite of a warm rain event in late 2013 that partially melted the snowpack. Soil VWC in 2014 peaked in March with 1 springtime storm making an impact to delay maximum drying by roughly 2 weeks. Summertime rains arrived in late July, and recharged soil VWC significantly. Unfortunately, no snowfall occurred in late 2014 and temperatures stayed elevated starting in early 2015 to make for a dry winter with only a single storm in March, besides very low-yield events. The soil VWC outlook for 2015 was very bleak until a wet May materialized; however, a very warm June with persistently high VPD quickly erased gains and a relatively hot and dry summer followed until a large rainstorm in October. Overall, precipitation at this site occurs at moderate to low intensity (<5 mm/h) that improves soil VWC in all seasons, with high-intensity

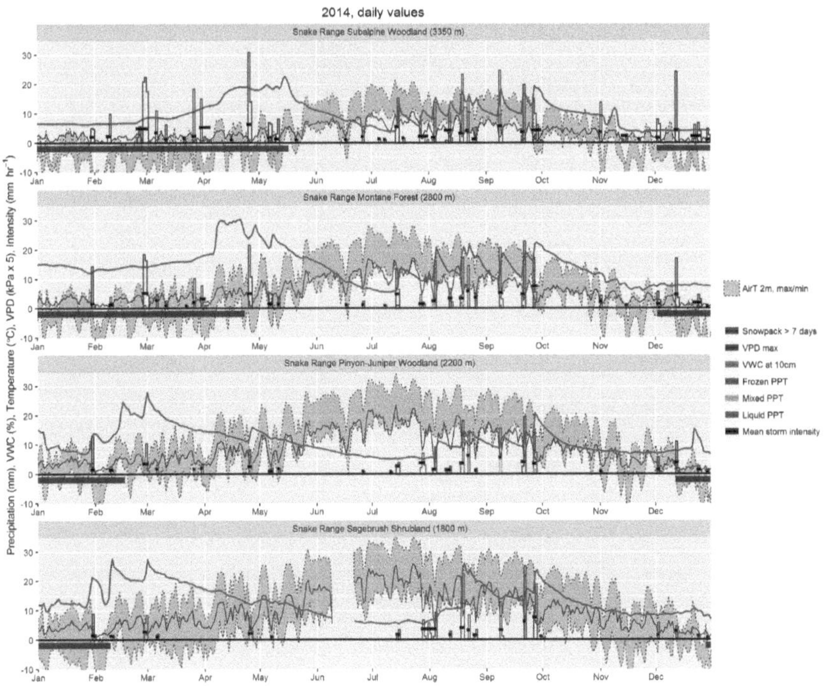

FIGURE 5.11 Daily ecohydrologic observations across the elevation gradient of the Snake Range for the year 2014.

(>10 mm/h; two present in 4 years) events that are not particularly effective due to runoff rather than infiltration. For instance, the August 20, 2013 storm totaled 39.4 mm (31% liquid, 19.5 mm/h weighted mean intensity), which was 19% of the annual total (203 mm) of classified events. That storm raised the 10 cm soil VWC from 5% to 9% in 2 days, whereas the very next storm (23.9 mm total, 91% liquid, 6.3 mm/h) raised the VWC from 8% to 19% over 3 days (Fig. 5.14).

Pinyon-Juniper Site (2200 m)

During 2012, the Pinyon-Juniper site started the year with several winter and springtime storms that kept soil VWC elevated until April. A dry April–June sequence ended in July when the soil VWC was approaching the apparent VWC_{PWP}. Weekly storm events maintained soil moisture throughout the late summer and fall, and low temperatures allowed snow cover to remain until March 2013. Drying in spring 2013 happened more slowly than in 2012, with several minor storm events and reduced VPD prolonging bottoming out of the soil VWC curve. A wet late summer continued into a wet fall, with soil VWC remaining high until a warm, dry January 2014. A cold snap in February 2014 prolonged snowmelt in an otherwise warm winter, and a warmer March storm

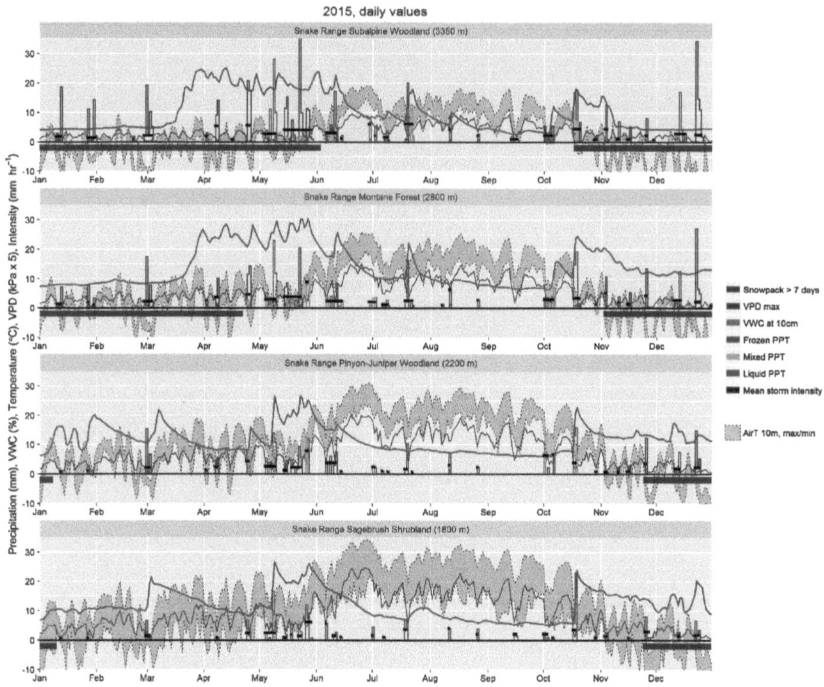

FIGURE 5.12 Daily ecohydrologic observations across the elevation gradient of the Snake Range for the year 2015. Air temperature and VPD calculations were made using 10 m data, rather than 2 m because of a sensor malfunction at one of the sites.

FIGURE 5.13 A mixed rain/hailstorm event at the Sagebrush site on August 20, 2013 deposited over 30 mm of precipitation in a short period of time. In this image, pooling and runoff are clearly visible, indicative of the "flash-flood" nature of some Great Basin high-intensity storms. The GDM classification procedure correctly flagged this event as "mixed" precipitation (Figs. 5.10 and 5.14), not entirely "frozen", and not entirely "liquid", demonstrating the usefulness of the method in capturing unusual precipitation events even during the warm seasons.

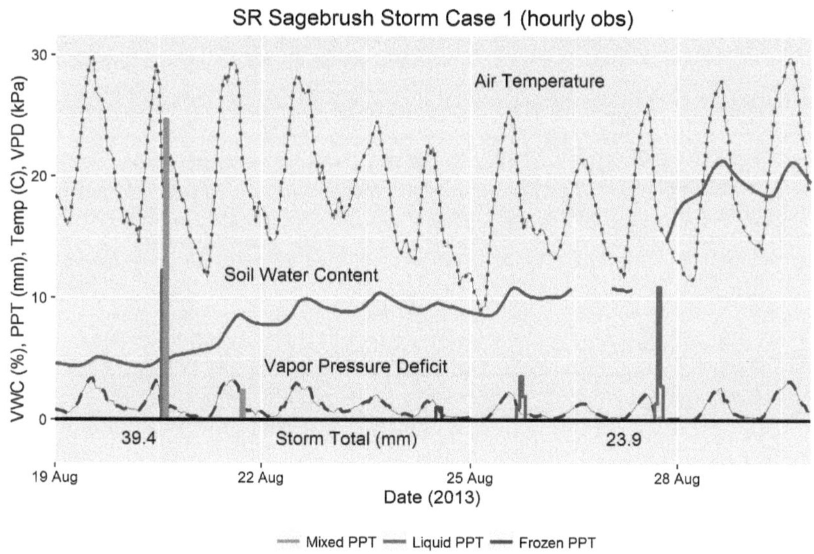

FIGURE 5.14 Two subsequent storms at the Sagebrush site had vastly different impacts on soil moisture, primarily due to the differences in intensity.

set the soil VWC peak for the year. Small storms mitigated the drop in soil VWC during the April–June window, and it was not until August 2014 that significant precipitation arrived. High temperatures and VPD quickly dried shallow soils between large, moderate-intensity events in August–September, and a dry October–November rapidly allowed the soil VWC to approach the apparent VWC_{FC} point. Only one snow event of significant volume/intensity occurred in December 2014 to February 2015, although the large soil VWC response and lower VPD sequence in late January 2015 indicates that perhaps a persistent low-intensity event occurred between January 14 and January 20 that was not picked up by the precipitation data processing procedure (Fig. 5.15). A substantial snowstorm in March 2015 was followed by elevated temperatures that rapidly dried soils to levels unseen in the previous three March–April time frames. Nearly continuous storm activity in May peaked the soil VWC in advance of the driest summer of the 4 years observed, which was not mitigated substantially until mid-October.

Montane Site (2800 m)

Assessment of precipitation/VPD/VWC/vegetation interaction and storm effectiveness is greatly enhanced at the Montane site, as several trees on site were instrumented to monitor sap flow in the outer xylem stem tissues, with data spanning 2013–15. Species included in this study were *P. flexilis* (PIFL; limber pine), *Cercocarpus ledifolius* (CELE; mountain mahogany), *Abies concolor* (ABCO; white fir), and *Pseudotsuga menziesii* (PSME; Douglas-fir).

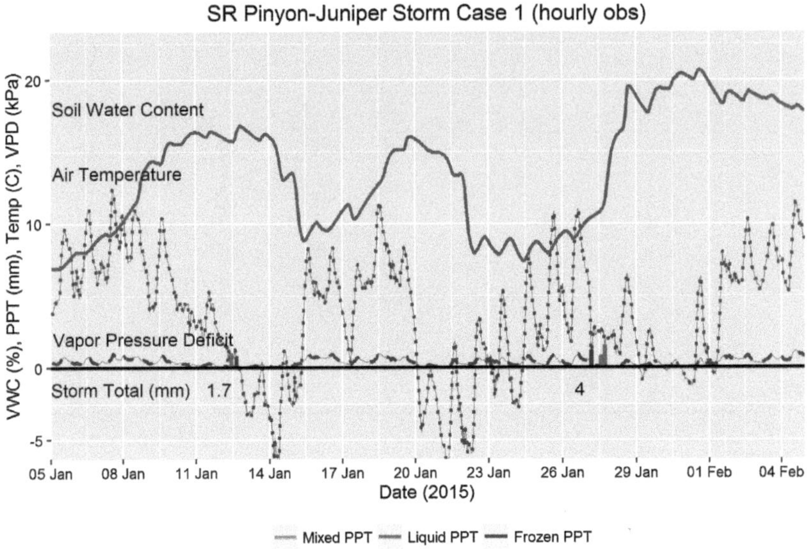

FIGURE 5.15 During the dry winter of 2014–15, unmeasured or dramatically underestimated, low-intensity precipitation events (<0.5 mm/h) may have combined with low VPD to provide additional soil VWC recharge, as measured storm events were very small and no snow cover melted at the time to otherwise recharge the shallow soil column.

The first two species are commonly found at mid and high elevations in major Great Basin mountain ranges, whereas ABCO and especially PSME are generally restricted to distributions in the central-eastern portions of the region, associated with Rocky Mountain ecoclimatic influences (Charlet, 1996).

The Montane site differs strongly from the two lower elevation sites in the sense that wintertime snow cover persisted in all years during the drought period. Soil VWC peaks in the springtime during snowmelt, and drops precipitously as temperature and VPD are already quite high in June. Snowstorms occur through May in all years, although 2012 was limited in this regard. Each summer within the observation period had very different precipitation and shallow soil moisture regimes. Overall, the sap flow trends closely follow this soil moisture curve (Fig. 5.16), but the trees do not always have the same seasonal response. For example, in 2013 and 2014, sap flow ramps up in the springtime as soil VWC from the melting snowpack peaks, declines as the warm season progresses and soil dries out, and then has a secondary response to influx of late summer precipitation that persists into October in accordance with the available soil moisture. In 2015, sap flow is generally lower, but has initial responses early in the season that are suppressed by the cool, wet storms during May. Consistent flow begins after May, but tapers off for the rest of the year and does not respond significantly to the moisture arriving in October, even though in previous years active sap flow was present in the same month.

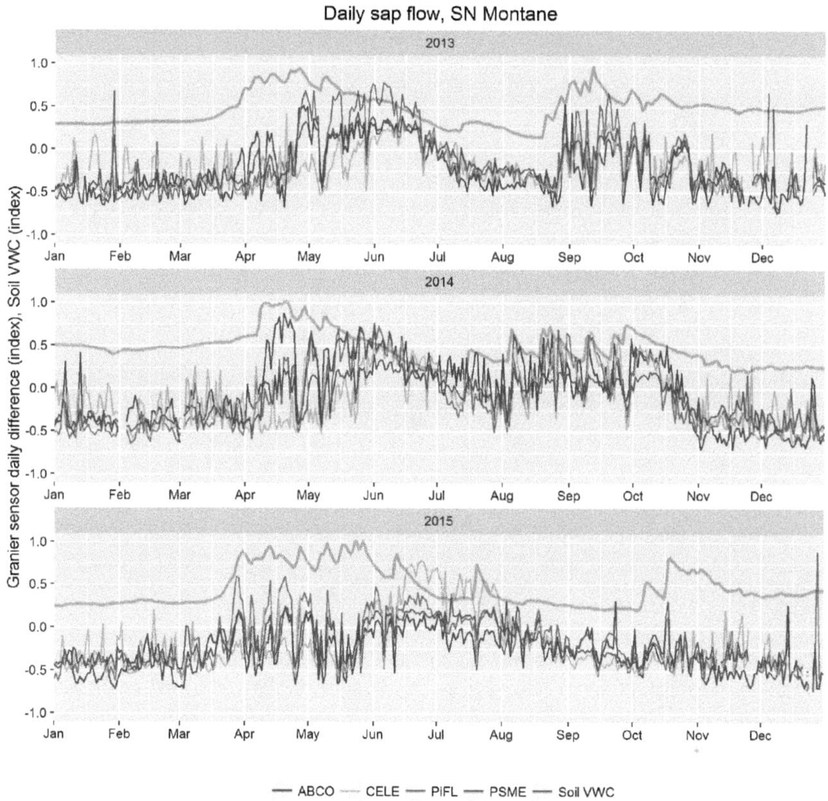

FIGURE 5.16 Maximum differences in daily temperature in the Granier probe data are indicative of higher sap velocity. The measurements from individual probes were combined as means for each species (ABCO n = 2, PIFL n = 2, CELE n = 3, and PSME n = 2), converted to daily differences (Granier probe differential daily $°C_{MAX} - °C_{MIN}$), and standardized for visual comparison. Soil VWC is also shown as an index. ABCO hourly data were particularly noisy due to problems with sensors, and were not used in further evaluations.

A detailed examination of processes occurring during the late summer of 2013 demonstrates relief of drought stress by storms that improve soil VWC and subsequent restart of sap flow by the local trees (Fig. 5.17). Prior to the storm on August 20, 2013, the shallow soil VWC was at the lowest point in the 4-year record (4.0%; August 18, 2013), one indication of the highest measured drought stress. Successive storms brought rain to the site, which increased the VWC to presummer levels. Sap flow prior to the storm of August 27 to August 28 does not appear to have been occurring, whereas afterwards a substantial increase of sap velocity took place during daylight hours throughout the stormy September period. This sap flow restart was coincident with initial reductions in daily VPD and increases in soil VWC above 15%.

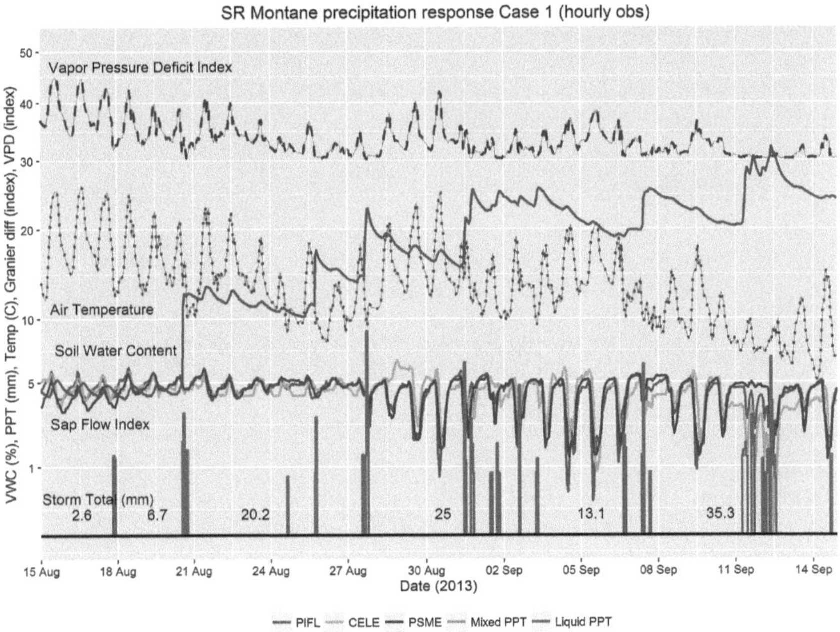

FIGURE 5.17 Flow response of tree sap to late summer storms is shown after soil VWC is brought up to springtime levels following the lowest point of VWC observed on the site in the 2012–13 period. VPD and sap flow are shown as dimensionless indices. Lower values of the sap flow index are representative of higher sap velocity. The y-scale is nonlinear to improve visual clarity.

A second case of interesting tree activity was in October 2014, where active sap flow occurred well into the fall season and then precipitously dropped off by November 1. Following a very wet summer, sap flow remained elevated near springtime levels, although storm events and associated low VPD (and cloudiness) temporarily reduced daytime sap velocities. After the final storm in September, October remained warm and dry, with temperatures rarely approaching 0°C (Fig. 5.18). Soil VWC tapered steadily down during this time to 10%, although nowhere near maximum drought levels of 4%. Sap flow is seen to have remained high until mid-October, where it began to reduce in diurnal amplitude following 2 days of colder temperatures. By the end of the month, significant sap flow had ceased for the 3 species shown after another 2-day cold event, even though subsequent temperatures remained above freezing, VPD was still elevated, and soil VWC was above 10%.

Finally, May 2015 is examined a month that in the 2 previous years experienced high sap velocities. A sustained cold storm sequence interrupted early season sap flow occurring in the conifers but not in the *Cercocarpus* (Fig. 5.19). Once the stormy period was passed, sap flow commenced immediately with the rise of VPD and air temperatures. This continued until

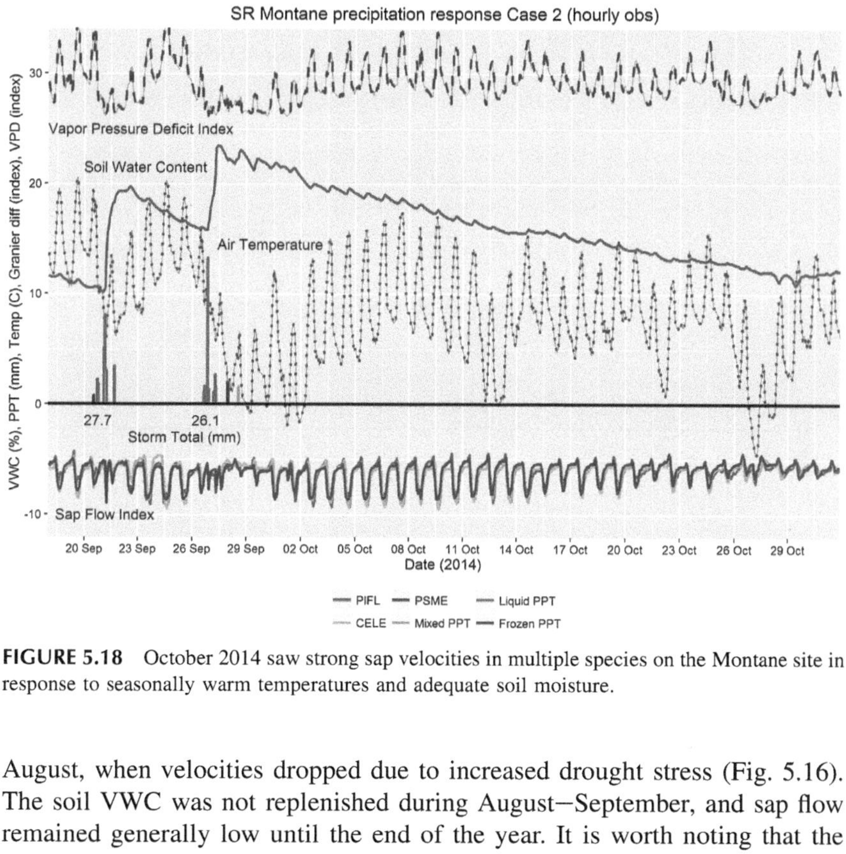

FIGURE 5.18 October 2014 saw strong sap velocities in multiple species on the Montane site in response to seasonally warm temperatures and adequate soil moisture.

August, when velocities dropped due to increased drought stress (Fig. 5.16). The soil VWC was not replenished during August–September, and sap flow remained generally low until the end of the year. It is worth noting that the relative velocity of sap flow in CELE was sustained at a higher level than in previous years during the June 2015 to August 2015 interval.

Subalpine Site (3350 m)

The Subalpine study site in the Snake Range represents a unique set of scientific observations, in that it provides comprehensive, high-resolution sensor data for several seasons from a Great Basin bristlecone pine (*P. longaeva*; PILO) woodland (Fig. 5.20). It is one of very few instrumented high-altitude sites in the region, and has a wider range of sensors and instruments than probably any other subalpine zone climatological research site in the Great Basin.

Trends in this site's precipitation and soil moisture record are similar to the Montane site, as it has persistent snow cover in all winters during the observation period, experienced significant summertime moisture in 2012–14, and had a dry late summer in 2015. The nature of precipitation events appears to be more dynamic at this site compared to the others in the Snake Range, due to the length of the snowfall season, notable summertime hail events, and the fact

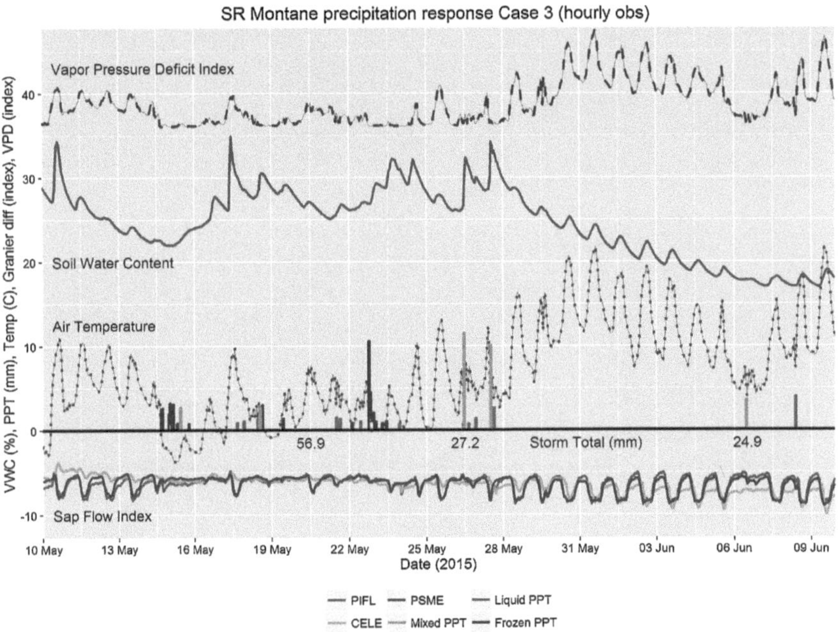

FIGURE 5.19 The stormy May 2015 interval halted springtime sap flow of conifers for nearly 2 weeks. At the end of the storm sequence, sap flow for mountain mahogany started for the year, along with the resumption of activity in limber pine and Douglas-fir. This was to be the peak of the conifers' sap flow season for the year, as no substantial precipitation relief of summertime dryness would materialize, contrary to the previous years (Fig. 5.16).

FIGURE 5.20 Ancient bristlecone pine inhabit the Snake Range in Nevada, with some individuals exceeding 4000 years in age (Currey, 1965). Strategies used by this species to survive large changes in climate may include adaptation to shifts in precipitation seasonality and reduction in snowfall.

that some storms appear to deposit significantly more precipitation here than at the other sites (Figs. 5.9–5.12).

Sap flow monitoring used in this study was installed in the summer of 2012 on three *P. flexilis* and four *P. longaeva*, commonly co-occurring east-central Great Basin high-elevation conifers (Charlet, 1996). The response of these two species to precipitation under regional drought conditions is of particular interest for conservation and biogeographic research purposes, and indeed the sap flow from each season seems to be closely tied to soil moisture conditions during the May–October time frame (Fig. 5.21).

Between the sap flow and precipitation regimes, several observations emerge during the drought period that improve our knowledge of high-elevation systems in the Great Basin. First, is the phenomenon of summertime precipitation events that occur as hailstorms. Two substantial events of

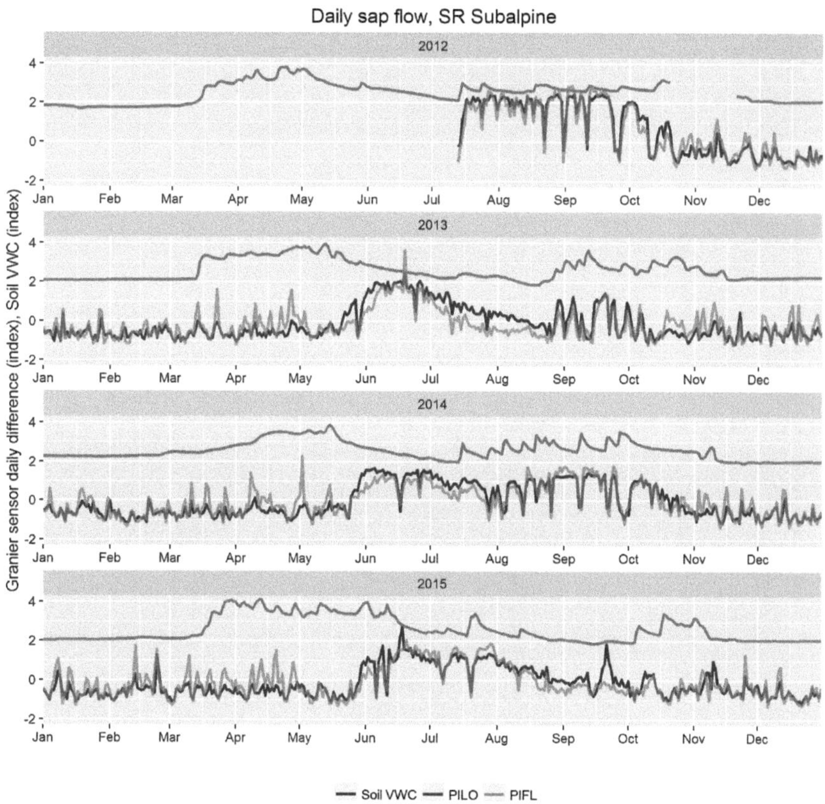

FIGURE 5.21 Relative sap velocity is shown for two subalpine conifer species, *P. flexilis* (PIFL; n = 3) and *P. longaeva* (PILO; n = 4), during the 2012–15 drought. Both species appear to directly follow trends in soil moisture even in this high-altitude environment, within 100 m of the upper tree line.

this nature occurred during the 4 years of observation: the first in August 2012 and the second in July 2015. Although hailstorm was probably associated with other summertime events during the time period, these stand out because of the volume and amount of the storm total that fell as ice rather than rain. The storm in August 2012 was particularly intense, with nearly 40 mm of precipitation falling within a 2-h window (Figs. 5.22 and 5.23). This single event raised the soil VWC from 10% to 14%, but did not immediately impact sap flow, as previous storms in July had already raised the moisture from the seasonal low point (Fig. 5.9). However, this recharge amount extended elevated summertime soil moisture (and high sap flow rates) until the next large event in mid-September, alleviating stress during a year that set high-temperature records across the country.

Another prominent feature of these data is the persistence of sap flow in the conifers until well into the fall season, as long as September soil VWC was elevated along with air temperatures. Both species responded to late summer precipitation and warm temperatures even after cold storms in September–October (Fig. 5.24). Several centimeters of snow fell in

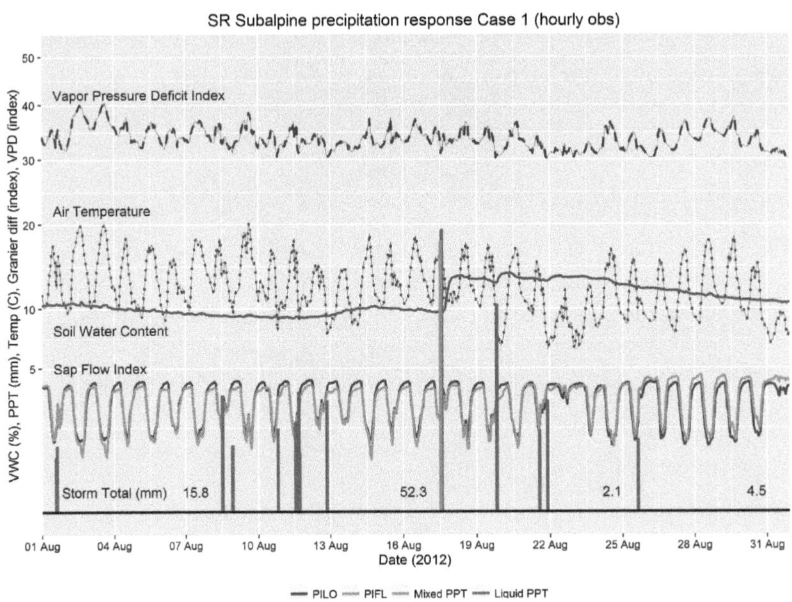

FIGURE 5.22 The highest intensity precipitation event (from an hourly perspective) during the observational period at the Subalpine site occurred as a mixed rain–hailstorm in August 2012. Although the 5-day storm total was 52.3 mm, the initial event deposited almost 40 mm of liquid equivalent in 2 h. The impact to the soil moisture for this event was much higher than adjacent events (note: nonlinear scale), presumably because even though this intensity–duration–amount combination should have resulted in substantial runoff, the form of the precipitation as ice allowed for additional infiltration as it melted over several hours (see Fig. 5.23).

FIGURE 5.23 A camera view from the Subalpine site shows a strong summertime hailstorm that deposited nearly 40 mm of liquid equivalent. The ice did not immediately melt and run off, but persisted for several hours and provided a much greater proportion of effective soil VWC recharge than if this same intensity–duration event had been rain. The elevated VWC persisted for multiple weeks afterwards during the middle of a record-setting drought year, with sap flow of subalpine conifer responding with seasonally high flow rates through September.

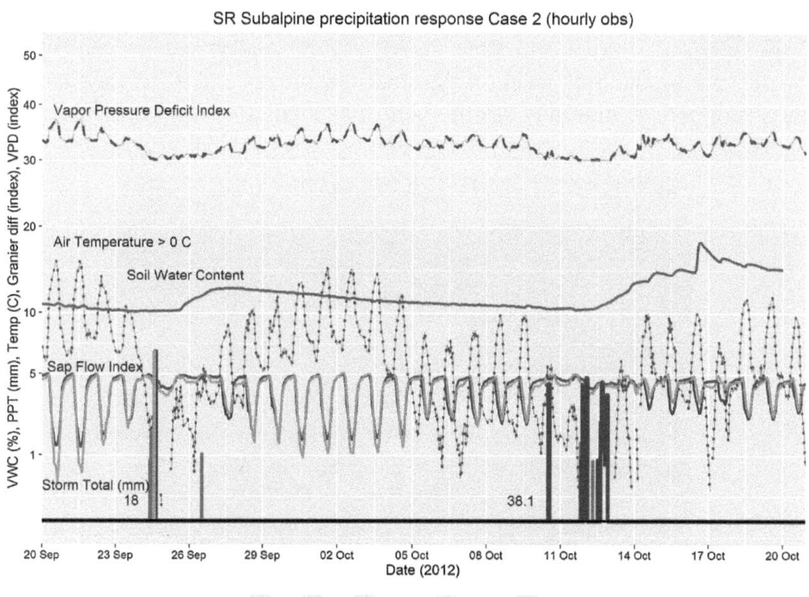

FIGURE 5.24 Late season sap flow at high altitude is observed in both PILO and PIFL, even between cold storms. Directly after this sequence, a short cold event occurred with daytime high temperatures near −10°C, which effectively ended elevated rates of xylem activity.

mid-October 2012, and in less than 1 week completely melted under warmer temperatures after the storm (Fig. 5.25). Although sap flow slowed considerably at this time, the diurnal changes in velocity persisted for another week before temperatures dropped substantially.

An interesting interaction between climatological inputs, soil VWC, and sap flow is the pattern that occurs at the very minimum shallow soils VWC during the growing season. Even though these years were considered regionally to be part of a major "drought" sequence, demonstrable moisture stress on the conifers only occurred on the Subalpine site twice—during August 2013 and September 2015. It is also possible that this happened in July 2012, but sap flow data used in this study did not begin until just after this time. In both cases, as drying, shallow soil VWC passed through the 5–6% threshold, a noticeable decoupling of diurnal sap flow occurred between limber pine and bristlecone pine (Figs. 5.26 and 5.27). In the 2013 case, high VPD, coupled with rapid soils drying is associated with a reversal of diurnal amplitude of sap flow in PIFL during an approximately 3-week period in August. PILO, on the other hand, slowed the daytime sap velocity during this period but never full ceased. A series of rain events recharged the soil moisture, and by August 29 both species were exhibiting elevated sap velocities even while VPD and high temperature in daytime was lower than previously (Fig. 5.26).

The water stress sequence in 2015, however, lasted for approximately 4 weeks. Similarly, elevated VPD and air temperatures accelerated soil drying, and a small storm event of 3.7 mm precipitation failed to mitigate decoupling of the species' diurnal sap flow activity. PIFL daily patterns once again reversed during the final week of the stress sequence, and PILO daytime flow nearly stopped. A multiday storm event that totaled 4.5 mm only raised the

FIGURE 5.25 A fall season snowstorm melted within a few days as temperatures remained between 0°C and 10°C afterwards. The daytime sap flow of the conifers continued on October 15th and persisted for another week.

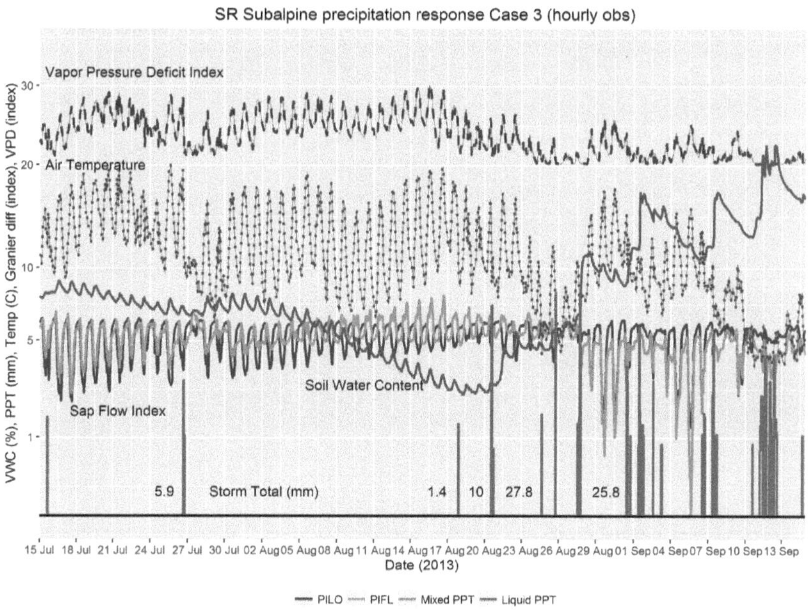

FIGURE 5.26 Data from August 2013 indicates high-stress levels on both conifer species, with daytime sap flow slowing significantly in bristlecone pine and in limber pine coming to a complete stop during the day.

shallow VWC by 2 percentage points, but it was enough for PILO to restart sap flow and actually increase velocity for a few days before tapering off again, while PIFL sap flow still indicated zero or negative daytime flow. A sudden cold storm brought 19.5 mm of snow, after which both species resumed an abbreviated diurnal pattern for a short time (Fig. 5.27).

One common theme in the 3 complete years of data is the onset timing of sap flow, which occurs at approximately the same point each year (late May). This time frame is somewhat coincident with rises in temperature and the disappearance of snow cover in open, generally unshaded areas (Figs. 5.9–5.12 and 5.21).

SUMMARY AND DISCUSSION

Four years of subhourly ecohydrological observations were presented from study sites which represent the valley to mountain gradient of vegetative zones in the central Great Basin region of the United States. These data were analyzed to extract the characteristics of precipitation events, which ultimately control the seasonal water budgets of these zones. Subsequent responses of rooting depth soil moisture and sap flow activity in dominant vegetative species at the two upper-elevation sites were compared to these precipitation events, and seasonal patterns described. Specific interactions at subdaily

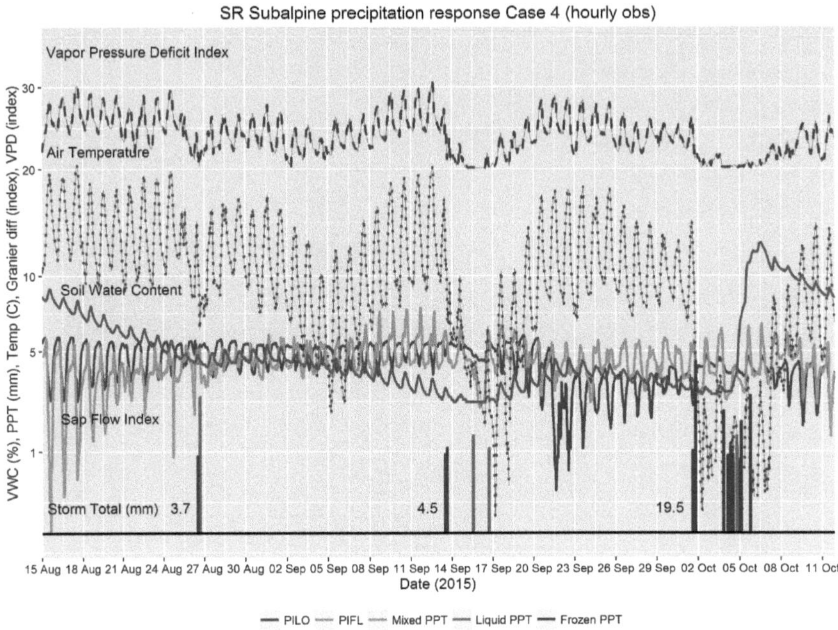

FIGURE 5.27 Another high-stress period on the Subalpine site was observed during September 2015. Similar to the 2013 sequence, PIFL daytime sap flow halted prior to PILO as soil VWC decreased.

resolution were highlighted, showing examples of cause and effect processes that had not been directly observed in this region. These processes have implications for ecological studies of these vegetative species, working concepts of seasonal water balances, and projections of climate adaptation. Lessons are to be learned about the adaptability of vegetation within semiarid regions, which already experience frequent moisture stress and interannual variability.

Seasonality of Precipitation Under Regional Drought

Because the 4 years of observation were coincident with prevailing regional drought, this provided an opportunity to examine processes with short-term changes in seasonal climate. Indeed, monthly distributions of precipitation during the period proved to be quite different than long-term "normals." In the Great Basin, interannual variability of climate is quite significant, both during the modern period (Redmond and Koch, 1991) and inferred from palaeorecords (Hughes and Funkhouser, 2003; Strachan et al., 2012). This partially due to the region functioning as a transition zone between longer term oscillations in large-scale circulation (Cayan, 1996; Wise, 2010), as well as the fact that the relatively few storms that occur have proportionally large impacts on any given season.

Each year of the study period exhibited slightly different seasonal patterns; however, the dominant theme in 3 of the 4 years (2012, 2013, and 2014) proved to be a 2-season mode of precipitation with significant summertime storm events balancing relatively light winters. Summertime storm events in the south-central Great Basin are most commonly associated with the seasonal pattern of the NAM, which transports high amounts of water vapor into the American Southwest during the July–September period (Higgins et al., 1997; Means, 2013). Convective activity from daytime solar insolation during these months generates active thunderstorms, which serve as the main precipitation mechanisms. Consequently, precipitation events during the summer would be expected to be spatially disjunct and highly variable in intensity and duration. Analysis of storm intensity by season and site location confirms that storms in winter exhibit lower hourly precipitation rates than spring and summertime events, and that summer intensities are more variable (Fig. 5.28). A one-way analysis of variance (ANOVA) on log-transformed, storm intensity by season, indicates that the difference between seasonal storm intensities (in particular, summer and winter) during drought conditions is significant [$F(3, 458) = 11.14$, $p < .0001$].

Influence of Topography

Topographic gradients play a critical part in the spatial distribution and phase of precipitation during most storm events. This is particularly evident during nonsummer seasons, and is most likely due to a combination of orographic enhancement, height of precipitable water in the large-scale air mass, and temperature lapse rates. Year-by-year comparisons of daily values across sites show that the relationship between daily precipitation and elevation varies with storm (Figs. 5.9–5.12). The percentage of precipitation as "frozen" during the observational period was significantly higher for the two upper-elevation sites (58.7% and 68.7%), than the two lower sites (19.0% and 36.7%).

In addition, the weighted mean hourly, storm intensity is directly related to elevation during nonsummer seasons. A one-way ANOVA on log-transformed storm intensity (for nonsummer seasons) by site indicates that the difference in hourly precipitation intensity between elevational zones is significant [$F(3, 306) = 7.85$, $p < .0001$]. As the spatial separation of the sites is minimal (8.9 km total) and all are located on the western gradient of the same mountain range, the primary geographic dissimilarity metric is elevation (1500 m total separation; Fig. 5.4).

Storm intensity in summertime does not differ significantly by site, and the monthly precipitation totals appear to vary by elevation only in a minor fashion during this season, with considerable variation. The GDM classification flagged multiple instances of high-intensity summertime "mixed" events, which were confirmed to be hailstorms. None of these events (during daylight hours) at the lower elevation sites left visible ice on the ground within 1 h after

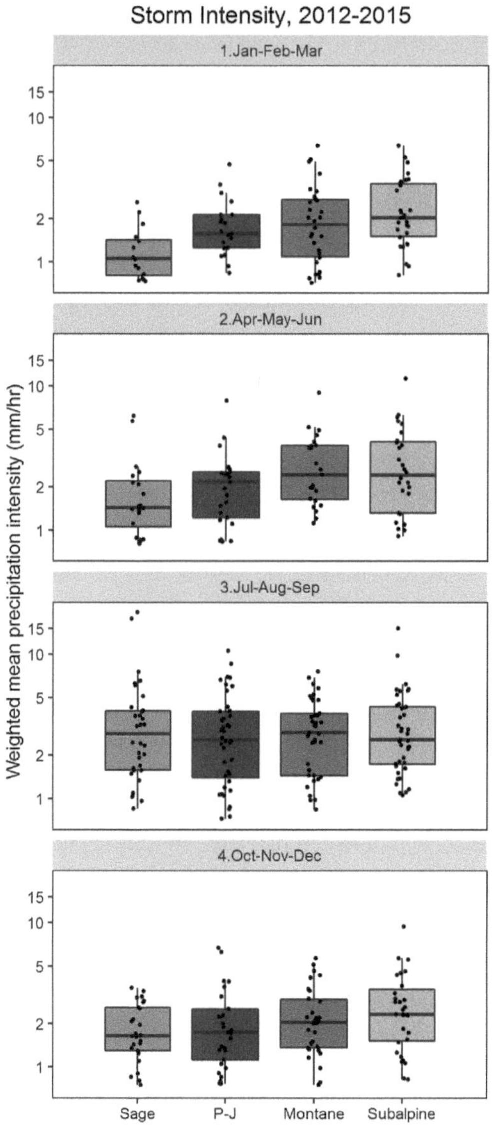

FIGURE 5.28 Distributions of seasonal storm intensities for all years are shown for each site (log scale). A clear relationship exists with elevation during the colder seasons.

the event, whereas the events at higher elevations had hailstones which persisted for more than 1 h after falling. Cooler temperatures and perhaps greater percentages of ice rather than liquid during high-altitude hail events assist in longer melting times and thus higher infiltration rates in shallow soils.

Opportunistic Behavior in Great Basin Trees

A significant ecohydrological observation of this study was the seasonally opportunistic xylem sap flow (as a proxy for transpiration) activity of the dominant trees on upper-elevation sites in response to effective precipitation inputs, and very late season activity even above 3000 m. The departures in precipitation seasonality from a winter-dominated regime during 2012, 2013, and 2014 resulted in two distinct "seasons" of peak soil moisture at rooting depth in each year. Tree transpiration responded accordingly for continuous time periods (weeks) for PILO, PIFL, ABCO, PSME, and CELE in the late summer, extending elevated sap velocities well into October in most cases. By contrast in 2015, a wet May–June suppressed VPD and shortened the transition from spring into a dry summer. Sap flow activity declined from a peak in June and, except for very short instances in some species, did not exhibit the dual season behavior of the previous 3 years. These observations highlight the importance of summertime monsoon activity in the central Great Basin in relieving winter season drought for mountain ecological communities, as well as impacts of July–September soil moisture on tree transpiration activity and the balance of resources that influence the following year's growth.

Mountain conifers in the western United States are generally associated with the wintertime seasonality of snow-dominated precipitation, and primary growth processes occur during the June–July window (Fritts, 1976), although photosynthesis and associated activities can occur across a range of seasons. This is particularly true in the semiarid regions of the Southwest. In a study by Adams and Kolb (2005), an intermountain assembly of conifer species similar to the Snake Range showed growth sensitivity to spring and summer water stress at both high- and low-elevation sites near the southern edge of their distributions in a dry, monsoon-influenced mountain environment, although the effects of previous year conditions were not evaluated. Williams and Ehleringer (2000) found that trees adapted to predictable monsoonal precipitation (such as *Pinus edulis*; PIED) make significant use of shallow soil moisture in September as long as they are located in a geographical zone where the monsoon makes a regular contribution, but that use of this seasonal water drops off as the geographic predictability of the monsoon is reduced. This geographic shift in response begs the question of phenotypical adaptation, and a study of *P. ponderosa* (PIPO) genetics (Potter et al., 2015, 2013) showed that in the central and southern Great Basin a gradient of differences occurred in the genetic markers analyzed. Similar studies have not been performed for the conifers in this study, but it would be interesting to see if PILO or PIFL had similar differentiating markers across geographical space. In one of the only examples of direct instrumentation and continuous phenology observation of high-altitude conifers in the Great Basin, Hal Fritts observed that *P. longaeva* in the White Mountains of California experienced radial increment growth in the basal stem during the June–July period, and that significant diurnal oscillations in dendrograph readings (a proxy for sap flow activity) persisted into August during 1962, 1963, and 1964 (Fritts, 1976, 1969). Unfortunately, the study only covered these summer months, so tree activity and response to environmental conditions outside of this window remained a mystery. The Fritts study was repeated in 2007 and 2008, with similar results, but once again the observations were only made during the June–August time frame (Hallman and Arnott, 2015). Similar early dendrograph work on mid- and low-elevation conifers in the southwest suggested that significant diurnal sap flow can occur in PIPO, PSME, and PIED from March to October on semiarid sites (Fritts et al., 1965).

Observation of this same type of seasonally opportunistic biologic activity near the tree line and at higher latitude is critical to our understanding of the adaptability of these central Great Basin conifer communities. Peaks of late summer sap flow activity were observed in August–September, indicating the use of monsoonal moisture to maintain respiration and photosynthetic activities. High sap flow rates extending into October, even at a near-tree line site such as the Subalpine location, seem to run the risk of cell cavitation due to freezing conditions, which is a real limitation for high-altitude tree growth (Körner, 2012). LaMarche and Mooney (1972) suggested that the Snake Range populations of *P. longaeva* were not as adapted to summer and fall drought as those in the White Mountains, and thus more susceptible to winter desiccation. Much more work on the physiology and biogeographic distinctions of Great Basin conifers remains to be done, but these contrasts bring into question if the Snake Range trees have rooting and water use strategies that are adapted to mitigate wintertime precipitation failures and springtime drought by exploiting frequent monsoon activity, even at the risk of extreme cold events in the early season. It is also possible that the southern topographic exposure of the Subalpine study site buffers this season-extending biologic activity due to higher amounts of direct solar radiation. This hypothesis is partially supported by examples of lower elevation PILO populations in seemingly warm and minimal snowfall environments across the south-central Great Basin, with a distinct absence of PIFL in these same stands. Of particular interest in this study's observations from the Subalpine site are the empirical differences between the relative VWC_{PWP} thresholds for PILO and PIFL. Diurnal sap flow activity between the two species is very similar overall (Fig. 5.21); however, at times of greatest water stress the PIFL appear to close stomata during the day at an earlier point in time than the PILO do. This is visible in Figs. 5.26 and 5.27 as a reversal of the diurnal pattern in sap flow. In fact, none of the peak water stress points (lowest soil VWC during elevated atmospheric VPD) have a complete daytime shutdown in PILO sap flow. PIFL, on the other hand, exhibits a greater sensitivity to moisture stress under these same conditions. Bristlecone pine possess characteristically smaller growth increments and generally denser xylem structures than limber pine (LaMarche, 1969), which would improve cavitation resistance and aid in the adaptation to water stress

(Hacke and Sperry, 2001) that characterizes the species and its geographic distribution. What is currently understood about the biogeographic history of conifers in the south-central Great Basin certainly points to a long time frame of regional adaptation (Charlet, 2007; Cole et al., 2013; Potter et al., 2015). The distribution of PILO, ABCO, PSME, and PIPO in particular across this region strongly suggests a tie to monsoonal geography and warmer temperatures (Charlet, 1996; Potter et al., 2013; van Devender and Spaulding, 1979). Making direct observations of these opportunistic processes across diverse seasonal patterns assists in interpretation of both genetic and paleoclimatic records.

Roles of Snow and Frozen Precipitation

Clearly, snow plays a significant role in annual ecohydrological processes in Great Basin mountains, even in times of regional drought and very low snowpack. In the Snake Range, the three mountain-block sites (Pinyon-Juniper, Montane, and Subalpine) underwent some period of snow cover longer than 7 days during the wintertime in each year in spite of regional conditions. The percentage of monthly precipitation classified as "frozen" was at or near 100% for the November–April time frame for the two upper-altitude sites in all 4 years. The subsequent delay in soil moisture peak as the snowpack melted each year provided the primary conifer growing season (April–June at the Montane site and June–July at the Subalpine site) with soil moisture levels above the apparent VWC_{PWP} for almost the entire season in each case. Snow at the Pinyon-Juniper and Sagebrush sites was much more variable in monthly percentages of total precipitation during the wintertime (Fig. 5.8), although when it did occur it seemed to impact soil VWC more substantially than rain events of similar magnitude. It is difficult to speculate on how warm season water balances would change if wintertime snow events were reduced to a minimum, as summertime precipitation dramatically augmented the light winters in this case. It seems, however, that even "low snow" years such as 2012, 2013, and 2014 brought enough water to maintain local conifer populations at mid-to high-altitude sites. This study does not explore the nature of vegetative competition and relative rooting depth of the species on each site, but the these conclusions from seasonal and sap flow measurements in case studies indicate that the instrumented species can have different times of sap flow peaks relative to one another, somewhat driven by the soil VWC, and likely other variables such as soil temperatures and remnant shaded snow.

Frozen precipitation during the summertime (i.e., hail) had a demonstrable effect on conifer sap flow at the Subalpine site compared to 100% liquid events of comparable magnitude. Although these events would seem to be rare, the fact that this 4-year study observed multiple examples of this phenomenon indicates that, during seasons of elevated convective activity, high-intensity hail events may be an effective mechanism at relieving summertime drought stress on mountain sites. It is important to note that current methods of automated weather monitoring and historical weather data from such sites do not differentiate between rain, snow, and hail, and thus the difference in "effective" precipitation between a 50 mm/h hail event versus the same falling as rain. Performing storm-by-storm analysis of precipitation using historical records in the context of water-limited ecohydrology is not as effective as when storm totals, durations, intensities, and most importantly, precipitation phase can be quantified and compared with ecological observations.

CONCLUSIONS

This 4-year study of precipitation across a Great Basin mountain gradient reveals concurrent ecohydrological processes at a level of spatiotemporal detail that has not previously been possible in the region. Combining precipitation phase, storm characteristics, winter snowpack, rooting depth soil moisture, air temperature, VPD, and tree sap flow at subdaily resolution into a single study over several continuous seasons provides new insights into the relative impacts of hydroclimatic inputs on Great Basin mountain communities. A more complete understanding of what "seasonal variability" actually looks like in the Great Basin will aid environmental research from a conceptual standpoint, as well as an operational one if future study designs are implemented with these results in mind. During regional drought conditions, the seasonality of precipitation in the Snake Range was altered from the long-term normals, with summertime (July—September) storms contributing much more to the annual totals during 2012, 2013, and 2014, than the more "normal" year of 2015. All 4 years were considered to be regional "drought" years, but water stress at middle and upper elevations on the mountains did not appear extreme if soil VWC and tree sap flow are reasonable indicators. This demonstrates the buffering effect of high altitude on semiarid ecological populations from drought due certainly to lower temperatures, but perhaps even more because of large changes in precipitation characteristics within Montane and Subalpine zones, compared to lower elevations. Two dominant modes of precipitation seasonality were observed during the study period: first mode that includes mild winters from the snowfall perspective but with significant summertime effective moisture, and second mode where springtime precipitation is very high and the summertime is dry. The long-term effects of each of these seasonal modes are difficult to gage, but these observations reinforce the concept of the Great Basin as a seasonally dynamic region with opportunistic and adaptable ecology. In addition, the role of mountain topography in the context of precipitation is highlighted here, which contributes to concepts of ecological refugia during extended periods of drought or increased temperatures. Because high-resolution measurements of climate variables in mountains of this region have not been made until relatively, it is difficult to say if the conditions observed are sustainable for the sizable conifer

FIGURE 5.29 Snow clearly plays a role in growing season ecohydrology in the Snake Range. The bristlecone pine tree line shown here may be resistant to changes in snowfall as long as minimal rainfall occurs during the spring or summer, but the responses of the other species may be questionable.

populations there, especially at the lower extents of the Montane conifer distribution. However, it is clear that the few snowstorms that occurred during the study period had a substantial role in elevating shallow soil moisture to levels useful to the conifers (Fig. 5.29).

Future studies should examine the spatial distributions, depths, and patterns of soil moisture from a microsite perspective. This should be done to examine whether conifers at these locations exhibit different seasonal strategies tied to individual tree rooting depths, local competitive dynamics, shading/persistence of snow, and moderation of soil temperatures.

REFERENCES

Adams, H.D., Kolb, T.E., 2005. Tree growth response to drought and temperature in a mountain landscape in northern Arizona, USA. Journal of Biogeography 32, 1629–1640.

Andrews, E.D., 2012. Hydrology of the Sierra Nevada Network National Parks: Status and Trends (Fort Collins, CO).

Austin, A.T., Yahdjian, L., Stark, J.M., Belnap, J., Porporato, A., Norton, U., Ravetta, D.A., Schaeffer, S.M., 2004. Water pulses and biogeochemical cycles in arid and semiarid ecosystems. Oecologia 141, 221–235.

Bell, J.E., Leeper, R.D., Palecki, M.A., Coopersmith, E., Wilson, T., Bilotta, R., Embler, S., 2015. Evaluation of the 2012 drought with a newly established national soil monitoring network. Vadose Zone Journal 14.

Beven, K.J., 2011. Rainfall-Runoff Modelling: The Primer. John Wiley & Sons.

Billings, W.D., 1951. Vegetational zonation in the Great Basin of western North America. In: Les Bases Écologiques de La Régénération de La Végétation Des Zones Arides. International Union of Biological Sciences, Series B: Colloquia, pp. 101–122.

Brekke, L., Thrasher, B.L., Maurer, E.P., Pruitt, T., 2013. Downscaled CMIP3 and CMIP5 Climate Projections: Release of Downscaled CMIP5 Climate Projections, Comparison with Preceding

Information, and Summary of User Needs. US Department of the Interior, Bureau of Reclamation, Technical Services Center, Denver, Colorado.

Burgess, S.S.O., Adams, M.A., Turner, N.C., Beverly, C.R., Ong, C.K., Khan, A.A.H., Bleby, T.M., 2001. An improved heat pulse method to measure low and reverse rates of sap flow in woody plants. Tree Physiology 21, 589–598.

Burgess, S.S.O., Dawson, T.E., 2008. Using branch and basal trunk sap flow measurements to estimate whole-plant water capacitance: a caution. Plant and Soil 305, 5–13.

Castelli, R.M., Chambers, J.C., Tausch, R.J., 2000. Soil-plant relations along a soil-water gradient in Great Basin riparian meadows. Wetlands 20, 251–266.

Cayan, D.R., 1996. Interannual climate variability and snowpack in the western United States. Journal of Climate 9 (5), 928–948.

Charlet, D.A., 1996. Atlas of Nevada Conifers: A Phytogeographic Reference. University of Nevada Press, Reno.

Charlet, D.A., 2007. Distribution patterns of Great Basin conifers: implications of extinction and immigration. Aliso: A Journal of Systematic and Evolutionary Botany 24, 31–61.

Chesson, P., Gebauer, R.L.E., Schwinning, S., Huntly, N., Wiegand, K., Ernest, M.S.K., Sher, A., Novoplansky, A., Weltzin, J.F., 2004. Resource pulses, species interactions, and diversity maintenance in arid and semi-arid environments. Oecologia 141, 236–253.

Cole, K.L., Fisher, J.F., Ironside, K., Mead, J.I., Koehler, P., 2013. The biogeographic histories of *Pinus edulis* and *Pinus monophylla* over the last 50,000 years. Quaternary International 310, 96–110.

Comstock, J.P., Ehleringer, J.R., 1992. Plant adaptations in the Great Basin and Colorado Plateau. The Great Basin Naturalist 52, 195–215.

Cook, B.I., Seager, R., Smerdon, J.E., 2014. The worst North American drought year of the last millennium: 1934. Geophysical Research Letters 41, 7298–7305.

Currey, D., 1965. An ancient bristlecone pine stand in eastern Nevada. Ecology 46, 564–566.

Daly, C., Neilson, R.H., Phillips, D.L., 1994. A statistical-topographic model for mapping climatological precipitation over mountainous terrain. Journal of Applied Meteorology 33, 140–158.

Dascalu, S., Harris Jr., F.C., McMahon Jr., M., Fritzinger, E., Strachan, S., Kelley, R., 2014. An overview of the Nevada climate change portal. In: Proceedings of the 7th International Congress on Environmental Modelling and Software (iEMSs 2014), vol. 1, pp. 75–82.

Denmead, O.T., Shaw, R.H., 1962. Availability of soil water to plants as affected by soil moisture content and meteorological conditions1. Agronomy Journal 54, 385.

DeRose, R.J., Bekker, M.F., Wang, S.Y., Buckley, B.M., Kjelgren, R.K., Bardsley, T., Rittenour, T.M., Allen, E.B., 2015. A millennium-length reconstruction of Bear River stream flow, Utah. Journal of Hydrology 529, 524–534.

Dettinger, M.D., Cayan, D.R., Diaz, H.F., Meko, D.M., 1998. North-south precipitation patterns in western North America on interannual-to-decadal timescales. Journal of Climate 11, 3095–3111.

Dettinger, M.D., Cayan, D.R., Meyer, M.K., Jeton, A., 2004. Simulated hydrologic responses to climate variations and change in the Merced, Carson, and American River basins, Sierra Nevada, California, 1900–2099. Climatic Change 62, 283–317.

Douglas, M.W., Maddox, R.A., Howard, K., Reyes, S., 1993. The Mexican monsoon. Journal of Climate 6, 1665–1677.

Dunne, T., Zhang, W., Aubry, B.F., 1991. Effects of rainfall, vegetation, and microtopography on infiltration and runoff. Water Resources Research 27, 2271–2285.

Fritts, H.C., 1969. Bristlecone Pine in the White Mountains of California, Growth and Ring-Width Characteristics. Papers of the Laboratory of Tree-Ring Research No.4, Tucson.

Fritts, H.C., 1976. Tree Rings and Climate. Academic Press, London.

Fritts, H.C., Smith, D.G., Stokes, M.A., 1965. The biological model for paleoclimatic interpretation of Mesa Verde tree-ring series. Memoirs of the Society for American Archaeology 19, 101–121.

Gleick, P.H., Chalecki, E.L., 1999. The impacts of climatic changes for water resources of the Colorado and Sacramento-San Joaquin river basins. Journal of the American Water Resources Association 35, 1429–1441.

Goodison, B.E., Louie, P.Y.T., Yang, D., 1998. WMO Solid Precipitation Measurement Intercomparison. World Meteorological Organization.

Graham, N.E., Hughes, M.K., Ammann, C.M., Cobb, K.M., Hoerling, M.P., Kennett, D.J., Kennett, J.P., Rein, B., Stott, L., Wigand, P.E., Xu, T., 2007. Tropical Pacific – mid-latitude teleconnections in medieval times. Climatic Change 83, 241–285.

Granier, A., 1987. Evaluation of transpiration in a Douglas-fir stand by means of sap flow measurements. Tree Physiology 3, 309–320.

Granier, A., Biron, P., Lemoine, D., 2000. Water balance, transpiration and canopy conductance in two beech stands. Agricultural and Forest Meteorology 100, 291–308.

Grayson, D.K., 2011. The Great Basin: A Natural Prehistory, second ed. University of California Press, Berkeley.

Hacke, U.G., Sperry, J.S., 2001. Functional and ecological xylem anatomy. Perspectives in Plant Ecology, Evolution and Systematics 4, 97–115.

Hallman, C., Arnott, H., 2015. Morphological and physiological phenology of *Pinus longaeva* in the white mountains of California. Tree-Ring Research 71, 1–12.

Hamlet, A.F., Mote, P.W., Clark, M.P., Lettenmaier, D.P., 2005. Effects of temperature and precipitation variability on snowpack trends in the western United States. Journal of Climate 18, 4545–4561.

Higgins, R.W., Yao, Y., Wang, X.L., 1997. Influence of the North American monsoon system on the U.S. summer precipitation regime. Journal of Climate 10, 2600–2622.

Hoerling, M., Eischeid, J., Kumar, A., Leung, R., Mariotti, A., Mo, K., Schubert, S., Seager, R., 2014. Causes and predictability of the 2012 great plains drought. Bulletin of the American Meteorological Society 95, 269–282.

Horton, R.E., 1933. The Rôle of infiltration in the hydrologic cycle. Eos, Transactions American Geophysical Union 14, 1689–1699.

Houghton, J., 1979. A model for orographic precipitation in the north-central Great Basin. Monthly Weather Review 107, 1462–1475.

Howell, T.A., 1995. Comparison of vapor-pressure-deficit calculation methods—southern high plains. Irrigation and Drainage Engineering 121, 191–198.

Hughes, M.K., Funkhouser, G., 2003. Frequency-dependent climate signal in upper and lower forest border tree rings in the mountains of the Great Basin. Climatic Change 59, 233–244.

Huntington, J.L., Niswonger, R.G., 2012. Role of surface-water and groundwater interactions on projected summertime streamflow in snow dominated regions: an integrated modeling approach. Water Resources Research 48, 1–20.

Knowles, N., Cayan, D.R., 2004. Elevational dependence of projected hydrologic changes in the San Francisco Estuary and watershed. Climatic Change 62, 319–336.

Körner, C., 2012. Alpine Treelines: Functional Ecology of the Global High Elevation Tree Limits. Springer, Basel, Switzerland.

Köstner, B., Granier, A., Cermák, J., 1998. Sapflow measurements in forest stands : methods and uncertainties. Annals of Forest Science 55, 13–27.

LaMarche, V.C., 1969. Environment in relation to age of bristlecone pines. Ecology 50, 53–59.

LaMarche, V.C., Mooney, H.A., 1972. Recent climatic change and development of the bristlecone pine (*P. longaeva* Bailey) krummholz zone, Mt. Washington, Nevada. Arctic and Alpine Research 4, 61–72.

Lanner, R.M., 1984. Trees of the Great Basin: A Natural History. University of Nevada Press, Reno.

Loik, M.E., Breshears, D.D., Lauenroth, W.K., Belnap, J., 2004. A multi-scale perspective of water pulses in dryland ecosystems: climatology and ecohydrology of the western USA. Oecologia 141, 269–281.

Lundquist, J.D., Minder, J.R., Neiman, P.J., Sukovich, E., 2010. Relationships between barrier jet heights, orographic precipitation gradients, and streamflow in the northern Sierra Nevada. Journal of Hydrometeorology 11, 1141–1156.

Maule, C.P., Chanasyk, D.S., Muehlenbachs, K., 1994. Isotopic determination of snow-water contribution to soil water and groundwater. Journal of Hydrology 155, 73–91.

McCabe, G.J., Wolock, D.M., 1999. General-circulation-model simulations of future snowpack in the western United States. Journal of the American Water Resources Association 35, 1473–1484.

McRoberts, D.B., Nielsen-Gammon, J.W., 2011. A new homogenized climate division precipitation dataset for analysis of climate variability and climate change. Journal of Applied Meteorology and Climatology 50, 1187–1199.

Means, J.D., 2013. GPS precipitable water as a diagnostic of the North American monsoon in California and Nevada. Journal of Climate 26, 1432–1444.

Mensing, S., Strachan, S., Arnone, J., Fenstermaker, L., Biondi, F., Devitt, D., Johnson, B., Bird, B., Fritzinger, E., 2013. A network for observing Great Basin climate change. Eos, Transactions American Geophysical Union 94, 105–106.

Mensing, S.A., Smith, J., Burkle Norman, K., Allan, M., Norman, K.B., 2008. Extended drought in the Great Basin of western North America in the last two millennia reconstructed from pollen records. Quaternary International 188, 79–89.

Miller, R.F., Chambers, J.C., Pyke, D.A., Pierson, F.B., Williams, C.J., 2013. A Review of Fire Effects on Vegetation and Soils in the Great Basin Region : Response and Ecological Site Characteristics, General Technical Report RMRS-GTR-308. Rocky Mountain Research Station, United States Department of Agriculture, Forest Service.

Mock, C.J., 1996. Climatic controls and spatial variations of precipitation in the western United States. Journal of Climate 9, 1111–1124.

Nadezhdina, N., Cermák, J., Ceulemans, R., 2002. Radial patterns of sap flow in woody stems of dominant and understory species: scaling errors associated with positioning of sensors. Tree Physiology 22, 907–918.

Nieto, R., Gimeno, L., de la Torre, L., Ribera, P., Gallego, D., Garcia-Herrera, R., Garcia, J.A., Nunez, M., Redano, A., Lorente, J., 2005. Climatological features of cutoff low systems in the Northern Hemisphere. Journal of Climate 18, 3085–3103.

Nowak, C.L., Nowak, R.S., Tausch, R.J., Wigand, P.E., 1994. Tree and shrub dynamics in northwestern Great Basin woodland and steppe during the late-Pliestocene and Holocene. American Journal of Botany 81, 265–277.

Oakley, N.S., Redmond, K.T., 2014. A climatology of 500-hPa closed lows in the northeastern Pacific Ocean, 1948–2011. Journal of Applied Meteorology and Climatology 53, 1578–1592.

Overpeck, J.T., 2013. The challenge of hot drought. Nature 503, 350–351.

Parish, T.R., 1982. Barrier winds along the Sierra Nevada mountains. Journal of Applied Meteorology 21, 925–930.

Pataki, D.E., Oren, R., Smith, W.K., 2000. Sap flow of co-occurring species in a western subalpine forest during seasonal soil drought. Ecology 81, 2557–2566.

Potter, K.M., Hipkins, V.D., Mahalovich, M.F., Means, R.E., 2013. Mitochondrial DNA haplotype distribution patterns in *Pinus ponderosa* (Pinaceae): range-wide evolutionary history and implications for conservation. American Journal of Botany 100, 1562–1579.

Potter, K.M., Hipkins, V.D., Mahalovich, M.F., Means, R.E., 2015. Nuclear genetic variation across the range of ponderosa pine (*Pinus ponderosa*): phylogeographic, taxonomic and conservation implications. Tree Genetics and Genomes 11.

Potts, D.L., Huxman, T.E., Cable, J.M., English, N.B., Ignace, D.D., Eilts, J.A., Mason, M.J., Weltzin, J.F., Williams, D.G., 2006. Antecedent moisture and seasonal precipitation influence the response of canopy-scale carbon and water exchange to rainfall pulses in a semi-arid grassland. New Phytologist 170, 849–860.

Redmond, K.T., Koch, R.W., 1991. Climate and streamflow variability in the western United States and their relationship to large-scale circulation indices. Water Resources Research 27, 2381–2399.

Reinemann, S.A., Porinchu, D.F., MacDonald, G.M., Mark, B.G., DeGrand, J.Q., 2014. A 2000-yr reconstruction of air temperature in the Great Basin of the United States with specific reference to the Medieval Climatic Anomaly. Quaternary Research 82, 309–317.

Reinking, R.F., Boatman, J.F., 1986. Upslope precipitation events. In: Mesoscale Meteorology and Forecasting. American Meteorological Society. Springer, pp. 437–471.

Reynolds, J.F., Kemp, P.R., Ogle, K., Fernandez, R.J., 2004. Modifying the "pulse-reserve" paradigm for deserts of North America: precipitation pulses, soil water, and plant responses. Oecologia 141, 194–210.

Robinson, D.A., Campbell, C.S., Hopmans, J.W., Hornbuckle, B.K., Jones, S.B., Knight, R., Ogden, F., Selker, J., Wendroth, O., 2008. Soil moisture measurement for ecological and hydrological watershed-scale observatories: a review. Vadose Zone Journal 7, 358–389.

Rodriguez-Iturbe, I., 2000. Ecohydrology: a hydrologic perspective of climate-soil-vegetation dynamics. Water Resources Research 36, 3–9.

Roe, G.H., 2005. Orographic precipitation. Annual Review of Earth and Planetary Sciences 33, 645–671.

Schwinning, S., Sala, O.E., 2004. Hierarchy of responses to resource pulses in arid and semi-arid ecosystems. Oecologia 141, 211–220.

Shukla, S., Safeeq, M., AghaKouchak, A., Guan, K., Funk, C., 2015. Temperature impacts on the water year 2014 drought in California. Geophysical Research Letters 42, 2015.

Sperry, J.S., Hacke, U.G., 2002. Desert shrub water relations with respect to soil characteristics and plant functional type. Functional Ecology 16, 367–378.

Stewart, I.T., 2009. Changes in snowpack and snowmelt runoff for key mountain regions. Hydrological Processes 23, 78–94.

Strachan, S., Biondi, F., Leising, J., 2012. 550-year reconstruction of streamflow variability in Spring Valley, Nevada. Journal of Water Resources Planning and Management 138, 326–333.

Strachan, S., McAfee, S., McCurdy, G., (in prep). The Gage-Difference Method: simple sensor-based detection of precipitation phase at fine timescales for energy-limited deployments.

Swain, D.L., 2015. A tale of two California droughts: lessons amidst record warmth and dryness in a region of complex physical and human geography. Geophysical Research Letters 42, 9999–10003.

Swain, D.L., Tsiang, M., Haugen, M., Singh, D., Charland, A., Rajaratnam, B., Diffenbaugh, N.S., 2014. The extraordinary California drought of 2013–2014: character, context, and the role of climate change. Bulletin of the American Meteorological Society 95.

van Devender, T.R., Spaulding, W.G., 1979. Development of vegetation and climate in the southwestern United States. Science 204, 701–710.

Wang, S.Y., Gillies, R.R., Reichler, T., 2012. Multidecadal drought cycles in the great basin recorded by the Great Salt Lake: modulation from a transition-phase teleconnection. Journal of Climate 25, 1711–1721.

Wang, S.Y., Hipps, L., Gillies, R.R., Yoon, J.H., 2014. Probable causes of the abnormal ridge accompanying the 2013–2014 California drought: ENSO precursor and anthropogenic warming footprint. Geophysical Research Letters 41, 3220–3226.

Welch, A., Bright, D., Knochenmus, L., 2007. Water Resources of the Basin and Range Carbonate-Rock Aquifer System, White Pine County, Nevada, and Adjacent Areas in Nevada and Utah. US Geological Survey Scientific Investigations (Report 5261).

Weltzin, J.F., Loik, M.E., Schwinning, S., Williams, D.G., Fay, P.A., Haddad, B.M., Harte, J., Huxman, T.E., Knapp, A.K., Lin, G., Pockman, W.T., Shaw, M.R., Small, E.E., Smith, M.D., Smith, S.D., Tissue, D.T., Zak, J.C., 2003. Assessing the response of terrestrial ecosystems to potential changes in precipitation. BioScience 53, 941.

Wigand, P.E., Nowak, C.L., 1992. Dynamics of northwest Nevada plant communities during the last 30,000 years. In: The History of Water: Eastern Sierra Nevada, Owens Valley, White-Inyo Mountains. White Mountain Research Station Symposium, pp. 40–62.

Williams, D.G., Cable, W., Hultine, K., Hoedjes, J.C.B., Yepez, E.A., Simonneaux, V., Er-Raki, S., Boulet, G., de Bruin, H.A.R., Chehbouni, A., Hartogensis, O.K., Timouk, F., 2004. Evapotranspiration components determined by stable isotope, sap flow and eddy covariance techniques. Agricultural and Forest Meteorology 125, 241–258.

Williams, D.G., Ehleringer, J.R., 2000. Intra- and interspecific variation for summer precipitation use in pinyon – juniper woodlands. Ecological Monographs 70, 517–537.

Wilson, K.B., Hanson, P.J., Mulholland, P.J., Baldocchi, D.D., Wullschleger, S.D., 2001. A comparison of methods for determining forest evapotranspiration and its components: sap-flow, soil water budget, eddy covariance and catchment water balance. Agricultural and Forest Meteorology 106, 153–168.

Wise, E.K., 2010. Spatiotemporal variability of the precipitation dipole transition zone in the western United States. Geophysical Research Letters 37, 1–5.

WMO, 2008. Guide to Meteorological Instruments and Methods of Observation. World Meteorological Organization, Geneva.

Yevjevich, V., 1967. An Objective Approach to Definitions and Investigations of Continental Hydrologic Droughts, Hydrology Papers 23. Colorado State University, Fort Collins.

Chapter 6

Impact of Hydropower on Mountain Communities in Teesta Basin of Eastern Himalaya, India

G. Choudhury
University of North Bengal, Siliguri, West Bengal, India
E-mail: govinda.choudhury@gmail.com

Chapter Outline

Mountain Ice and Water. http://dx.doi.org/10.1016/B978-0-444-63787-1.00006-8

INTRODUCTION

The development of hydropower in the Himalayan Mountains has been mired in controversies as distributed benefits do not add up to the social and environmental cost of such projects. The race for hydropower development in the Himalayas has serious implications for the well-being of the mountain community. Although the major share of the electricity produced from hydropower projects (HEPs) are transferred for consumption and production to the plains, the mountain communities bear the cost arising from environmental damages and socioeconomic changes. HEPs have been reluctantly agreed by the mountain communities and were never accepted by choice. The proliferation of the HEPs in the Himalayas during last couple of decades has been a direct outcome of the changes in economic and energy policy of India since the 1990s.

In 1991, India adopted the new economic policies, and since then the demand for energy for consumption and production purposes have increased manifold. The increase in energy consumption in India is directly related to growth in economic activities in all three sectors of the economy, namely, the primary, secondary, and tertiary sectors. In addition, the growth in household income during the last three decades has also increased demand for consumption of goods such as air conditioners, coolers, and other electrical home appliances in India. With increasing temperature over the years, increase in consumption demand for electricity has increased exponentially. The cumulative annual growth rate of electricity consumption for the past decade has been increasing almost at 10% per annum in India. The major source of energy in India has been thermal power generated from fossil fuels, followed by other nonconventional sources of energy. The higher dependence on fossil fuel has been associated with increased level of pollution and carbon emission in the last couple of decades. Also, increase in consumption of fossil fuel and coal has put immense pressure on the balance of foreign trade of India. The high cost of import price of energy has serious implications for the Indian economy, making the country's product uncompetitive. As a result, alternative sources of cheaper, clean renewable energy have become a priority in the country's energy plan. The emphasis of the government policy has been to increase efficiency in energy use, reduce dependence on fuel imports, and increase production of clean renewable energy.

The expansion of hydropower sector in India is shown in a Central Electricity Authority, India (GoI) 2016 report that hydropower accounts for 14.35% of the total electricity generation capacity in India. The total installed hydroelectricity capacity is 42.78 GW and another 4.27 GW of capacity has been developed through small hydropower units (Government of India (GoI), 2016, 4). More than half of this capacity is installed in the Himalayas, where the natural gradient of the rivers makes it convenient to harness the energy. The immediate effect of the construction of HEPs is the loss of biodiversity due to inundation of the valley and change in hydrological flow of the river, displacement, and loss of livelihood of the people. These losses are estimated to an extent based on available information at the time when projects are initiated. The environmental damages are addressed through compensatory afforestation programs for loss of forest cover, identifying a minimum environmental flow of the river to ensure aquatic life in river, pollution abatement measures, and technology to reduce the damage from pollution. But the cost imposed on people due to displacement, land acquisition, and loss of livelihood is compensated through financial packages, the amount of which has always been contested. In lieu of this, the project always promises development packages to the neighborhood community to ensure that people provide support for the project. But financial compensation has remained confined to the directly project affected people only who are either displaced or who lose their assets. The compensation mechanism fails to include those who are affected during the course of project construction and operation due to diminishing natural resource base and loss of livelihood. As a case study, this chapter enquires into the socioeconomic impact of two HEPs in Teesta River on the mountain community. The two HEPs considered here are the Teesta Low Dam Projects (TLDP) III and IV in Darjeeling district of India. In the mountains where the population density is low, displacement is not a major issue for enquiry for such projects. Given that only 20 households were identified as project-affected families (PAFs) in the TLDPs, this study therefore focuses on the impact of hydropower on development experiences of the mountain community living in the project area.

Objectives

The main objective of this chapter is to study the socioeconomic impact of TLDP-III and IV on rural mountain communities living in the lower region of Teesta valley. The study examines the impact on livelihood, infrastructure, culture, and riparian environment due to construction of HEPs in TLDP area. The chapter also explores the factors which have led to proliferation of hydroelectricity plants during the last couple of decades in India. In this context, the chapter examines the debate on the nexus between economic growth and consumption, and what its implication is for India's energy policy. Based on these findings, conclusions and recommend policy prescriptions are drawn.

The chapter is divided as follows: The first is the introductory section that introduces the research problem and elaborates the objectives of the chapter. The second section explains the methodology followed in this chapter. The third section is on economic development and energy demand where the debate on the role of energy in economic development has been briefly explored. In the following section, the growth of hydropower in the Himalayan region has been discussed in the context of the trade-off between conservation and development. In the fifth section, the political economy of hydropower development in the Teesta River basin are discussed. In the sixth section, the impact of TLDP on the mountain communities are explained, followed by the project's environmental impact assessments (EIAs) to understand the baseline assessment of the impact on the environment and socioeconomic conditions. During the course of focus group discussions (FGDs), we took these issues to find out to what extent these problems have been addressed by National Hydro Power Corporation (NHPC). In the following section, we study the demographic and livelihood changes in the TLDP-affected area. In the absence of good-quality baseline data, the census data for 2001 and 2011 are considered to see the changes in demographic pattern and corresponding livelihood changes in the affected region. The next section analyses the information collected through FGDs to understand the impact on five well-defined parameters. The last is the concluding section in which suggestions to mitigate the adverse impact of the project are made and policy prescriptions are recommended.

METHODOLOGY

The purpose of this chapter is not to estimate the socioeconomic impact in some quantitative value. This is because without well-defined household level, baseline data on socioeconomic variables, any present estimation of the variables will not be comparable with any earlier data. As a result, we have considered the interdecadal changes in the demographic and socioeconomic variables for the census data from 2001 to 2011 to compare the pre- and post-TLDP development scenarios. The development of hydropower in the Himalayas in general and in Teesta basin in particular has been studied in the context of economic growth and energy demand debate. To understand the socioeconomic impact of TLDPs on mountain community, five broad thematic issues were identified from the review of EIA studies of the two TLDPs. On the basis of secondary data the two villages for FGDs were identified. The purpose of the FGDs was to study the community experience and their understanding of the impact of HEPs on their well-being. In addition, we interviewed a few regional activists and researchers to understand the impact of TLDPs on the mountain communities. From our analysis of secondary data and case studies, conclusions were drawn for this study.

ECONOMIC GROWTH AND ENERGY DEMAND

The role of energy, or for that matter any natural resources in economic growth, has been debated in economics ever since the publication of Adam Smith's *Wealth of Nations*. The main point of the debate was whether energy produced from natural resources would be a major constraint in economic growth of a nation. Economics has always been concerned about immediate need and well-being of the people, ignoring the effect of a decision on the well-being of the future periods. Depletions of natural resource stocks were not considered an obstacle to economic growth because of the possibility of substituting one input for another (Solow, 1974, 41). In other words, technology and capital formation is the panacea for all future progress and economic growth. However, this does not mean that the role of energy has been totally ignored in economics. Economic models incorporating the role of energy in growth and development have mostly remained confined in the field of resource economics.

During the last two centuries, the growth of the world economy was achieved through exploitation of cheap fossil fuels. Other factors such as division of labor, technological innovation, and increased world trade also played significant roles in economic growth. But without concentrated and controllable sources of energy such as oil, coal, and natural gas, the transition of agrarian society to modern economy would not have been possible. Even when growth theorists have ignored the role of natural resources such as fossil fuel, they were able to largely explain the economic experiences of the developed countries for the postwar period. Abundant and cheap prices during this period in the developed countries were never considered to be a challenge to economic growth. However, when we include essential inputs such as energy into these growth models, any depletion of the resource stock implies that they must be substituted by capital to sustain economic growth. This means constant accumulation of capital is required for sustainable growth. These models were designed to either maximize the sum of net discounted welfare over some meaningful time horizon, or the ability to sustain the level of social welfare over time.

Assuming that nonrenewable resources have zero extraction cost and a constant capital stock, Solow (1974) showed that sustainable growth is achieved even under these models when factors of production are perfectly substitutable and saving rates are sufficiently high. But what happens when factors of production are not perfectly substitutable? In such a case, sustainable growth is not feasible when the stock of nonrenewable resources decreases without a possibility of substitution. Can technological progress ensure economic growth or constant consumption level over time given a finite stock of nonrenewable resources? The answer to this question is not that simple. In endogenous models with essential nonrenewable resources, it is not possible to attain an unbounded growth in consumption. On the other hand, ecological economics argue that technological progress has a limited role in substituting capital for essential resources (Stern, 1997).

Although writing on energy issues in the development literature has proliferated since the oil crisis of 1973, concerns were raised as early as 1865 by Jevons about finite stocks of coal and the challenges it posed to economic growth in his book *The Coal Question*. Malthus (1798), however, argued differently concerning the limits posed by nature and the need for restraint, and not to exceed the subsistence level of consumption. Since then, the importance of natural resources has been revisited in the development literature and political ideology many times.

During 18th and 19th centuries, the conservation of resources was highly neglected under the laissez-faire system, and management of natural resources were left in the hands of private owners. The concern for the environment was observable only in those conservation policies introduced in the colonies; for example, the 1865 forest conservation policy in the Indian subcontinent. Such conservation policies were highly skewed in favor of the elite class, and the cost of conservation was actually borne by the rural communities. John Muir, the founder of *Sierra Club*, argued that conservation policies were never strong enough to protect the natural world because at the core of such policies is economic interest. In the United States, the argument of a political, ideology-driven Conservation Movement (1890–1920) was for restrictive use of nonrenewable resources. The movement considered laissez-faire approaches as wasteful and inefficient. Economic competition and monopolies were the major challenge to judicious use of natural resources. Such conservation movement enhanced the government role in distribution of natural resources to ensure inter- and intrageneral equity. Thus, irrigation projects and dam building were promoted by the federal government to provide irrigation to small-scale farmers. Millions of acres were earmarked for conservation (national parks, bird reserves, game reserves) to be kept for posterity. However, conservation programs and environmental regulations have provoked debate on whether environmental regulations impede economic growth.

Two distinct views have emerged: first those who consider that regulations are needed for protection of the environment as this will promote efficient use of natural resources and second those who suggest that technological changes brought about by market forces can protect the environment even in the absence of regulation. A closer look into these two views shows that the implication of conservation and environmental regulation on economic growth can only be sustained through more efficient use of inputs. This means economic production must use less input per unit of output, or in energy terms, less energy per unit of output (X-efficiency). This will reduce pollution and waste produced in economic production and consumption. Even when zero pollution and waste are desirable, can we reduce energy consumption to a level at which the desirable outcome is achievable? To answer this question we do not have to understand the laws of thermodynamics. But it will suffice to understand that all production involves work and will therefore, require energy; and also limits exist to substitute other inputs for energy (Stern, 1997, 199–200). So the alternative is not only to use clean form of energy (like electricity), but also to reduce the pollution and waste in the production of energy itself. The solution to the environmental problem is considered to lie in higher use of renewable energy sources such as hydro, solar, and wind, which are collectively being promoted as clean energy.

Economic Growth and Energy Consumption Nexus in India

The nexus between economic growth and energy have intrigued researchers for quite some time, without any definite causal relationship being established. One extreme view is that price of energy plays a vital role in economic growth and has been a crucial factor for industrial revolution. The industrial revolution has been generally considered as a technical revolution in economic production. But, such inventions became absolutely necessary when the labor wages were relatively much higher than the price of energy (coal) and capital (Allen, 2009). Technology breakthrough in production made it possible to substitute capital and energy for labor. The objective of such technology was to achieve growth of production by using more of a cheaper factor. So more and more of the cheaper fossil fuels were used (ignoring their consequences on the environment) to achieve higher growth. Does this imply that energy consumption and economic growth feed each other? In other words, do the variables have a bidirectional relationship? Since Kraft and Kraft (1978) pioneering empirical work on the relationship between energy consumption and growth, numerous studies have been done to understand the causal relationship between the two variables. However, no conclusive evidence has been produced from these studies, as multiple causal relationships have been reported.

At least four alternative hypotheses have been produced on the causal relationships between energy consumption and economic growth (Abosedra et al., 2015, 2). First is the growth hypothesis, which considers that energy plays an important role in economic growth as a complement to capital and labor. Any energy conservation policy will reduce capital formation and negatively affect economic growth. Second is the conservation hypothesis, which does not consider energy conservation policies to have any adverse impact on economic growth. In other words, energy conservation policy does not adversely affect the economic growth. Third is the feedback hypothesis, which considers that energy consumption and economic growth are interrelated and are complementary to each other. Fourth is the neutrality hypothesis, which implies that no causal relationship exists between energy conservation and economic growth.

Without analyzing the details of the econometrical work on energy consumption and gross domestic product (GDP) in India, studies have found either a unidirectional relationship (Mohanty and Chaturvedi, 2015, 199) or a bidirectional causality (Gollagari and Rena, 2013, 19), and that the neutral hypothesis is satisfied between the two variables (Asghar, 2008, 174). What are the implications of these findings for energy policy in India? First, a unidirectional causality from energy consumption to GDP implies that to sustain high economic growth, consumption of energy must be kept at a high level. This means economic growth will be arrested if energy conservation policy is followed. Second, the bidirectional causality implies that increase in energy consumption will lead to more damage of the environment, and if the problem is not addressed this will act as a constraint for achieving higher economic growth. The neutrality hypothesis implies that any energy conservation policy will not adversely affect the economic growth of the nation.

To what extent conservation policies can be implemented in a country like India is highly debatable, when the demand for energy is continuously increasing due to population growth and rapid urbanization. From 2005–06 to 2013–14, the compound annual growth rate (CAGR) for conventional energy consumption has been 5.28%, whereas for electricity alone it was 9.95%. During this period, the CAGR of per capita energy consumption was 4.53%. The energy intensity has decreased by −1.30% CAGR for the same time interval, implying an increase in efficiency of energy use in India (GoI, 2015a, 44–45). This does not mean that the carbon emission in India will decline if the country continues to rely on coal for production of electricity, and 83.5% of rural and 22.9% of urban households continue to depend on firewood and chips for cooking purposes (GoI, 2014, 30). Although 96.7% of the villages are electrified (GoI, 2015b), the number of rural households electrified is not more than 60% in India. Economic growth and ensuring basic amenities to people has led to a huge increase in demand of electricity both for industrialization and household consumption. Therefore, the policy thrust of India's energy policy has been to ensure access and energy security to its billion-plus population.

To sustain the engine of growth in economic activities after 1991 in India, environmentally sensitive conservation zones in the mountains and coastal regions have also been opened up for economic activities. Although the coastal ecosystems are being increasingly exploited for aquaculture, tourism, and real estate business, the mountains on the other hand are being exploited for its natural resources. As a result, the Himalayan have during the last couple of decades witnessed rapid growth of unmindful construction for tourism, mining, hydropower, and other factors. Most of these investment projects pay little heed to environmental norms or blatantly flout environmental regulations to make immediate economic gains at a huge cost and risk to the people of the region. Capital invested by HEPs is much larger than other industries in the Himalayan region. Although the benefits of such huge investment are enjoyed by the society at large through increased supply of electricity and reduction in carbon emissions, but how such projects affect the local communities living in these ecologically sensitive zones is a little-studied subject.

HYDROPOWER DEVELOPMENT IN THE HIMALAYAN MOUNTAINS OF INDIA

The Himalaya–Hindu Kush mountain range and the Tibetan Plateau region is also known as third pole because its ice covers contains the largest reserve of freshwater outside the two Polar regions. The Himalayan region is also the source of 10 major river systems which has for past several centuries provided freshwater for agriculture and domestic consumption, and also supported aquatic and terrestrial biodiversity. The big river basins have also been used for navigation, and the sedimentation brought from the mountains has fertilized the agriculture lands in the plains. For centuries, the water of Himalayan rivers flowed unobstructed through the Indian subcontinent into the sea. The use of water by upstream riparian communities for agriculture and consumption purposes was never considered as an infringement on the downstream riparian rights. After the independence of India, the upstream river storage projects and barrages were considered as means of reducing damage from floods and to rationalize downstream water use through better distribution of the river water. However, rivers such as the Indus, Ganga, and Teesta, which have acquired transboundary characteristics after the partition of the country, have witnessed rising conflicts between the upper- and downstream riparian countries.

In the past few decades, the debate on conservation and development has evolved into a new discourse on political economy in India. States with large, ecologically sensitive zones share the responsibility of conservation, and are deprived of development. This has especially affected the mountain communities because their cost burden for nonuse of natural resources has not been matched by transfer of equivalent benefits to them (Verma et al., 2014, 2). The benefit from nonuse of resources such as water and forest by the mountain communities is transferred to the plains below through enhanced hydrological and environmental flows. This has been acknowledged in India and the Twelfth Finance Commission for the first time made a recommendation for the transfer of fund from central government to the state government (Verma et al., 2014, 9). For the first time a weight of 7.5 has been given to land under forest cover of a state as a criterion for transfer of central government funds to the state by the Fourteenth Finance Commission of India. The amount of the transfer has been estimated by an index for opportunity cost foregone due to conservation of natural resources. The opportunity cost of development foregone has two components. The first component of opportunity cost is the loss in income that otherwise would have accrued had forestlands and wetlands been developed into agriculture, industry, township, and economic zones. The second component of opportunity cost is the loss of tax revenue as the ecosystems are always undervalued in any market transaction. As a result, the Indian states in the Himalayan region always have an urge to develop the natural resource into economic good to increase tax revenue and generate livelihood opportunities. The total electricity demand of the Himalayan states of India put together is much less than what will be produced from all HEPS being developed in the mountains. The question is why do the Himalayan states agree to build a large number of HEPs? A common argument that has been put forward by the states in favor of the dams and HEP is that they will bring in much needed revenue to the state exchequer for development. Apart from producing clean energy, revenue consideration has been a major cause for the Himalayan states as being facilitators of HEP development. However, cost–benefit of an HEP project based entirely on capital-accounting methods may not reflect the true cost of HEPs. Even when individuals loosing from the HEPs receive fair compensation, inclusion of the cost of environmental damage would make many of these projects uneconomical.

Direct compensation is paid only to those identified as PAFs. Therefore, the question remains whether people living in the ecologically sensitive Himalayan states stand to benefit from these development projects? Does obstruction of the free flow of water violate the riparian rights of the downstream communities? Whether loss of forest cover and biodiversity due to submergence of land leads to reduced flow of hydrological and environmental benefits to downstream communities? Because upstream development has implications for the downstream states, such development has been the bone of contention between the upstream and downstream states in India. Thus, consent of the downstream states is required for any development in the upstream. As the Indian federal system has been maturing over the years, the states in India are becoming more assertive about their rights. Downstream states only agree to any upstream HEPs proposal when they are assured that the project will not reduce the flow of benefit adversely, and their downstream riparian rights will be protected. It is for this reason that the HEPS are being projected as the run-of-the-river (RoR) projects, whereas in practice they become storage projects. The much-touted clean and green HEPs have consequences for the regional environment and ecology, thereby affecting the local community directly. The EIA of the biotic and abiotic environment of the proposed HEPs is a precondition for environmental clearance (EC) by the Ministry of Environment & Forest and Climate Change, GoI. To understand the entire process of building HEPs and their socioeconomic impact on the mountain community, we have selected the Teesta basin as our case study. We have considered the Teesta basin as it has the highest density of HEP among all river basins being developed for HEPs in India (Pandit and Grumbine, 2012, 1065).

DEVELOPMENT OF HYDROPOWER IN THE TEESTA BASIN OF THE EASTERN HIMALAYA

Hydropower in India has been promoted as environmentally benign and a cheap source of electricity. The hydropower potential estimated for the country by CEA is 148.7 GW, which works out to an economic potential of 84 GW at a 60% load factor. At present, the total installed capacity is 42.78 GW, which is 28.77% of the total hydropower potential estimated for the country (GoI, 2013, 4). The estimated hydropower potential of Teesta is 3.9 GW. To tap this potential, numerous HEPs were proposed in the Teesta valley in Sikkim and Darjeeling districts. The history of hydropower development is not new to the region. The earliest HEP in India was the Sidrapong Hydel Power Station built in 1898 and was located some 12 km from Darjeeling town. The main objective of the project was to supply electricity to the Darjeeling municipal area. In the eastern Himalayan region, HEPs were also later built at Jaldhaka in 1967 through state investment. In India until 1991, the entry of private investors in hydropower generation was totally restricted. It was even much later and not until 1998 that private-sector investment was permitted for transmission. In 1995, the Power Trading Company was established to facilitate sale of electricity from power companies to State Electricity Boards. By enactment of the Electricity Regulatory Commission Act in 1998, the Central Electricity Regulatory Commission (CERC), and State Electricity Regulatory Commissions were constituted in India to fix the electricity tariffs. However, all this did not encourage private investment in power generation in general, and hydropower in particular. Unlike the state-owned, public-sector power companies such as National Thermal Power Corporation and NHPC, private-sector, power-generating companies did not have sovereign guarantees available to them for financing of huge debts. This was a major obstacle for private investment in the hydropower sector.

In the 2000s, quite a number of sweeping changes occurred in power generation, transmission, and distribution policy by the government. First, the Electricity Act of 2003 allowed the state governments in India to provide subsidies to power-generating companies if they agreed to lower the electricity tariff more than the regulated tariff. Second, through this Act, the National Electricity Policy of 2005 was framed to recover the cost of electricity generation (Sec. 61d). In addition to the above statutory legislations, the National Tariff Policy 2006 introduced the multiyear tariff policy (Sec. 5.3h), CDM benefits (Sec. 5.3i), and a guaranteed return on capital investment to be declared by the CERC from time to time. To resolve the problems related to losses and costs imposed on the PAFs, a National Policy on Resettlement & Rehabilitation Policy was formulated in 2003. It was found that the policy did not address the PAFs problem in a time-bound manner, and as a result, a more comprehensive policy on R&R was framed in 2007. The Mega Power Policy 2006 also provide for tax incentives for the proliferation of HEP in India in general and in the Teesta valley in particular is closely related to these policy changes in the power sector.

The River Teesta originates from the Kanchenjunga range at an elevation of 5280 m in Sikkim (GoI, 2006, 5), flows south through Darjeeling district in West Bengal, and then enters the plains at Sevoke to travel for another 80 km before it enters Bangladesh. The river descends sharply at 4–35 m/km, making it one of the fastest flowing rivers in the Himalaya. The Teesta watershed receives an average annual rainfall of almost 2218 mm. This makes it an ideal river for harnessing of hydroelectricity. Since Sikkim is included in the North-East Development Council of India, it has additional development grants and tax incentives to attract capital investment for HEPs.

The numbers of HEPs in different phases of completion in the Teesta watershed are 26 in Sikkim, and only 2 HEPs are being commissioned in the Darjeeling region. In Sikkim, a number of private investors along with public-sector hydropower companies are developing HEP. In West Bengal, the government has both the HEPs and NHPC—a public-sector hydropower company of the GoI. In addition, four more HEP projects in the Teesta River have been allotted to NHPC (Sarkar, 2014), for which the initial feasibility study has been completed. The two HEPs in Darjeeling are at Rambhi (TLDP-III) and in Kalijhora (TLDP-IV), with installed capacity of 132 and 160 MW, respectively. All the four units at Stage III are in operation since 2013, whereas only unit number 2 in Stage IV in the downstream is operational from March 2016 (EAC, 2016, 8). This study will consider the impact of two HEPs in the Darjeeling district on the immediate downstream community of the two projects. Here we must mention that the impact of Teesta valley HEPs on the environment in Darjeeling cannot be studied by ignoring the impact of numerous HEPs further upstream in Sikkim.

IMPACT ASSESSMENT OF TEESTA LOW DAM PROJECTS IN DARJEELING

To understand the impact of the TLDPs on the riverine environment and community, it would be appropriate to review the two EIA studies for the HEPs. The review will help us to compare the present scenario with the baseline data. In addition, it would help us to understand the deviation from the anticipated outcomes as projected in the two studies. In India, the statutory provision for development of HEPs makes it binding on the part of the developers to get an EC from the appropriate authority. Following EIA guidelines, the two EIA studies for the respective projects broadly identified land environment, water environment, aquatic environment, terrestrial ecology, and socioeconomics as the criteria for impact analysis. Both the HEPs are an RoR project and hence have been projected as environmentally benign. Technically, an RoR does not require the storage of water for production of electricity. Instead it uses the kinetic energy of the flowing river water to generate electricity, and is therefore assumed to cause little or almost no damage to the aquatic and terrestrial ecology of the watershed.

The Teesta watershed region is seismically unstable and prone to landslides. Earthquakes measuring 7 or more in the Richter scale in close vicinity of the two HEPs are common. The occurrence of landslides increases during the rainy season in the Teesta valley due to steep slope and loose soil and rock formation. It has also been argued that the landslide occurrence has been aggravated due to deforestation, construction of roads, and faulty agriculture practices. The developer's argument for the two HEPs was that as these will not be a storage project, therefore, it will not aggravate the problem and increase risk and damage to the people. Because these are RoR projects and the water diverted for producing electricity will be immediately discharged to the river, the damage to aquatic life can be avoided (EIA, 2002, 3.11). Any change in the availability of water in the river may be due to change in precipitation in the Teesta watershed, and not due to construction of HEP barrages on the river (EIA, 2002, 4.8). Moreover, the chance of river water getting polluted due to low discharge is also ruled by these assessments as water will not be stored. Waste from the labor colony will be treated and no wastewater will be discharged into the river (EIA, 2003, 2.8–2.9).

The Teesta watershed is rich in biodiversity and has a large area under forest cover. Large areas of forestland along the Teesta valley in the Darjeeling district are plantation forests and the rest are natural forests, mostly along the ridges. The only wildlife identified by the EIA (2002) study along the river is the monkeys. The study assumes that since the monkeys live on food given by the travelers, any construction on the river will not threaten the species. But in the downstream of the TLDP-IV, the territorial wildlife is more diversified due to the Mahananda Wildlife Sanctuary. A rich stock of mammal, avian, and reptile fauna were reported to have been found during the EIA study. Also to ensure the conservation of aquatic life in the river water, 1.25 cumecs would be continuously released through the fish ladder (EIA, 2003, 5.12). The report also concluded that because no commercial or organized fishing occurs in the Teesta, the construction of dams will not adversely affect fishing activity of the people. Instead construction of diversion dams were considered to create opportunities for fishing activities once the projects became operational.

The forestland permanently required for the construction of TLDP-III project was respectively 183.49 and 119 ha temporarily during the construction phase for quarrying and dumping of waste. Besides 25.70 ha of government land was permanently required for construction labor colony, approach road, store, and workshop (EIA, 2002, 8.1). The extent of land that was required for the Stage IV project was 308 ha of permanent acquisition of forestlands, and another 29 ha of temporary acquisition of forestlands for road, dumping of waste, and quarry land for mining construction aggregates (EIA, 2003, 3.18). In addition, TLDP-IV acquired 353.85 ha of government land (EIA, 2003, 7.12). The study also considered that damage to the riverine forest cover will be negligible due to the V shape of the river valley. However, the influx of a large number of construction workers will put additional demand for fuel for cooking and heating purposes. This could be taken care by arranging alternative fuel to the workers. Seven quarries measuring 253 ha are in the river, which are within 30–40 km of the dam construction site, were leased for the purpose of quarrying construction aggregates; and two quarries for clay mining measuring 1.15 ha in private land were leased for construction purposes. The estimated summary of total construction aggregates required for TLDP-III is presented in Table 6.1. The report also concluded that the enhanced quarry activities in other rivers of the region will not have any effect on the flora and fauna, as the species would suitably find alternative habitat for themselves (EIA, 2002, 5.8; EIA, 2003, 5.21).

The EIA study had anticipated that during the period of construction some temporary effect on downstream environment may likely happen. There will

TABLE 6.1 Estimated Amount of Construction Aggregate for TLDP-III and TLDP-IV

Construction Aggregates	TLDP-III	TLDP-IV
Coarse aggregate (nonwearing surface concrete)	454,000 cumecs	820,000 cumecs
Coarse aggregate (wearing surface concrete)	15,700 cumecs	30,000 cumecs
Fine aggregate	235,000 cumecs	410,000 cumecs
Clay	110,000 cumecs	90,000 cumecs
Rock (fill material)	–	195,000 cumecs
Fly ash	–	21,888 MT

Reproduced from EIA (2002, 5.1) and EIA (2003, 7.5).

be some deposition of construction material in the river, an increase in turbidity, and the normal flow of water will be hampered during the construction phase of the projects. Both the EIA study identified six types of impacts due to construction of the two HEPs in the Darjeeling Teesta region. The environmental impact was likely to happen due to large influx of labor, expansion in quarry activities, use of heavy machineries, and transport vehicles, dumping of waste, and construction and expansion of roads for the project (EIA, 2002, 8.2; EIA, 2003, 7.2). The reports acknowledge that there will be an increase in air and noise pollution during the construction period, which the developers would mitigate through various means.

The impact on socioeconomic and cultural aspects considers that the advantage of a large influx of workers will lead to exchange of ideas and cultures. It also contemplates that during the construction phase the livelihood of the local people will improve from increased employment and business opportunities. The report also identifies improvement in transport facilities, education, health, and market facilities. In addition, the developer would improve the water-supply infrastructure, sanitation, and communication facilities. A large influx of the migrant workers would have some temporary effect on the local cultural values and lifestyles. The projects did not require any displacement of people because of acquisition of land, but had identified 9 PAFs household downstream of the dam in TLDP-III and another 10 PAFs downstream of TLDP-IV. The report also anticipated an improvement of the esthetic environment in the project site due to landscape designing and beautification program to be undertaken by NHPC. Picnic spots such as Kalijhora were anticipated to gain because the dams would add to the recreational value of the locality. In addition, the dam would create connectivity of the two banks and this would increase the accessibility of the left bank of the Teesta.

Map showing TLDP III & IV and the project affected villages

***Source**: Map developed using Google map, available at: https://www.google.com/maps/d/edit?hl=en&authuser=0&mid=1Vn1Taymadk35dNXHycXCfkQA_RI*

The focus of EIA studies in practice has been mainly confined to estimate the impact of development projects on the surrounding ecological environment, and has a cursory reference to the socioeconomic impact of the project for the purpose of identifying the amount of compensation to be given to affected families. A socioeconomic impact study can be an important tool to understand the cost imposed on society from any project and also to help plan mitigation of the negative impacts on community. In India, social impact assessment for certain categories of projects has been made compulsory under the National Rehabilitation and Resettlement Policy 2007. Social impact assessment is also a perquisite condition put forward by multilateral and private financing agencies. However, after more than a decade since the commencement of construction for the two projects and operationalization of TLDP-III and IV, this study enquires about the socioeconomic impact of the projects on the community. The socioeconomic impact of the project and the demographic and livelihood changes in the TLDP area have been studied using secondary data, and also through interview with key informants like local activists, researchers, and non-government organization (NGO) working with the affected community.

DEMOGRAPHIC AND LIVELIHOOD CHANGES IN TEESTA LOW DAM PROJECTS AREA

Rural livelihood in the villages on the bank of the Teesta River has been dependent on agriculture and raising livestock. After the merger of Sikkim with India as its 25th state in 1975, the volume of traffic increased on the highway (NH-31A) connecting the Himalayan states with the rest of India. This increased the livelihood opportunities and led to the growth of settlements along the highway. The highway runs along the western bank of Teesta River following the course of the river. Some of the junctions on the highway have over the years evolved into market places, creating business opportunities for trading goods, transport, construction work, and so forth. All along the river at regular intervals people from the neighboring villages would quarry stones, boulders, and sand. The construction aggregate would then be immediately loaded onto trucks waiting on the highway. Thus people living in the mountains in the Teesta valley depended on agriculture, quarrying, and livelihoods along the highway.

With the commencement of construction of TLDP-III and IV in 2003 and 2005, respectively, a huge flow of capital investment from both the projects has occurred, which had its immediate effect on the livelihood of the local community. The initial approved cost of TLDP-III was INR 7.69 billion (USD \$118.31 million), which was later approved to INR 18.08 billion (USD \$278.16 million) by CERC order (CERC, 2015, 15). The cost of TLDP-IV has escalated from CCEA approved INR 10.62 billion (USD \$163.39 million) and has escalated to INR 20.26 billion (USD \$310.77 million) until March 2016 (NHPC, 2016, 2). This gave a boost to the local economic activities and people moved out of agriculture to new livelihood activities which have increased during the construction work of the projects. To understand the impact of the TLDPs on the neighboring settlements, we considered all the 14 villages along the 2 banks of Teesta from Tista Bazar District Improvement Fund (DIF) in the upstream to Sevoke Forest to the downstream where Teesta enters the plains of West Bengal. Most of these villages have a large geographical area, and have several rural settlements dispersed from each other because of the difficult physical terrain and dense forest cover. The population density of this area has increased from 88.22 to 121.59 km^2, and the interdecadal growth of population during 2001–11 was 1.04% the CAGR, which is much below the country's CAGR for the same period. This implies that there has no major inflow of migrant settlers in this region, other than TLDP workers who have settled in the region.

The author, being a frequent traveler along this road, has witnessed the expansion in economic activities during the past 10 years. Quarrying had expanded along the river basin as private construction in the region has increased. Employment, which is mostly temporary wage work, petty contracting, rent through leasing of rooms, transport work, trading in goods and services, has expanded in settlements along the Teesta valley because of numerous HEP and other government and private projects in Sikkim and Darjeeling. However, the livelihood opportunities have not increased equally all along the valley, because instead economic activities have become concentrated mostly along the right bank of the river in those places that are rapidly acquiring an urban character due to better infrastructure suitably located at some nodal junctions. The density of population on the right and left bank are 236 and 20 km^2, respectively. The interdecadal (2001–11) increase in population on the right and left bank of the river was 904 and 177 persons, respectively. As a result, settlements such as Tista Bazar, Rambi Bazar, Riyang, Kalijhora, and Sivok have undergone a very high growth rate of population on the right bank of Teesta. On the left bank, the increase in population is seen more at Mangpong Forest, which is on the opposite bank of Sivok Forest. The increase in population of Mangpong Forest does not relate to an improvement of work opportunities. On the other hand, due to the lack of, or diminishing livelihood opportunities in villages such as Riyong Forest, Sittong Forest, and Sivok Hill Forest, a considerable decline in population occurred during 2001–11 (Table 6.2). Sivok at the foothills of the mountain has the highest growth of population at 6.51% per annum. The quarrying activities at Sivok have increased manifold, which provided livelihood opportunities to the rural community in Teesta valley. A huge increase in construction aggregates has occurred due to the construction of HEPs and the demand for construction of the Asian highway. Also because Sivok is the nodal junction of the road going to Sikkim and North-East India, it provides an ample scope for eatery businesses. The large increase in the number of eateries in Sivok provides ample employment opportunities to the youth of the neighboring villages.

In India, the scheduled cast (SC) and scheduled tribe (ST) categories of people are identified as vulnerable groups and they are entitled to various state supports for their development and livelihood security. During 2001–11 while the SC population in the region has increased by a little over 2% per annum, the ST population had increased by more than 12% per annum. The huge increase in ST population during the decade was due to inclusion of two more communities, namely, the Tamang and Limbo people onto the ST list. In Sittong Forest, the increase in the number of the ST population was the highest among all villages within the TLDP region. Even when an increase in number of ST occurred in Sittong Forest, the total population had decreased during the decade (Tables 6.2 and 6.3). The loss of population in some villages in the region corroborates with the all-India scenario of rural occupational patterns. Agriculture is no longer able to support rural livelihood and hence people move out of agriculture in search of nonfarm work.Interdecadal change in literacy rate shows an interesting pattern within the TLDP region. The right-bank villages have a higher literacy percentage as compared to the left-bank villages (Table 6.3). But the growth in literacy rate in the left bank is appreciable, given that during the decade 2001–11, the literacy has increased more than 27%. The literacy rate of a village bears a strong relation with access to school. The TLDP-IV that is located at Kalijhora has at least 15 schools within close distance. In the hills, more than the distance from school, it is the connectivity and road condition that determine access to school. The two villages on the left bank of Teesta, namely, Mazeok Forest and Mangpong

TABLE 6.2 Interdecadal Change in Population of the TLDP Area, 2001–11

	Area (ha)	Household 2001	Household 2011	Person 2001	Person 2011	Male 2001	Male 2011	Female 2001	Female 2011
Right-Bank Villages									
Tista Bazar DIF	31	528	658	2527	2953	1333	1530	1194	1423
Riyong Forest	11	49	43	229	185	123	97	106	88
Riyang Railway Station	7	53	83	237	371	132	188	105	183
Turzam Forest	107	87	93	391	406	210	216	181	190
Rambi Bazar DIF	3	156	200	738	1000	381	510	357	490
Birik Forest	516	17	24	80	93	42	48	38	45
Sittong Forest	389	584	632	3043	2930	1518	1468	1525	1462
Sivok Hill Forest	1480	129	76	632	321	341	164	291	157
Sivok Forest	1395	119	224	556	1045	296	557	260	488
I. Right-bank total	**3939**	**1722**	**2033**	**8433**	**9304**	**4376**	**4778**	**4057**	**4526**
Left-Bank Villages									
Ringking Pong Forest	662	2	3	7	10	5	9	2	1
Mazeok Forest	911	46	53	218	282	106	139	112	143
Gulling Forest[a]	453	–	–	–	–	–	–	–	–
Panbu Forest	830	1	5	4	20	2	13	2	7
Mangpong Forest	4173	195	246	1017	1111	533	598	484	513
II. Left-bank total	**7029**	**244**	**307**	**1246**	**1423**	**646**	**759**	**600**	**664**
Total (I+II)	**10968**	**1966**	**2340**	**9676**	**10727**	**5022**	**5537**	**4657**	**5190**

DIF, District Improvement Fund; *TLDP*, Teesta Low Dam Projects.
[a]*Gulling Forest is an uninhabited village.*
Source: Census of India 2001 and 2011.

TABLE 6.3 Interdecadal Change in Literacy Level of the TLDP Area, 2001–11

	Schedule Caste 2001	Schedule Caste 2011	Schedule Tribe 2001	Schedule Tribe 2011	Person Literate 2001	Person Literate 2011
Right-Bank Villages						
Tista Bazar DIF	422	501	86	443	1734 (77.38)	2053 (78.18)
Riyong Forest	5	0	0	34	103 (54.21)	109 (66.46)
Riyang Railway Station	68	83	5	101	121 (57.08)	252 (76.13)
Turzam Forest	6	6	338	366	167 (50.61)	224 (60.87)
Rambi Bazar DIF	107	148	18	86	504 (77.42)	775 (86.50)
Birik Forest	5	4	9	43	50 (73.53)	55 (67.07)
Sittong Forest	154	192	186	1057	2126 (79.39)	2223 (82.18)
Sivok Hill Forest	90	173	17	40	329 (59.07)	227 (83.15)
Sivok Forest	78	87	32	203	414 (86.79)	734 (83.31)
I. Right-bank total	**935**	**1134**	**691**	**2373**	**5548 (74.93)**	**6652 (79.58)**
Left-Bank Villages						
Ringking Pong Forest	6	3	0	4	3 (50.00)	8 (80.00)
Mazeok Forest	0	0	13	34	109 (45.61)	194 (80.50)
Panbu Forest	4	2	0	11	4 (20.00)	11 (78.57)
Mangpong Forest	58	27	219	541	430 (44.75)	668 (69.58)
II. Left-bank total	**68**	**32**	**232**	**590**	**546 (44.43)**	**881 (71.92)**
Total (I+II)	**1003**	**1226**	**923**	**2963**	**6094 (70.59)**	**7533 (78.60)**

Note: Values in parentheses indicate the percentage of literate people for the population age group 7 years or more.
DIF, District Improvement Fund; *TLDP*, Teesta Low Dam Projects.
Source: Census of India 2001 and 2011.

Forest, have a substantially improved literacy rate during the decade because of better access across the river. Mazeok Forest now has better road connectivity across Teesta because of TLDP-III, and the increased flow of public transport on the highway has increased access to the village. From the table, it is seen that the percentage increase in literacy rate only may not be the good indicator for the real changes taking place. In the case of Birik and Sivok Forest, even with an increase in the absolute number of literate population, the percentage shows a decline over the decade (Table 6.3). Sivok Hill Forest shows a major increase in literacy rate even when the decline in absolute number of literate people is substantial for this village. With the increase in literacy rate, young people tend to move out of agriculture to nonfarm employment. The Census data of India include all nonfarm activities in the other-worker category. On the left bank, the experience of Mangpong Forest shows that there has not only been a substantial increase in literacy rate but the work participation rate (WPR) has also improved. On the left-bank villages of Teesta, the shift has not been toward nonfarm activities, rather dependence on agriculture has increased.

As compared with the left bank, the right-bank settlements of the Teesta show a huge increase in the other-workers category (Table 6.4). This means the opportunities from nonfarm activities such as quarrying, construction work, trade, commerce, business, transport, government, or nongovernment jobs have expanded on the right bank. By abandoning agricultural work, villagers have moved to nonfarm work on this side of the Teesta. The decline in the number of cultivators, agriculture labor, and household industry workers during 2001–11 was 171, 423, and 145, respectively. But the expansion in the other-worker category has not led to an increase in WPR of all the villages on the right bank of TLDP region. Barring the two villages of Sittong and Sivok Hill Forest, all the villages in the TLDP show an increase in the number of nonworkers during the decade. In the Sittong Forest, the number of cultivator, agriculture labor, and household industry workers has been drastically reduced during the period. If we broadly consider the area between Tista Bazar to TLDP IV as the upstream and below it as the downstream area, then the WPR has declined in the upstream villages, whereas it has improved in the downstream. But our main concern is that the increase in the number of total workers during 2001–11 has been achieved through a shift of the workforce from the main- to marginal-work category (Table 6.5). This means that people do not find adequate work throughout the year, although there have been such huge investments in the TLDP area. To understand what challenges are faced by the community to find work for more than 6 months, we took up this issue for discussion with the community. We also approached the community to understand what the implication of these changes is on their well-being and culture. These issues are discussed in the following case study section.

TABLE 6.4 Interdecadal Change in Occupational Pattern of the TLDP Area, 2001–11

	Cultivator 2001	Agriculture Labor 2001	Household Industry Workers 2001	Other Workers 2001	Cultivator 2011	Agriculture Labor 2011	Household Industry Workers 2011	Other Workers 2011
Right-Bank Villages								
Tista Bazar DIF	9	5	73	720	8	5	23	773
Riyong Forest	27	40	14	10	2	1	1	52
Riyang Railway Station	0	0	14	75	2	1	0	101
Turzam Forest	159	0	8	15	45	105	1	56
Rambi Bazar DIF	4	0	42	195	2	0	10	257
Birik Forest	1	0	0	23	1	0	0	31
Sittong Forest	45	572	65	313	4	70	5	1042
Sivok Hill Forest	0	0	10	143	3	0	17	110
Sivok Forest	0	1	1	183	7	13	25	349
I. Right-bank total	**245**	**618**	**227**	**1677**	**74**	**195**	**82**	**2771**

Continued

TABLE 6.4 Interdecadal Change in Occupational Pattern of the TLDP Area, 2001–11—cont'd

	Cultivator 2001	Agriculture Labor 2001	Household Industry Workers 2001	Other Workers 2001	Cultivator 2011	Agriculture Labor 2011	Household Industry Workers 2011	Other Workers 2011
Left-Bank Villages								
Ringking Pong Forest	0	0	2	1	0	4	0	2
Mazeok Forest	1	0	0	23	4	69	1	15
Panbu Forest	0	0	0	1	5	6	0	1
Mangpong Forest	61	26	2	210	62	111	10	205
II. Left-bank total	**62**	**26**	**4**	**235**	**71**	**190**	**11**	**223**
Total (I+II)	**307**	**644**	**231**	**1912**	**145**	**385**	**93**	**2994**

DIF, District Improvement Fund; *TLDP*, Teesta Low Dam Projects; *WPR*, work participation rate.
Source: Census of India 2001 and 2011.

TABLE 6.5 Interdecadal Change in Workforce of the TLDP Area, 2001–11

	Total Workers 2001	Total Workers 2011	Main Worker 2001	Main Worker 2011	Marginal Worker 2001	Marginal Worker 2011	Nonworkers 2001	Nonworkers 2011	WPR 2001 (%)	WPR 2011 (%)
Right-Bank Villages										
Tista Bazar DIF	807	809	684	601	123	208	1720	2144	31.94	27.40
Riyong Forest	91	56	90	44	1	12	138	129	39.74	30.27
Riyang Railway Station	89	104	82	59	7	45	148	267	37.55	28.03
Turzam Forest	182	207	177	28	5	179	209	199	46.55	50.99
Rambi Bazar DIF	241	269	213	206	28	63	497	731	32.66	26.90
Birik Forest	24	32	20	29	4	3	56	61	30.00	34.41
Sittong Forest	995	1121	669	693	326	428	2048	1809	32.70	38.26
Sivok Hill Forest	153	130	80	59	73	71	357	191	24.21	40.50
Sivok Forest	185	394	167	74	18	320	447	651	33.27	37.70
I. Right-bank total	**2767**	**3122**	**2182**	**1793**	**585**	**1329**	**5620**	**6182**	**32.81**	**33.56**

Continued

TABLE 6.5 Interdecadal Change in Workforce of the TLDP Area, 2001–11—cont'd

	Total Workers 2001	Total Workers 2011	Main Worker 2001	Main Worker 2011	Marginal Worker 2001	Marginal Worker 2011	Nonworkers 2001	Nonworkers 2011	WPR 2001 (%)	WPR 2011 (%)
Left-Bank Villages										
Ringking Pong Forest	3	6	3	3	0	3	4	4	42.86	60.00
Mazeok Forest	110	89	69	12	41	77	108	193	50.46	31.56
Panbu Forest	1	12	1	6	0	6	3	8	25.00	60.00
Mangpong Forest	299	388	211	151	88	237	718	723	29.40	34.92
II. Left-bank total	**413**	**495**	**284**	**172**	**129**	**323**	**833**	**928**	**33.15**	**34.79**
Total (I+II)	**3180**	**3617**	**2466**	**1965**	**714**	**1652**	**6453**	**7110**	**32.86**	**33.72**

DIF, District Improvement Fund; *TLDP*, Teesta Low Dam Projects; *WPR*, work participation rate.
Source: Census of India 2001 and 2011.

SOCIOECONOMIC IMPACT OF THE TEESTA LOW DAM PROJECTS ON MOUNTAIN COMMUNITIES

Among the upstream villages we selected Rambi Bazar DIF and in downstream, Sivok Forest (see Map). The two villages are located to the downstream of TLDP-III and TLDP-IV, respectively. The growth of population in the two villages during 2001–11 is the highest among all villages in the TLDP area. Interdecadal growth of population per annum for Rambi Bazar DIF and Sivok Forest was 3.08% and 6.51%, respectively. These villages also have the highest increase in literacy rate in the TLDP area. The highest growth rate in the number of workers in upstream and downstream of the TLDP area was at Rambi Bazar DIF and Sivok Forest, respectively. The per annum increase in workers was 7.85% at Sivok Forest. To understand the impact of HEPs on the mountain communities, we formally approached the *Samaj* of the two places, namely, Rambi Gram Samaj Sudhar Samity at Rambi Bazar and Sivok Gram Samaj Sudhar Samity at Sivok Bazar to help us to conduct FGDs with the villagers. The Rambi Bazar association has been formalized under Indian law, but the latter association in Sivok continues to function informally. We requested both the associations that the participants of the FGDs should include village leadership, youth, elders, and women members from the locality.

From our earlier discussion on the EIA documents of TLDP-III and IV, and our analysis of the secondary data of the villages in the TLDP area, we identified five broad thematic issues for inquiry to understand the socioeconomic impact of the HEPs in the Teesta basin. The five thematic issues are livelihood of the people, education and health of the community, infrastructure of the region, culture of the villagers, and regional resource endowment. These five issues were brought up for discussion with the *Samaj* members who participated in the FGDs. To guide the discussion, we put a number of leading questions related to the different issues and helped steer the discussion. The participants of the FGDs then narrated the community experience and also explained the reason for the changes.

Rambi Bazar is on the NH-31A on the right bank of the Teesta River, and is only 2 km downstream of TLDP-III. The village is bounded by Teesta Valley Forest in the north, Teesta River in the south, Riyong Forest in the east, and Sunthali and Mongpoo Cinchona Plantation in the west. The FGD was arranged at the *Samaj* community hall. The discussion started with 10 participants who were all males of different age groups. Later it was joined by a few more members of the *Samaj*. The participants of the meeting were engaged in different livelihood activities. The participants included office bearers of the *Samaj*, casual workers of TLDP-III, daily wage laborers, petty contractors and traders, school teachers, elderly retired workers, and staff from the health department.

Sivok Bazar is in the foothills of the mountain, and it is here that the Teesta River enters the plains of India. The village is bounded in the north by the

Mahananda Wildlife Sanctuary, Teesta River in the south, Mangpong Forest in the east, and Samardanga Forest in the south. In this village, the FGD was held in the community hall attended by 16 people, of which 7 were women. The participants included ex-army personnel, daily wage laborers, housewife, shopkeepers, and teachers.

In the following sections we compare the experience and impact of HEPs on the two communities located at different altitudes.

Hydropower Projects Impact on Livelihood in Teesta Low Dam Projects Region

The livelihood of the people in the Rambi Bazar has been mostly dependent on transport work. People work as taxi drivers or drive freight vehicles. After the NHPC hydropower project commenced in 2003, a gradual in-migration has occurred of people from the neighboring regions to Rambi. As a result, one-third of the population is now constituted by the migrant households. These people have made their houses in the DIF land. The growth in population is a direct consequence of enhanced livelihood opportunities in Rambi as people got work as wage laborers in HEP. People also worked as quarry workers as demand for construction aggregates increased for the construction of the project. In addition, they also got work in transport mainly carrying construction material for the hydropower project. A few of them also started to work as subcontractors for the project. Renting out of rooms for project workers and staffs gave a boost to private construction activity also. Business increased and people started to enjoy the benefits of increased income. After the completion of the construction work in 2013 and full operationalization of the project, a sudden decline occurred in all economic activities. The demand for labor was no longer there, hiring of vehicles has almost become zero, thus making many transport workers jobless. At present, people no longer demand rented premises for living, as workers have either migrated out or have shifted to the workers quarter of NHPC. Business has fallen and the subcontractors have shifted to other businesses.

The main effect was on the quarry workers who were poor people who made a living by mining in the river basin. As the project became fully operational, availability of the construction aggregates in the basin has declined due to submergence of the valley, and also it has become risky to work in the river. People reported in the FGD that no certainty existed that stone and sand collected downstream of the dam could be sold to contractors, as it was highly likely that the aggregates might get washed away once the water is released from the dam. The few people who earlier caught fish in the river could not do it now as the river has dried up and it is being reduced gradually to a *nala* (drain). The place where TLDP-III has been constructed was where the Samco Ropeway operated. The place was admired for its scenic beauty and has drawn attention of filmmakers. This gave a boost to recreational activities in the area. Picnickers and adventure sport enthusiasts visited the place and created income–earning opportunities in the locality. Flooding of the river basin upstream of the dam and releasing of water downstream without any warning has stopped all such activities in the region.

The decrease in water in the springs had been noticed much earlier during the dam-building stage. But many springs in the region had immediately dried up after the 2011 Sikkim earthquake with 6.9 magnitude on the Richter scale that struck the valley. This has made it difficult to cultivate agricultural lands in the neighboring villages such as the Turzam Forest. Most of them grow vegetables in small plots of land as the wages of labor have become higher during construction of dams and irrigation water is scarce.

The experience of the people in the Sivok Bazar is quite different from those in the Rambi Bazar, as they never got any direct earning opportunities from TLDP-IV. Instead, a huge increase occurred in the quarrying activities in the Teesta at Sivok to meet the construction demand of the upstream dams. This has increased livelihood opportunities, not only in river quarrying activities, but also in eatery businesses that got a boost. Because Sivok was a junction on the NH-31A, it has become a convenient place for travelers to rest, before they set out for the long journey to Sikkim or Dooars. Sivok also has an old Hindu temple that attracts many followers each year. Many families from the village have small outlets in front of the temple selling goods to the worshippers.

Agriculture has decreased on the right bank of the TLDP-IV area, such as in Sittong and Sivok. The main reason is the lack of irrigation and animal intrusion. The loss or damage to crops is not compensated as villagers do not have land rights. Sivok Bazar has grown up on DIF land and forestland. People living on DIF land have a 30-year lease agreement with the government. The lack of agricultural activities has pushed people to start their own eatery businesses. Around 15% are engaged in trading, a few as casual workers in the eateries, a handful of people have government jobs or are in the army. Many of them sell corn that is brought from the neighboring state of Bihar, and almost 60% of the villagers work as quarry workers.

The increase in quarrying activity required a large number of workers, who either have moved close to the quarrying site in Sivok, or commute daily by train from neighboring places. The decrease in demand from dam construction has been more than compensated for by the demand from other activities. The new Asian Highway has created a huge demand for stone and boulders from the Teesta. On an average, about 100 trucks carry stones and boulders from the quarry. The place also attracted large numbers of people for its scenic beauty and has become an important picnic sport. Tourists also visited the place for its scenic beauty and under a joint forest management program, facilities for picnics and tourism were developed. As a result, the business flourished in the Sivok Bazar. But the construction of dams had created a fear among people of getting drowned when water is released from them. The result has been a drastic fall in the number of people who came here for recreational purposes.

The impact of these changes on the community has been an increase of income during the construction period, and joblessness after the projects became operational. The complaint of the community is that the NHPC has not employed local youths in the projects. In the Rambi Bazar, only a few young people have found work as casual workers with no job security under service contractors through some outsourcing arrangement. These young people have to pay a part of their wage to the supervisor to continue working on the project. The rest who have been out of work either migrated for jobs or are left behind with few work opportunities. Under the joint forest management scheme, the people in Rambi were trained in mushroom cultivation and INR 17,000 (USD $261.5) was given as seed money to each household to start the cultivation. Also, the enterprising households were provided water-supply connections by the forest department. This had become an important activity now. But more and more people prefer to buy mushroom from neighboring places at a low price and sell it in the Bazar at a high price, as there is good reputation of Rambi for mushrooms.

Hydropower Projects Impact on Education and Health in Teesta Low Dam Projects Region

The Rambi Bazar has couple of schools and also a few more are within the settlement. The participants reported that all children from the village are enrolled in school and the literacy level is high. We found young people who have university degree in engineering, a master in social science, and vocational degree holders in the village. When we inquired about the role of NHPC for expanding education and health services, the response was mixed. The participants reported that initially after the public hearing was held some goodwill gesture came from the NHPC. Scholarships were given to four students, and some work for the local school was also done. An NHPC-KV school of up to eighth standard had been opened in 2007 mainly for the wards of the company employees. The expansion of the transport system has helped the children to travel and attend school and colleges. But the impact of the dam has been detrimental to schooling of children from the poor households, as a large number of them had dropped out of school and started working in the TLDP-III. The administration did nothing to check this social problem, and the poor families had no means to stop this trend. As a result, a large number of young boys did not even complete their elementary schooling.

The basic condition for good health of the people depends on the availability of nutritious food, safe drinking water, good sanitation, and a clean environment. In Rambi Bazar, the people have to depend on spring water for drinking and all other consumption purposes. The spring from where the community collected drinking water earlier was 1.5 km from the settlement, but it has receded almost to 3 km from Rambi Bazar. The administration from its DIF, and also later the Forest Department had laid pipelines to supply water to the community, which then failed. Now the community on their own initiative, used plastic pipes to bring water to the village. It is not difficult to understand what the quality of water they have for consumption is likely to be. Sanitation is not universal and is a serious health issue of the villages in TLDP-III area. The people also complained that it was after the project became operational that there has been outbreak of dengue in the region. The destruction of the biodiversity has led to invasion of alien species such as the multiple disease-spreading, Aedes mosquito in the region.

In Sivok Bazar when we asked people about their health condition, they said that the village was facing difficulty in fetching water from the aquifer, as the groundwater level has decreased after the dams were built. The problem was all attempts to bore a deep tube well have failed and the villager has to rely on dug wells made by the *Panchayat*. To solve the problem, a drinking water project was sanctioned by the district administration and a pipeline was to be constructed from Chabok Khola, a stream to the north of the Sivok Bazar. The work could not be started as the stream was in the Wildlife division and no work is permissible in such land. Thus the problem of drinking water remains in the settlement. The problem of sanitation is similar to that in Rambi Bazar. Unlike Rambi, Sivok Bazar does not have any primary health center, so they have to depend on Siliguri town, which is 20 km away. The village has also been affected by dengue in the past couple of years.

Even for higher education, students have to travel to Siliguri. The elementary school in Nepali language is there in the village. The cost of commuting daily to school is INR 100 (USD $1.54) at least, which few households in the village can easily afford to pay for their children. This is not the only cost parents pay, they also have to pay for their own travel for safety while they accompany small children to the distant school. An Integrated Child Development Services center also exists that provides food, preschool education, and primary health care to children. This scheme motivates the family to educate their children. With an increase in literacy level, the youths of Sivok Bazar do not want to work in the quarry and instead move to nearby towns and at times to far-away cities in search of jobs. To meet the demand of labor for quarry work, large numbers of illiterate workers from outside the village are settling in and around the Bazar to work as quarry workers. As a result, although the numbers of literates are increasing in the village, the percentage of literate people shows a decline.

Impact of Teesta Low Dam Project on Infrastructure

Development of an area depends on the availability of appropriate infrastructure facilities. In the previous section, we discussed the educational and health infrastructure of the region. In this section, we see the impact of HEPs on the physical infrastructure of TLDP area. Economic activities, offices, and settlement have all grown linearly along the NH-31A. Any damage to the highway will, therefore, affect the livelihood of the people. The road is now under threat as submergence of the valley by the HEPs dams is making the soil below the highway loose and soft. Heavy vehicles plying the road are causing the loose soil to slide down. Landslides have increased in the valley after the HEPs in TLDP area started to submerge the valley to hold water for generation of electricity. Construction of an alternative road bypassing the sink zone of the highway was attempted, but it has failed miserably as the new road has been damaged beyond repair by landsliding. The threat of losing their homestead and livelihood due to damage to the road will undoubtedly increase the number of projected affected people. The question is how will these people be compensated for their losses? The upstream Rambi Bazar community who stood by NHPC now feel that they have been taken for a ride by the hydropower company. In the EIA public hearing, the NHPC had promised that the locality would be developed. The company would take up the task of electrification of the street, development of drinking water pipelines to the locality, a community hall with good amenities, expansion of the health center, and so forth. However, to date no street electrification has been done, the community hall is without electricity or water, and the school constructed by the NHPC has been damaged by the earthquake of 2011. Instead when the people demanded what was promised to them, a coil of wire with a few electrical lamps was given to them for lighting of the street. No authorization was given to use electricity from the poles and anybody who would do so would be prosecuted under the law. The participants of the FGD admitted that prior to the coming up of HEPs in the valley, they had little knowledge about HEPs and their impact. Also during the public hearing no information was provided by anyone to the community in Nepali language. They believed what the agents of the NHPC told them, and they agreed to support the project because the company promised many things to them. However, all these promises were never official and were only a verbal assurance from the company.

In downstream Sivok Bazar, the community has been a victim of administrative indecision for many years. Its administrative jurisdiction has been shifted from one police station to another. For many years, the village had an elected gram panchayat (GP) member under the Matigara GP. However, after the village had been included under the Gorkha Territorial Administration (GTA) area, no representation in the panchayat occurred and all development work in the village has stopped. The village has electricity in all houses and shops, but no high school or health center. The villagers were assured by the NHPC in the public hearing on EIA for TLDP-IV at Kalijhora that the company would help to transform the symbolic Coronation Bridge into a heritage structure. It also agreed to fund construction of schools and hospitals at Sivok Bazar. But all these have not yet happened, and any representation of the villagers on this matter is no longer entertained.

Development of upcoming infrastructure projects in Sivok has already increased a sense of threat as people fear eviction due to the upcoming new projects. The proposed railway to Sikkim will require acquisition of land and also land will be required for the new road and bridge to be constructed downstream of the Coronation Bridge. The villagers do not have any land rights as these are forest or DIF lands, only some of them settled in DIF land have a lease of 30 years. The lease does not guarantee that the land cannot be revoked by the state. The threat of eviction is a common phenomenon all along the TLDP settlement, as there is no security of land tenure in DIF, forest, and *khasmahal* government lands for lease. As a result, people are reluctant to invest in construction of their houses. It is only after many years that people have developed a sense of right and are determined to fight any threat of eviction. Our understanding is that the increase in population in the TLDP area has given them enough political space and voice in decision-making. Earlier alignment of the people of Sivok and Rambi with HEPs was a well thought out, strategic decisions of the community to engage with the developers. The objective was to make the maximum possible gains from the project. At the household level, the gains from employment and opportunity of doing business with NHPC was a priority, and at the community level, it was the assurance for development of village infrastructure that people considered as their reason for supporting the project.

Impact of Teesta Low Dam Projects on Mountain Culture

The young people in the Teesta valley who were employed in various activities during construction of TLDP-III are now unemployed and frustrated, as there are no alternative livelihood options in the area. Rise in unemployment has led to the growth of crime in the locality. Smuggling of drugs and illegal extraction of timber from forests has increased in the valley. The immediate local community at TLDP-III and IV is strongly opposed to people from outside the locality being employed in the projects. The mistrust between different settlements in the TLDP area is high as they consider the others as rivals vying for employment in the projects. The Teesta valley communities have lived in close harmony with each other, but at present intercommunity relationships have been affected to a large extent because of the HEPs. The mountain communities had an unwritten code of conduct and rules, which people observed and obeyed. The social and cultural fabrics have been in flux because of the radical changes in economic activities in the Teesta watershed region.

Teesta is worshipped as a deity and is known as *Teesta Maharani*. People also worship in the river during *Chaat Puja* and in other festivals. The decrease in water and also the release of water from the dams make it difficult to perform any religious and cultural practices in the Teesta River. The decline of agricultural activities and out-migration from the region has made the younger generation detached from their culture and religion. Youth no longer participate in local cultural and religious activities. Due to the lack of participation, many cultural and religious activities are no longer observed. The younger generation is under pressure as they are uncertain about their future. The older generation feels that the youth idolizes people who have made a fortune for themselves, irrespective of the manner how they acquired their wealth. This they feel, has been due to flow of investment to the region, and the old and traditional values such as honesty, sincerity, and integrity of the person are no longer idolized. Conspicuous consumption and imitation of the urban lifestyle by the youth has led to alcohol and drug abuse. People have earned money but only few knew how to invest the money. However, participants also felt that economic changes due to HEPs have made villagers more outward looking, and they do not hesitate to migrate to distant places for livelihoods.

Impact of Hydropower Project on Mountain Environment

Large number of trucks-loading stones and boulders from the riverbed in Sivok Bazar are polluting the river due to leakage of oils and lubricants in the riverbed. Drying of the water downstream is creating pools of stagnant water suited for breeding of mosquitoes. Also due to insufficient water, the river is becoming polluted and during the dry period the riverbed stinks. Due to the decreased flow of water and increasing pollution, different species of fish in the Teesta River are fast declining. The loss of forest habitat due to clearing of forest and submergence of the land, along with loss of aquatic life in the Teesta, has threatened different birds living in the valley. The population of birds has declined and hardly any birds are seen flying in the locality. The loss of forest has reduced the flow of minor products and people no longer can get fodder and grass for their livestock. Fuelwood has also become scare due to loss of forest, and alternative fuels such as kerosene and natural gas are available at a high price. After the construction of the dams, the river no longer brings down fuelwood during the monsoon. Also, the loss of forest cover and neighboring forests being earmarked as wildlife sanctuary makes it difficult for the people to collect fuelwood.

Although quarrying, which was an important livelihood activity in the TLDP area, is no longer possible in the upstream, the activity has increased in the downstream from the Sivok Bazar. The participants considered that the annual deposition of sand and boulders is gradually decreasing after the construction of the dams. As a result, the increase in quarrying has resulted in a lowering of the riverbed itself as more is extracted from the river than what is being deposited each year. About 60–70 truckloads of boulders from the Sivok Bazar is being exported to Bangladesh every day. The result has been a lowering of the riverbed by 15–20 feet in the last 5 years.

The major impact on health from the dam was an increase in pollution. Dust and noise pollution had increased during the construction phase of

TLDP-III. Also, an increase in population in the absence of civic services led to piling of waste and garbage. The problem has become more acute due to stagnation of water in the dams and creation of swamps around them, providing a good breeding place for the mosquitoes. All degradable, nondegradable, hazardous waste flow down the river from the upstream settlements accumulates around the dam. River water is highly polluted during the dry months, as the use of water during dry months leads to skin infections. The people of the Rambi Bazar considered that pollution in the region is detrimental for the revival of the tourism activities, and it is adding to the cost of health. People are burning the waste, which also has a detrimental impact on the environment. With increase in income comes change in the products being sold in the local market. More and more of plastic-wrapped items are being sold in the shops, adding to the quantity of nondegradable waste.

Discussion

The EIA process for the two TLDPs has been widely criticized in the regional media. The main criticism was that the report has been prepared in gross violation of rules for conducting the EIA study, deliberate exclusion of the geological report of the region, and the study has been poorly done. The EIA reports of the two TLD projects considered PAFs to be only those people who were to be displaced. The reports did not estimate the downstream and upstream population who are affected due to change in river hydrology and ecosystems as a result of the constructions. The identification of the PAFs is arbitrary, as the number of families facing threat of inundation is larger than 20 for the 2 projects. The threat of displacement due to more area submerging and large areas surrounding it, being subject to landslides, is increasing. The main objection to the projects was that even though when they are claimed as an RoR project, the height of the dams is much more than what is permitted under such schemes.

In the TLDP region, the major problem of the people is that they do not have secured land rights. The lands in which they live are either DIF leased lands, or forestlands or in *khasmahal* government lease lands in which they have no land title. The hydropower developers, as a result, have always projected them as encroachers and have issued veiled threats to evict them. The compensation package initially has been very low, but later increased by the R&R committee constituted by the district administration due to strong resistance and determination of the PAFs. However, the revised compensation of almost USD $13,000 in a lump sum, as mentioned in the TLDP-IV, a 6-month progress report for May 2016 is not enough to account for the loss of livelihood by the PAFs.

The threat of a disaster from the dams is looming in the region. The region is a highly seismic zone and prone to landslides. The incidence of landslides has increased after the operationalization of the projects due to submergence of the valley. Any major seismic activity can trigger a cataclysm disaster in the region, the cumulative impact of which will be felt by the downstream communities. NHPC has only conducted a few disaster awareness camps in the region. Although the 6-month progress report of the projects elaborates the disaster management plan of the company, the people in the region have complained that even warnings are not given regularly before the releasing of water from the dams.

The projects have not undertaken development work in the region as has been promised to the villagers during the public hearing of the project's EIA. The 6-month progress report shows that almost all budgetary allocation under corporate social responsibility have been exhausted, yet there has been no visible sign of development work in the region. The development work of the project is confined around the HEPs and only a few token works have been undertaken in places outside the construction site.

It is important to note that the projects with all their limitations cannot be put on hold now. Also, the need for expanding the production capacity of clean renewable energy cannot be ignored. The need is therefore to relook into the whole issue of hydropower development once again.

CONCLUSION

Studies on the socioeconomic impact of HEPs on communities have mostly focused on displacement of people due to project construction. In most of these studies, the downstream population has come to be identified with people living in the plains, and mountain settlements have been clubbed together as upstream communities. It is not that the upstream communities share the burden of cost without any benefits from the projects. Our study showed that in the upstream of Teesta the benefits and costs of the HEPs are not uniformly shared by the mountain communities. Within the mountain community, the cost imposed on the downstream population has been largely ignored while estimating the benefit–cost of the HEPs. As a result, the socioeconomic impact studies have discounted the cost imposed on the mountain communities.

Our study showed that in the absence of adequate livelihood opportunities, mountain communities had aligned with the developers, with an expectation to benefit from TLDPs. Their support for HEPs development was not based on any information and knowledge about HEPs. During public hearing to solicit community approval, other than the PAFs and local NGOs, none had opposed the two projects as the villagers in TLDP area were promised for better livelihood and development works.

Livelihood opportunities had increased during the construction phase, but reduced drastically after the projects became operational. The community expected that the NHPC would undertake a capacity-building program to enhance livelihood activities in the affected area. The company has not made any initiative to strengthen livelihood activities. At present, the people have expressed in unequivocal terms that they do not want any hydropower project, since they know the outcome of such projects. The experience of the community showed that the gains from HEPs are limited only to the construction phase of the project. It is only when the project becomes operational that the extent of the damage becomes pronounced.

The riverine community in Teesta does not want their problems being compounded by new construction. However, without paying any heed to people's voice, the government has announced approval of many more HEPs in Teesta. We suggest that before any further activities are undertaken in the river, cumulative impact analysis for the entire river basin should be done. The basin-level study should identify the carrying capacity of the river. In this context, we should mention that a carrying capacity study has been limited to the Sikkim portion of Teesta River. However, in all cases of EIA in Teesta, including the carrying capacity study in Sikkim, participation of the communities was never ensured.

Following WCD (2002) we suggest that the participation of the stakeholders must be statutorily ensured at every stage of the project work. The stakeholders should not only include the riparian community, the developers, and the government, but also regional experts on river hydrology, geology, ecology, and biodiversity. The information provided by the stakeholder will help the HEPs to generate electricity, while at the same time ensuring ecological and social sustainability in the river basin. The result will be a better level of cooperation between upstream and downstream communities. At present in the TLDP area no communication occurs between the communities, and each community considers the others as their rival. As a result, the objective of each community is to get a larger share of the benefit from the HEPs, which is a manifestation of any rent-seeking behavior. The outcome of such behavior has been a skewed distribution of benefits associated with a high socioeconomic and environmental cost on the mountain communities.

Once the project has been designed, monitoring of the progress in work should be done through the stakeholder committees. This will ensure that the progress of the projects adheres to the ecological and socioeconomic parameters set by the stakeholders. Also, it is important to consider that a river is a dynamic system and any change in the ecological and socioeconomic parameters must be immediately addressed. In the case of the TLDPs, the cost of construction has been almost double that which was allocated in the DPRs. The NHPC argument has been that this escalation is due to geological surprises and delay caused by political disturbances in the region. But the same company denies any further claim for compensation when people have suffered losses caused by landslides due to submergence of the river valley.

However, as an immediate measure we suggest that the NHPC undertakes capacity-building programs to not only enhance livelihood opportunities for the people, but also to make the local youth to be more capable of being absorbed as employees in the project. This will reduce the distrust and grievances of the people toward the TLDPs. The projects should develop methods to minimize the pollution caused by stagnation and decreased flow of water in the river. This will not only help to address the health problems such as dengue, but also the improved environment will help to expand tourism activities in the region.

Tourism infrastructure should be developed at the earliest under the environmental management plan of the two projects, and its benefit must be distributed equitably between the communities. This can be done by forming a committee for coordination between the mountain and riparian communities. Through tourism, regional culture and activities can be promoted and this will orient the regional youth toward their culture and social activities. Also, it is essential that the community be given secure land rights, which will reduce their fear of eviction. This will also ensure that in case of any displacement or acquisition of property, they are entitled to a fair compensation for their losses.

But, before any further expansion of HEPs is sanctioned, we need to ask ourselves do we need hydropower by imposing high cost on environmental and mountain societies? Is the risk from hydropower development commensurate with the benefits generated by the projects? Are these benefits and risks equitably distributed? Are renewable energy sources such as wind and solar better alternatives as compared to hydropower? The answer to these questions lies in revisiting our renewable policy and developing a sustainable energy plan.

ABBREVIATIONS

CAGR Compound annual growth rate
CCEA Cabinet Committee on Economic Affairs
CEA Central Electricity Authority, India
CERC Central Electricity Regulatory Commission
cumecs Cubic meters per second
DPR Detailed project report
EAC Expert Appraisal Committee
EC Environment clearance
EIA Environmental impact assessment
FGD Focus group discussion
GNP Gross national product
GoI Government of India
GP Gram panchayat
GW Giga Watt
Ha Hectare
HEP Hydro electricity project
ICDS Integrated Child Development Services
INR Indian rupee
MNREGA Mahatma Gandhi National Rural Employment Guarantee ACT
MoEF Ministry of Environment & Forest, Government of India, Government of India
MoEFCC Ministry of Environment & Forest and Climate Change, Government of India

MW Mega Watt
NGO Non government organization
NHPC National Hydro Power Corporation
PAFs Project affected families
R&R Rehabilitation and Resettlement
SC Schedule Caste
ST Schedule Tribe
TLDP Teesta Low Dam Project
USD United States of America dollar
WCD World Commission on Dams

NOTES

1. WPR (Work Participation Rate) = (Total Workers for the year $\div$ Total Population for the year) $\times$ 100.
2. In this chapter, the exchange rate has been considered as USD 1 = INR 65.
3. Main worker are defined as the "workers who worked for more than six months (180 days) in the reference period" (Census Metadata, 2011, 16).
4. Marginal workers are defined as the "workers who have worked less than six months (180 days) in the reference period" (Census Metadata, 2011, 16).
5. *Samaj* is an association of the villagers with close cultural affinity. In Teesta Low Dam Project (TLDP) area, they are mostly an association of Nepali-speaking community. It evolved as an informal institution for ensuring financial and other help to their members when they require help. Some of them are registered and function formally.
6. *Panchayat* is the village-level local administration in India.

CONFLICT OF INTEREST

The author declares that there is no conflict of interests.

ACKNOWLEDGMENT

The experience of writing on a subject which requires an interdisciplinary focus has been quite challenging. The writing would have not been completed without the help and active support of professional associates, students and my family. I am deeply indebted to Arnab Bhattacharjee of NESPON for providing me the copy of the two EIA reports. I would like to acknowledge Sailesh Chettri and Sheela Karji who have helped to arrange the FGDs. I would also like thank to members of the Rambi Gram Samaj Sudhar Samity and Sivok Gram Samaj Sudhar Samity for agreeing to give their valuable time and for providing information and insight.

I would also like to thank everybody whose work I have cited to convey important ideas. However, the responsibilities of interpreting these works are entirely mine. A special thanks to my editor John F. Shroder and Gregory B. Greenwood, and also to Tasha Frank for her professional and supportive role.

Finally would like to thank my wife Debjani for her continued support and encouragement.

REFERENCES

Abosedra, S., Shahbaz, M., Sbia, R., 2015. The links between energy consumption, financial development, and economic growth in Lebanon: evidence from cointegration with unknown structural breaks. Journal of Energy 2015 (965825), 1–15. http://dx.doi.org/10.1155/2015/965825.

Allen, R.C., 2009. The British Industrial Revolution in Global Perspective. Cambridge University Press, Cambridge, UK.

Asghar, Z., 2008. Energy-GDP relationship: a causal analysis for the five countries of south Asia. Applied Econometrics and International Development 8 (1), 167–180. Available at: http://www.usc.es/economet/journals1/aeid/aeid8114.pdf.

Census Metadata, 2011. Census of India 2011 – Meta Data, Office of the Registrar General & Census Commissioner, India, New Delhi. Available at: http://www.censusindia.gov.in/2011census/HLO/Metadata_Census_2011.pdf.

CERC, 2015. Order Related to Petition No. 115/GT/2013 for Approval of Generation Tarrif of TLDP III for Period 1.4.2013 to 31.3.2014 Under Section 62(1)(a) of the Electricity Act, 2003. http://www.cercind.gov.in/2015/orders/SO115.pdf.

EAC, 2016. 'Minutes of the 92nd Meeting of the Expert Appraisal Committee for River Valley and Hydroelectric Projects Held on 28th & 29th March', Teesta Meeting Hall, 1st Floor, Vayu Wing, Indira Paryavaran Bhawan, Jor Bagh Road, New Delhi – 110003. Available at: http://environmentclearance.nic.in/writereaddata/Form-1A/Minutes/13042016WU3TMVJDMinutes92ndEAC.pdf.

EIA, 2002. Environmental Impact Assessment – Tista Low Dam Project, Stage III, West Bengal, Study Conducted by University of North Bengal for NHPC (A Government of India Enterprise), Sector-33, Faridabad.

EIA, 2003. Environmental Impact Assessment – Tista Low Dam Project, Stage IV, West Bengal, Study Conducted by University of North Bengal for NHPC (A Government of India Enterprise), Sector-33, Faridabad.

GoI, 2006. Carrying Capacity Study of Teesta Basin in Sikkim – vol. 1, Ministry of Environment & Forests, Government of India, New Delhi. Available at: http://www.actsikkim.com/docs/CCS_I_Introductory_Volume.pdf.

GoI, 2013. Forty Third Report, Standing Committee on Energy 2013–14 (Fifteenth Lok Sabha), Ministry of Power, Lok Sabha Secretariat, Government of India, New Delhi. Available at: http://164.100.47.134/lsscommittee/Energy/15_Energy_43.pdf.

GoI, 2014. 'Household Consumption of Various Goods and Services in India 2011–12', NSSO Report No 588-68th Round, Ministry of Statistics and Programme Implementation, Government of India, New Delhi. Available at: http://mospi.nic.in/Mospi_New/upload/Report_no558_rou68_30june14.pdf.

GoI, 2015a. Energy Statistics 2015, Central Statistics Office, National Statistical Organisation, MoSPI, Government of India, New Delhi. Available at: http://mospi.nic.in/Mospi_New/upload/Energy_stats_2015_26mar15.pdf.

GoI, 2015b. Open Government Data Platform India, Central Electricity Authority, Ministry of Power, Government of India, New Delhi. Available at: https://data.gov.in/catalog/progress-report-village-electrification.

GoI, 2016. Executive Summary Power Sector April 2016, Ministry of Power, Central Electricity Authority, Government of India. New Delhi. Available at: http://www.cea.nic.in/reports/monthly/executivesummary/2016/exe_summary-04.pdf.

Gollagari, R., Rena, D., 2013. An empirical analysis of energy consumption and economic growth in India: are they causally related? Studia Oeconomica 58 (2). Available at: http://repository.uwc.ac.za/xmlui/bitstream/handle/10566/1070/RamakrishnaEmpiricalAnalysis2013.pdf?sequence=5&isAllowed=y.

Jevons, W.S., 1865. The Coal Question. Macmillan and Co., London. Available at: http://www.econlib.org/library/YPDBooks/Jevons/jvnCQ18.html.

Kraft, J., Kraft, A., 1978. On the relationship between energy and GNP. Journal of Energy and Development 3 (2), 401–403.

Malthus, T., 1798. An Essay on the Principle of Population. J. Johnson, in St. Paul's Church-yard, London. © 1998, Electronic Scholarly Publishing Project. Available at: http://www.esp.org/books/malthus/population/malthus.pdf.

Mohanty, A., Chaturvedi, D., 2015. Relationship between electricity energy consumption and GDP: evidence from India. International Journal of Economics and Finance 7 (2). http://dx.doi.org/10.5539/ijef.v7n2p186. Also available at: http://citeseerx.ist.psu.edu/viewdoc/download?doi=10.1.1.665.973&rep=rep1&type=pdf.

NHPC, 2016. Six Monthly Progress Report for the Period Ending March 2016, in Respect of Environmental Standards of TLDP Stage – IV, West Bengal. Available at: http://environmentclearance.nic.in/writereaddata/Compliance/30_May_2016_130418660HYPOMLFG102-115.pdf.

Pandit, M.K., Grumbine, R.E., 2012. Potential effects of ongoing and proposed hydropower development on terrestrial biological diversity in the indian himalaya. Conservation Biology 26 (6), 1061–1071. http://dx.doi.org/10.1111/j.1523-1739.2012.01918.x. Also available at: http://du.ac.in/du/uploads/research/Best%20Publications/EnvSt3-PanditGrumbine,%202012.pdf.

Sarkar, D., 2014. Over burdened West Bengal government goes for new hydropower projects. The Economic Times. September 15, 2014, Kolkata. Available at: http://articles.economictimes.indiatimes.com/2014-09-15/news/53942791_1_tldp-iv-mw-teesta-low-dam-tlpd-iv.

Solow, R.M., 1974. 'Intergenerational equity and exhaustible resources', review of economic studies. In: Symposium on the Economics of Exhaustible Resources, vol. I, pp. 29–46. Available at: http://folk.uio.no/gasheim/zSol1974.pdf.

Stern, D.I., 1997. Limits to substitution and irreversibility in production and consumption: a neoclassical interpretation of ecological economics. Ecological Economics 21, 197–215. Available at: http://www.academia.edu.

Verma, M., Negandhi, D., Mehra, S., Singh, R., Kumar, A., Kumar, R., 2014. High Conservation Value Forests: An Instrument for Effective Forest Federalism in India, Report Submitted to Fourteenth Finance Commission of India, Government of India. Available at: http://fincomindia.nic.in/writereaddata/html_en_files/fincom14/others/28.pdf.

WCD, 2002. Dams and Development: A New Framework for Decision-making. The Report of the World Commission on Dams, Earthscan, UK. Available at: http://www.unep.org/dams/WCD/report/WCD_DAMS%20report.pdf.

Chapter 7

Climate Vulnerability, Water Vulnerability: Challenges to Adaptation in Eastern Himalayan Springsheds

R. Seidler,*[,1] G. Sharma[§] and Y. Telwala[¶]

**University of Massachusetts Boston, Boston, MA, United States;* [§]*The Mountain Institute-India, Gangtok, Sikkim, India;* [¶]*Global Mechanism, UNCCD, Rome, Italy*

[1]*Corresponding author: E-mail: reinmar.seidler@gmail.com*

Chapter Outline

Mountain Ice and Water. http://dx.doi.org/10.1016/B978-0-444-63787-1.00007-X

WHAT IS CLIMATE VULNERABILITY? CONTEXTUAL SPECIFICITY

Due to their vast size, height, and positioning across the head of the South Asian subcontinent, the Himalayan range and Tibetan Plateau are uniquely implicated in both generating and maintaining the large-scale weather patterns of the South Asian monsoon. As global climates distort over the coming decades, we can expect these patterns to change. But local weather patterns in the Himalaya are extraordinarily complex, and impacts on people of changes in these patterns will be difficult to predict. The great topographic and geologic complexity of the range itself and of its forehills where most people live, make hydrological systems particularly challenging to understand in more than broad strokes. Meteorological, geological, and orogenic drivers interact to determine characteristics, both of local groundwater supply and patterns of seasonality in temperature and rainfall. Yet, calls for climate-smart villages and climate-adaptive rural development have become common. Understanding these details to design appropriate and timely adaptive measures is becoming critical for mountain and hill people confronting climate change.

Partly due to research efforts directed at finding ways to "downscale" global climate models to administrative scales, the larger scale dynamics and teleconnections driving the South Asian monsoon are themselves becoming better understood (e.g., Sabin et al., 2013; Annamalai and Sperber, 2016). As moisture-laden monsoon winds approach the southern exposure of the Himalayan hills and begin to interact directly with the orogenic zone, however, they enter a "black box" where local predictions are hard to make (Houze, 2012). As rains fall—first, and principally, on the approximately south- and east-facing slopes—proportions of runoff and infiltration are determined by combinations of rainfall volume and timing, degree of slope, soil qualities, anthropogenic land use, and geological substructure, all of which may vary in any direction at fine scales (i.e., kilometers, hundreds or even tens of meters).

Good reasons exist to be skeptical of hopes of downscaling global climate models to administratively useful scales (Fatichi et al., 2016). What are administratively useful scales? In the Indian Himalaya, these would be the district level, because it is at this level that funds are distributed and government agencies commonly interact with community groups and individuals. At these scales, it is likely that patterns of precipitation, in particular, will remain irreducibly uncertain. And in high-relief mountain landscapes, spatial differences in temperature (often related directly to precipitation patterns) are also likely to be acute and unpredictable. Finally, the impact of weather extremes, which are the variables of most urgent interest to local communities, will often vary greatly from one side of a ridge to the other.

Fatichi et al., (2016) point out that scientific techniques for decision-making under uncertainty are highly developed. Such solutions, of course, may not be easily available or necessarily helpful to local decision makers.

Fatichi et al. also found that the envelope of current and historic precipitation variability is already wide enough to encompass many of the potential future trajectories. In short, people have always been dealing with tough weather challenges. On the other hand, because life in the Himalayan mid-Montane has always been somewhat marginal, especially as regards to agriculture, it is arguable that past and current experience may not prepare "climate vulnerable" communities adequately for future challenges. Much depends on the precise nature of those challenges, as they play out particular locations.

The overarching question motivating this chapter is how we can define "climate vulnerability" (especially water vulnerability) among rural mountain communities. A number of sets of "vulnerability indicators" already exist, but they have been criticized as both scientifically unsound and/or insufficiently policy relevant (e.g., Eriksen and Kelly, 2007; Klein, 2009; Hinkel, 2011). In this section, assumptions underlying concepts of climate vulnerability in the context of socially, topographically, geologically, and meteorologically complex mountain landscapes are examined.

In South Sikkim Case Study: Contextual Complexity section, a case study of a climate-adaptive project to regenerate springs in a watershed in the Indian state of Sikkim (Eastern Himalaya) is briefly described. In this watershed (or "springshed"), separate state- and district-level government agencies have initiated projects to diversify and strengthen rural village-based livelihood sources on the one hand, while maintaining and regenerating groundwater springs on the other. In the previous years, these projects have coincided and interacted in unexpected ways. The study region's geographical position in the lee of a high ridge creates specific local challenges for water supply and groundwater recharge, yet for the communities, the real problems were created by the interaction of geography with local economy.

This case study points to significant, inherent challenges in defining and then addressing climate vulnerability in complex mountain environments. In Water Vulnerability: Contextual Uncertainties and Water Vulnerability: Distribution and Equity sections, several dimensions of these challenges, including social, economic, and scientific uncertainties that complicate our ability to distinguish historical trends, measure current changes, and model future scenarios are described. The final section draws some inferences about the nature of climate vulnerability.

SOUTH SIKKIM CASE STUDY: CONTEXTUAL COMPLEXITY

Case Study Region

South Sikkim district is situated near the heart of the swathe of the Himalayas that receives the greatest average annual precipitation, most of it arriving during the summer monsoon season of June–September (Bookhagen and Burbank, 2010). The mountain state of Sikkim displays a complex topography and an

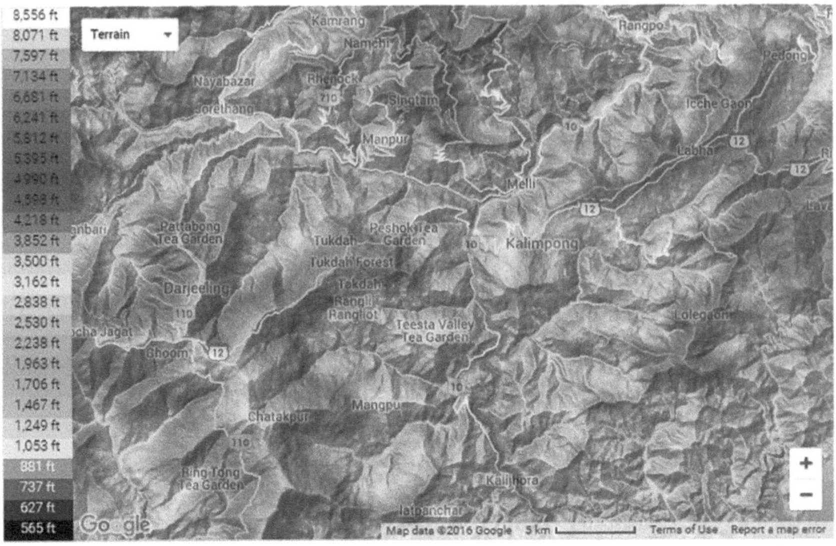

Relief map of Namchi subdivision of South Sikkim (upper central), separated from the Darjeeling Hills (lower left) by Rangit River and from the western margin of the Bhutan Himalaya (right) by the Tista.

extreme vertical gradient (220–8500 m a.s.l.). Annual rainfall varies across the state from as much as 4000 mm in temperate to subalpine zones (2000–4000 m a.s.l.), to 1250 mm or less in the lower elevations (300–1500 m) of South Sikkim and <500 mm in the dry plateaus of North Sikkim (see e.g., http://en.climate-data.org/region/779/). South Sikkim district is flanked by East Sikkim district, with some of the highest rainfall data in the state (e.g., Gangtok at 1650 m elevation receives 3494 mm annually), and West Sikkim district (e.g., Barsey Rhododendron Sanctuary at 1600 m receives >2500 mm). Indeed, most of the inhabited parts of the state get 2400–3600 mm of precipitation annually, but Namchi subdivision in south-central South Sikkim district is an exception. Namchi covers the roughly south-facing slopes overlooking Rangit River, a west-to-east flowing major tributary of the Tista, one of the large tributaries of Brahmaputra/Jamuna in Bangladesh. (The Rangit also forms the border between South Sikkim and Darjeeling District of West Bengal.) Namchi is characterized as a "drought-prone region" (Sharma, 2015), apparently due to a combination of southern exposure, low altitude (the deep Rangit valley), steep slopes, and especially the looming presence on the southern side of the Rangit of the high Darjeeling ridge. Much of this ridge is over 2000 m in elevation, with the highest point being 2566 m. To the east, just across the deeply incised, narrow valley of the north–south flowing Tista River, lies the western margin of the Bhutan Himalaya, also rising about 2000 m above its piedmont (Soja and Starkel, 2007).

The southern front of the Darjeeling Hills records annual rainfall of between 2000 and 4000 mm, in exceptional years exceeding 5000 mm (e.g., at Ambootia Tea Estate). The Bhutan Himalaya foreland receives 4000–6000 mm annually, occasionally reaching 8000 mm and perhaps more at the mountain front where there are no rain gauges (Soja and Starkel, 2007). Thus, Namchi subdivision falls in a classic "rain-shadow" vis-a-vis the prevailing weather systems ascending from the Bay of Bengal, to the south and east. And this relative deficit seems to be sharpening: Over the last decade, dry season (March–May) spring flow is perceived to have declined by nearly 50% in drought-prone areas of South and West Sikkim (Tambe et al., 2013).

Dhara Vikas Program

In response to this incipient crisis, the Sikkim Rural Management and Development Department (RMDD) initiated a research/action program in 2009 called *Dhara Vikas*, to "Revive Dying Springs." The hypothesis was that precipitation in the South Sikkim Gram Panchayat Units (GPUs) was failing to infiltrate the soils as it had in the past, due to fewer rain days, more intense rain events separated by longer drought periods, and changes in land use, including some deforestation and the expansion of agricultural fields. If it was true that rain water was running uselessly off into *jhoras* (seasonal streams) and eventually into the Rangit, it stood to reason that finding ways to slow down gravity's pull on it and redirect its flow beneath the soil surface might help recharge groundwater reservoirs and support below-surface flows.

Tambe et al. (2012) reported that over its first 3 years, Dhara Vikas targeted the revival of 50 springs and 4 lakes in 20 drought-prone areas in Sikkim. Nearly 700 springs were mapped, on each of which an average of 27 households depends for domestic water consumption. The first year of the program (2009–10) was spent on training Block-level RMDD staff as para-hydrogeologists and watershed managers. With the help of a number of non-governmental organizations, seven master trainers were certified and during 2010–11, the program provided a pilot group of five springs with groundwater recharge structures, primarily infiltration pits and trenches. Tambe et al. (2012) reported that dry season (winter) discharge increased from 4.4 to 14.4 L/min—an increase of 236% over just 1 year.

A crucial early success of Dhara Vikas was the formation of a collaborative relationship with the Mahatma Gandhi National Rural Employment Guarantee Scheme (MGNREGS). This nationwide program is the largest social employment program in the world. Administered by the Ministry of Rural Development, it began in 2005 with dual objectives of guaranteeing a minimum of 100 days of paid work per year to each rural household in the country, while "regenerating the natural resource base and creating productive assets" for rural communities (Ministry of Law and Justice, 2005). The program has not been uncontroversial, but previous research seems to show that in most places, MGNREGS projects have succeeded in generating useful community assets while also putting strong upward pressure on minimum local wages and bringing large numbers of women into the work force. Dhara Vikas, a state-level project, was able to partially fund its activities through MGNREGS funds, while also helping to usefully reorient the Sikkim MGNREGS program toward climate-adaptive and agricultural resilience measures (for further details on the MGNREGS collaboration, see Water Vulnerability: Distribution and Equity section).

Dhara Vikas Outcomes Assessments

Several published assessments have been produced of program outcomes. A Government of India document (NITI Aayog and UNDP, 2015) reported that through Dhara Vikas, 900 million liters of groundwater have been recharged annually over 4 years and over 400 ha had been treated, in return for a total investment of INR 25 million (US$372,000). These numbers are taken from Tambe et al. (2013).

A 2013 review of the MGNREGS program examined program outcomes in five districts across India, including the MGNREGS-Dhara Vikas collaboration in South Sikkim. This review found that in Sikkim, "beneficiaries reported an increase of 10–15% in the quantity of water collected from springs for domestic purposes, throughout the year" (Esteves et al., 2013).

In a previous (unpublished) report on the Dhara Vikas program, Sharma (2015) focused on the socioeconomic impact of the program in the following three GPUs (the local administrative unit, each containing several wards and perhaps a dozen individual villages) in South Sikkim district:

1. Mellidara Paiyong GPU in Melli block (population 6577) receives 1250 mm rainfall annually, about half that of the surrounding areas. The wards in this GPU range between 100 and 1600 m in elevation, and slopes are mostly steeper than 35 degrees with relatively impermeable, though fractured, underlying rock structures (phyllites, quartzites, black slate, and mica schists). Surface runoff may be as high as 80% (Tambe et al., 2012; Sharma, 2015).
2. Turuk Ramabong GPU in Melli block (population 2971) similarly averages 1250 mm rainfall—not extremely low, but people do report water stress and have given 7 reasons for their water sources drying up: (a) increase in the seasonality of rainfall (25%), (b) earthquakes (21%), (c) population growth and increase in demand (18%), (d) landslides (14%), (e) deforestation (11%), (f) road construction in the catchments (7%), and (g) rise in average temperatures (4%; RMDD, 2014). Wards in this GPU range between 600 and 2000 m in elevation.

Dhara Vikas initiated in forest land. *Photo Durga P. Sharma.*

3. Lungchok-Kameray GPU (population 1845) consists of 6 wards. Of these, Mungram and Lungchok-Rolu lie at the highest elevation, house the largest tribal and "Most Backward Caste" communities, and suffer the most severe water stress because they have no reliable, perennial, and accessible water sources and must rely on rooftop water harvesting during the dry winter months (RMDD, 2014).

According to the Block Development Officer in Sumbuk (South Sikkim), 140 ha of forest lands and private lands were treated under Dhara Vikas between 2011 and 2014 in these three GPUs, with an estimated investment of INR 2.62 million (US$39,000). Thus these three GPUs absorbed 10% of the overall program effort but contained over one-quarter of the treated land area.

Starting work on a Dhara Vikas initiative in agriculture lands (Sumbuk, South Sikkim). *Photo Ghanashyam Sharma.*

Water collects in infiltration trenches under the Dhara Vikas initiative. *Photo Suren Mohr.*

Treatments consisted mostly of staggered trenches dug along the slopes, mostly at high elevations. The trenches were designed to fill during rains, encouraging infiltration instead of surface flow.

Of the 3 GPUs, containing some 63 perennial springs, 8 wards that were situated along an elevational gradient between 260 and 2050 m were selected. A total of 339 household surveys were carried out, addressing 50% of the households in the 8 wards. Households were selected for surveys in consultation with village communities; the goal was to sample a wide range of the spring sources used by households for their domestic needs. Maps were prepared by an independent consultant developing water security plans for RMDD.

Of the 339 households surveyed, 187 (55%) are supplied with water under the government supply scheme. Here the water is supplied in galvanized (GI) pipes first to a common storage tank which acts as a distribution tank, and then to each of the households through GI pipes. About 40 (12%) households supplement the water they received through the government scheme with water from local sources, whereas 90 (27%) households have no government pipes and are entirely dependent on local springs from which they collect water either through their own rubber pipes or by carrying water from the sources. A larger number of households fetch water from springs during the dry season (January–May). In high-elevation villages, people often spend time filling their containers at night while there is lower demand, since during daytime long queues develop at springs.

Over 50% of surveyed households own a private water storage tank, but virtually all households (98%) report experiencing water scarcity for basic needs (drinking, hygiene, domestic use, etc.), whereas 42% report water scarcity for animal husbandry. Less than 20% of households use any water for irrigation.

Lack of water access (indicated by distance from house to source, frequency of water collection, and time spent collecting) is roughly correlated with village elevation, whereas household water consumption rates and annual income are inversely related with elevation. However, at the fine scale, *distribution* determines access to water, rather than geography alone: Even within individual wards, some households face water scarcity throughout the year while others do not.

How did beneficiaries themselves perceive the effectiveness of the Dhara Vikas initiative? The survey results presented in Sharma (2015) reflect a wide range of perceptions. Slightly more than half the survey respondents perceived no impact from the program on water flow, around 20% felt that dry season water availability had improved, 3–5% of respondents thought there had been other positive impacts from the program, and 4% did not know. In addition, a large number of surveyed households had not heard of the project at all. Awareness was highest in the high-elevation wards, reaching 100% in the wards situated on or close to the ridgeline, such as Manglabarey, Gupti, and Mungrang (Table 7.1). This was partly because the high-elevation communities are keenly aware of the water situation, but also because MGNREGA stipulates that projects must hire residents from villages very near to the work site. Thus, it was the high-elevation villagers who were directly involved in constructing infiltration pits under convergence with the MGNREGS. Villages farther downslope—although they might in fact have been benefiting from an increased water flow—may not have been aware that the Dhara Vikas-MGNREGS work could be at least partly responsible for the change.

In general, although respondents from higher elevation wards (Gupti and Mungrang) were aware of Dhara Vikas, they tended to feel that the program had brought little or no change, whereas people in some lower elevation wards (e.g., Manglabarey and Melli Dhara) did feel that water is now available for more months of the year. In the mid-elevation wards (Upper Payong, Melli Gumpa, and Beeling), more respondents said they did not know. Communities from high-elevation Mungrang felt that forest cover had increased.

The survey also asked respondents to rate the success of the Dhara Vikas program on a scale of 1–10. Such survey questions are fraught with problems, including strategic responding. In this case, mean response was quite low at 3.6.

Over its first 5 years, the Dhara Vikas project recognized several important lessons (Sharma, 2015). First, lack of public awareness was perhaps partly a public communications problem—a common challenge to many extension and service programs in difficult-to-access areas. Second, and importantly, the problem of inequitable distribution remains. Even within a small numeric sample and a small geographical area, the specific geographical locations of villages have historical and social implications. Higher elevation villages are more remote from roads and markets, relatively difficult to access, and located farther from reliable springs. Government programs have not yet succeeded in making water accessible to all wards and villages. Even within single wards

TABLE 7.1 Perceptions of Surveyed Households (HHs) of the Dhara Vikas Program in Eight Wards of South Sikkim

Relative Elevation	Elevation (m a.s.l.)	Ward	HHs Aware of Dhara Vikas (%)	HHs Receiving Benefits From Government Water Distribution Schemes (%)	Participation of HH in Government Livelihood Schemes[a] (%)	HHs Self-Initiated New Livelihood Options[c] (%)	HH Assessment of Program Impact (Average Scale 1–10[b])
High	1210–2050	Gupti	100	3	14	11	1.6
	1500–2000	Mungrang	97	3	11	16	3.5
Mid	930–1610	Upper Ramabong	93	89	26	18	3.0
	600–1210	Beeling	79	39	16	50	4.5
	840–1510	Manglabarey	100	77	25	26	4.8
Low	160–1600	Upper Payong	38	88	39	93	4.8
	160–930	Melli Gumpa	62	83	24	47	3.7
	500–1200	Melli Dara	90	95	29	57	4.8

[a]*New livelihood options initiated with government agency support: poultry, dairy, beekeeping, horticulture, and floriculture.*
[b]*New livelihood options initiated by HH without government support: poultry, dairy, beekeeping, horticulture, and floriculture.*
[c]*Respondents' ratings of Dhara Vikas as a positive influence, on a scale of 1–10 (averaged within each ward).*

and villages, individual households' access to water remains extremely variable, and this variability is nonrandom, being correlated with income, educational level, family background, life opportunities, social connectedness versus marginalization, and other factors. Each of these endogenous factors crucially affects climate vulnerability at the household and the individual levels.

Third, it is not yet clear whether the infiltration trenches have actually had a direct impact on subsurface water flow. In the following sections, we address several reasons for this:

1. Early evaluations based on only 1 or 2 years' data have not been able to account for substantial year-to-year variations in rainfall. Funding for posttreatment monitoring has been deficient (as is common). Simple monitoring urgently needs to be put in place to verify that early positive results are repeated.
2. In these geologically complex landscapes, it remains unclear whether the geomorphological "rule of thumb" methods used are adequate for consistently inferring aquifer boundaries to position the infiltration trenches (Mahamuni and Kulkarni, 2011). Currently, attempts are underway to test this technical question through isotope analysis and other means (Michener and Lajtha, 2008). We address several of the broad contextual uncertainties associated with these first two questions defined in the Water Vulnerability: Contextual Uncertainties section.

Beyond these as-yet unresolved technical issues, a series of social and institutional questions have arisen. Maintenance of some (perhaps most) trenches has been inadequate. Beneficiary communities downslope found they could not oversee maintenance of the trenches servicing their springs, whereas communities high on the mountainside in the vicinity of the trenches (and which in many cases they themselves had originally dug) did not stand to benefit since the water would only exit down below. Together, these technical and social challenges produce a classic economic challenge: How to ensure that resource management investments and benefits are distributed equitably across a nonhomogeneous social-geographical landscape. How can investments (e.g., of labor) and benefits (e.g., of increased water flows) be more equitably distributed among stakeholders? Is there a compensation mechanism that could account for and rebalance inherent inequities?

WATER VULNERABILITY: CONTEXTUAL UNCERTAINTIES

Challenges in Understanding Spring Discharge and Recharge Dynamics

Across much of the Eastern and Central Himalaya, spring discharge appears to be fairly closely correlated with rainfall, responding with roughly consistent time lags (Negi and Joshi, 2004; Tambe et al., 2012; Agarwal et al., 2014). But how predictable are the temporal lags? And, do these lags point to recurring, and therefore predictable, spatial patterns of infiltration? Is it possible to link infiltration points reliably with spring outflows hundreds or thousands of meters distant? And, can such information from one springshed be transferred reliably to another?

Andermann et al. (2012) attempted to address some aspects of this problem in central Nepal, in large river catchments with fractured basement aquifers whose diffusivity ($\sim 1\ m^2/s$) may or may not be comparable to those of South Sikkim. In that study, the typical residence time for water in the reservoir appeared to be about 45 days, and the infiltration–discharge cycle was confirmed to be seasonal. However, the study was carried out at much coarser spatial and temporal resolution and the dynamics of storage in smaller scale, presumably shallower aquifers remain unclear in relation to the large and deep river basin.

It is thought that isotope signature analysis could help to delineate the aquifers and identify their interconnectivity by establishing the chemical relatedness or difference—both of water samples taken from neighboring springs and of spring water versus rainwater—but few such analyses have been performed in the Eastern Himalaya (NITI Aayog and UNDP, 2015). In an unpublished report, Dhakal et al. (2015) analyzed isotopes from 12 springs in several "climate vulnerable" catchments close to the district boundary between South and East Sikkim districts (elevation 650–1850 m, annual rainfall 1610 mm). The authors concluded that their study area is likely honeycombed by a system of small aquifers which are independent during dry winter periods but become interlinked during heavy rainfall periods (Dhakal et al., 2015). However, it is true that the relationship of rainfall to spring discharge will presumably be characterized by nonlinear threshold effects. Electrical conductivity (and salt content) of springwater samples varied noticeably over the course of a year, being overall higher during premonsoon months (March–May) than during the monsoon (July). This was interpreted as indicating that during the drier months, water occupies soil pores and fractures near the soil surface where salts gradually dissolve, increasing electrical conductivity. During the monsoon, in contrast, fresh new rainwater enters, first driving out old water and then itself flowing relatively quickly through the aquifer to emerge little changed from low-conductivity rainwater. Relatively low conductivity, thus, indicates low underground residence time and freshwater. This, in turn, implies that the recharge points are "local" and that the recharge is dominated by the flushing process. However, this brief study was unable to conclusively identify the recharge area and help map the underlying linkages. For this, isotope signature analysis will need to be performed on a series of samples taken at regular intervals along an elevational gradient above the spring.

Shivanna et al. (2008) performed such a study in the Western Himalaya. The authors analyzed isotopes from a series of 11 springs located in 3 valleys in Chamoli district of Uttarakhand, all showing steep slopes and thin topsoil bedded on phyllites with some hard granites and quartzites. As in South Sikkim, the springs responded sensitively to precipitation. At the higher elevation springs, discharge was chemically identical to rainwater, whereas lower down, the chemistry revealed longer passage through rock. Based on a combination of conventional geomorphological methods and environmental isotope analyses, possible recharge zones were identified and a series of subsurface dikes, check bunds, and infiltration trenches were constructed. A single-year data seem to show a very considerable increase in discharge at all springs throughout the dry season, increases ranging from 56% to 464% and averaging 227%. Unfortunately, only a single-year data were reported; it would be interesting to know whether that experience was repeated.

To fully understand the commonly complex groundwater recharge process in mountain areas, information is needed on many aspects of the subsurface structural variables that control water infiltration rates, including "type of rock(s) in an area, their attitude (dip and strike), the presence of openings, and [other] structural features" (Tambe et al., 2012). Collecting this amount of location-specific geological information for underresearched areas such as the Eastern Himalaya is a challenging demand. Meeting it would certainly require considerable investment by government agencies, and perhaps an increase in the number of geologists working in the region. Para-hydrogeologists, as developed by the Dhara Vikas program, may be able to satisfy some of these needs but it is unclear that the structure of underlying aquifers can always be inferred from surface forms (Mahumani and Upasani, 2011). Larger scale studies inferring recharge areas from mixed geologic and isotopic methods (such as those by Harris et al., 2010 on Table Mountain in South Africa and Parisi et al., 2011 on volcanic hills in Italy) produce some surprising results and show the importance of long-term records and of integrated, multipronged analyses. Nonetheless, close linkages between quantity and timing of rainfall and spring discharge have been found for most of the springs in the Sikkim-Nepal Eastern Himalaya, perhaps due to the generally thin topsoils preventing deep infiltration. Therefore, the second principal group of variables influencing spring discharge rates is amount and timing of precipitation.

Challenges in Understanding Monsoon Rainfall

Research on precipitation in the Eastern Himalayan range suffers from a dearth of the historical data that could help in determining long-term trends. Even today, weather stations in the hills are relatively few, considering the great fine-scale variability of weather patterns. Sikkim state does have 18 new Automatic Weather Stations across the state, set up in 2010 with the help of the Indian Space Research Organization of the national Department of Space. But longer time series exist only at a coarse scale and contain large gaps and questionable data.

To varying extents, this problem applies to all across the South Asian subcontinent. Partly for this reason, much of the research on the Indian monsoon until found various aspects of the monsoon to be "highly variable in intensity from year to year" but overall "trendless and mainly random in nature" (Prokop and Płoskonka, 2014; Mooley and Parthasarathy, 1984). Over the past 5−10 years, research has responded to calls for examination of trends in weather *extremes*, since it is largely the extremes that directly influence people and ecosystems. Some degree of consensus seems to be emerging that there is a statistically significant trend of increasing spatial and temporal heterogeneity in monsoon season rainfall (at the subcontinental level). Ghosh et al. (2012), for instance, showed first that the various methods for defining and calculating extremes—for example, annual maxima, use of fixed/variable thresholds, or 30-year/100-year return levels—themselves produce differing results, and second that increasing spatial heterogeneity points to some combination of local/regional factors driving extremes, including topography, land-use change (LUC) and urbanization.

The influence of anthropogenic LUC on local precipitation patterns may still be commonly underestimated. Soil moisture and vegetation can both influence rainfall at local and regional scales (e.g., Pielke, 2001; Douglas et al., 2006; Basistha et al., 2009; Niyogi et al., 2010; Spracklen, 2012). This effect may be particularly strong in the case of high-relief mountain topography and dryland areas (Houze, 2012). Basistha et al. (2009) looked at precipitation trends in Nainital district in the foothills of Uttarakhand state (Western Himalaya) over the first 80 years of the 20th century, that is, in the period *before* the strongest effects of global climate trends became noticeable in regional data sets. They found an inflection point in rainfall during the mid-1960s. From 1901 to 1964, annual precipitation had been increasing slowly, but after 1965 the trend reversed and precipitation rates fell conspicuously. This district-level pattern does not occur in the plains immediately to the south, implying that it is unlikely to be due to global or larger scale regional trends. But the pattern coincides well with the rapid draining and conversion of 25% of the total area of Nainital district from marshy Terai forest to agricultural land, along with large-scale immigration and population growth in Dehradun and adjoining valleys (Basistha et al., 2009).

Indeed, the 1960s were a time of major LUCs all across the Himalayan forehills and mid-hills. Bhutiyani et al. (2007) examined a hundred-year data set for seasonal temperature and precipitation from three weather stations in the Western Himalaya (the only stations for which the whole data series exists, namely, Srinagar, Shimla, and Leh). Trends for temperature and precipitation seem closely correlated again until the mid-1960s, after which they diverge. What accounts for this apparent regional-scale decoupling of trends in temperature and precipitation? Bhutiyani et al. (2007) conclude that "the teleconnections between the variation of temperature and precipitation ... appear to have weakened gradually since the late 1960s" and more rapidly in the 1990s, perhaps implying that natural trends were being overwhelmed during this period by strengthening anthropogenic forcings.

Which anthropogenic forcings, specifically? Bhutiyani et al. (2007) point to the increasing global-scale atmospheric concentration of greenhouse gases, but the most latest research has shown that simple signal occurs across South Asia of increase or decrease in precipitation, but rather an increase in *variability* among years (interannual), between seasons (intraannual) and even within seasons (more frequent dry periods, or breaks, within individual monsoon periods; Duncan et al., 2013; Singh et al., 2014; Pai et al., 2015). This increasing multiscale temporal variability in rainfall is combined with high and increasing spatial variability of rainfall as well, especially in those regions of the subcontinent where orography strongly affects rainfall patterns (Goswami et al., 2006; Dash et al., 2009; Panthi et al., 2015).

A complication is that across India, increasing incidence of heavy local rain events seems to be correlated with rapid *urbanization* processes (Kishtawal et al., 2010). Urbanization is a particular (extreme) case of LUC, in which not only does the surface area undergo far-reaching changes in vegetation cover, albedo and other factors, but the occurrence of increasing localized outputs of heat, gases, and aerosols change the atmospheric column above the urbanizing area and eventually throughout the region (Mahmood et al., 2014). The predominant natural driver of precipitation in the tropics, including over peninsular India, is convection: the development of cumulus clouds as a response to the flux of heat and moisture from surface to atmosphere. When a saturation threshold is reached, potential convective energy is released in the form of rainfall. Urbanization is thought to stimulate the convection process by altering the vertical thermodynamic stratification of the local atmosphere. In India, incidence of heavy rain days (>70 mm) shows an increasing trend since the early 1970s, but this is not true either of moderate rain days or of total annual rainfall. According to the analysis of Kishtawal et al. (2010), it appears that so far only some urban areas seem to exhibit the trend (i.e., 6 of the 15 most populous Indian cities).

Furthermore, indications are that atmospheric aerosols may also be directly involved in these large-scale changes. The strength of the prevailing summertime (monsoon) winds may be constrained if aerosols dim the sun and reduce the surface temperature gradient between South and North Indian Ocean waters. Ramanathan and Carmichael (2008) and other workers describe this effect as a "weakening" of the monsoon—though in fact only some aspects of the monsoon system are weakening, whereas others are strengthening. In general, the temperature contrast between air over the Indian Ocean and the low-pressure field over the Tibetan Plateau seems to be moderating, which relaxes the north and westward movement of moisture-laden air. This may result in periods of "stagnation" of the monsoon's progress. However, the warmer atmosphere is also capable of absorbing more moisture, which may contribute to an increasing frequency of severe cyclonic storms and intense downpours.

Rosenfeld et al. (2007) applied these observations to mountain areas by showing that upwind aerosol concentrations can suppress precipitation locally by forming clouds consisting of small droplets that coalesce only slowly into full raindrops. Such clouds may form when aerosol-laden air rises over mountain ridges, but may then evaporate when forced downward on the other side. Where this pattern is common, for instance in the Eastern Himalayan forehills, it is likely to strengthen any inherent topographical tendency toward the development of "rain shadows." Such local processes may also contribute to a general weakening of atmospheric "teleconnections" and a corresponding strengthening of locally generated trends.

Returning the focus to Sikkim, Rahman et al. (2012) looked at a 30-year time series (1981–2010) of meteorological records from a single weather station located on the campus of ICAR Research Complex in Tadong, located at 1350 m in east-central Sikkim, just a little south of Gangtok. They compared decadal rainfall means for the 1980s (2956 mm/yr) to the 1990s (3258 mm/yr), and to the 2000s (3080 mm/yr). These means show first a rise and then a fall, but the range of variation remains within the historical envelope. Thus Tadong, at least, does not seem to have experienced a clear trend in precipitation over the 30 years (1981–2010).

Other Nonmeteorological Forcings

Finally, as we noted in South Sikkim Case Study: Contextual Complexity section, village communities throughout the Eastern Himalaya commonly suggest that reductions in spring discharge of the dry season may be linked to earthquakes. In Sikkim often a particular focus occurs on the 2011 Sikkim earthquake. This earthquake with magnitude 6.9 was centered on the Kanchenjunga Conservation Area, close to the Sikkim-Nepal border. Several villages in Sikkim were destroyed and over 100 people died in Sikkim and Eastern Nepal. Could such an event alter groundwater streamflow? Montgomery and Manga (2003) reported that for earthquakes with this magnitude, such effects have indeed been recorded at distances of up to 100 km from the epicenter. Although globally, most reported instances of postearthquake streamflow changes have referred to *increased* rather than *diminished* flow, it should be possible in principle for near-surface fractures to close and produce sustained reductions in stream discharge (Montgomery and Manga, 2003). Thus, sudden changes in below-surface structures cannot be discounted as possible contributors to altered spring discharge rates (although it seems somewhat unlikely that all the reported instances in the Eastern Himalaya would go uniformly against the global trend of earthquakes releasing groundwater flows rather than blocking them).

On the other hand, in the Western Himalaya where formal environmental research began earlier, documented reports of springs drying go back at least several decades. Many publications of the 1980s and 1990s attribute drying of springs to deforestation, replacement of oak forests by pine, and other local anthropogenic and vegetative changes (e.g., Myers, 1986; Singh and Pande, 1989; Haigh et al., 1990; Valdiya and Bartarya, 1991). Today, global climate change is more commonly cited as the driver of loss of springs (e.g., Tripathi, 2016). Yet it has been difficult to point conclusively to causative factors, or even to rigorously document trends in the absence of adequate historical data.

To summarize, it appears that multiple potential drivers of changing weather patterns occur that may affect weather outcomes and Himalayan mountain spring discharge rates at regional and local scales. These include LUC, deforestation and agricultural expansion, urbanization and heat-island effects, aerosol and particulate emissions (including both industrial and domestic biomass)—all of which must be presumed to be interacting at multiple scales. It is scarcely surprising when climate vulnerability researchers ask "Regional climate downscaling: What's the point?" (Pielke and Wilby, 2012). No matter how much policy makers would like to believe that the next downscaling model may provide a key to fit the lock at the administratively useful scale, the exogenous drivers of "supply-side" water dynamics in mountain landscapes introduce daunting levels of uncertainty (Hawkins and Sutton, 2009; Kerr, 2011). The situation is further complicated by uncertainty regarding the possible drivers of endogenous or "demand-side" change. In the next section, we briefly examine some of these challenges in the context of South Sikkim district.

WATER VULNERABILITY: DISTRIBUTION AND EQUITY

Supply- and Demand-Side Problems

Villagers commonly report reduced availability of water in comparison to previous decades, and they attribute the reduction to various causes. Most of these reasons focus on perceived changes in water supply, but in a sample survey, one in five respondents mentioned increasing demand and population growth as contributing factors. Demand for water (both per capita and in the aggregate) is probably growing in many places, but in regions with little historical documentation, demand-side changes can be tricky to assess. Southern Sikkim was until at least the later 19th century only thinly settled. Many of today's villages have existed over just a few generations, having been established by British colonial powers or later by Indian government agencies or private estates to supply labor for tea gardens, in forestry, and in cinchona plantations. For these communities, mostly deriving from Nepali immigrants, farming was originally a subsistence-oriented occupation focused on basic grains, potatoes, and small kitchen gardens, only later producing market crops, especially cardamom. Areas of relatively flat land with access to reliable springs were doubtless settled first. Today, on these steep hillsides, almost every cultivable section of the slopes supports small plots, many of them terraced. Over half the surveyed households own private water storage tanks, but only one-third of these regularly use water for crop irrigation. Little tradition of irrigation systems occurs, and few of the sophisticated techniques of efficient traditional agriculture exists in places where communities have been farming intensively for many generations. Many of the original settlers were brought in from parts of mountainous central Nepal in which cattle and yak herding form the core of traditional livelihoods. However, the communities of plantation labor were composed of a mosaic of groups from the Terai, the mid-hills as well as the mountains, in addition to Sikkimese and some other groups.

Today, very little opportunity exists for grazing cattle, and stall-feeding is essentially universal. Throughout the Indian East Himalaya, per capita land holdings have been falling as the population has grown and family holdings have fragmented. In South Sikkim, population grew 12% between 2001 and 2011, and today, average per capita land holdings are exceptionally low at 0.53 acre (Sharma, 2015). In addition, the number of independent households has grown faster than the population itself, as nuclear-family living arrangements have tended to replace the more traditional extended family. Because smaller households commonly consume relatively more water per capita, this again complicates estimation of current rates of water consumption. Certainly, demand for water is unlikely to be stationary with respect to historical baselines, especially where rural livelihoods are in the process of integrating into larger market economies.

Livelihood Diversification Schemes

In 1978, just 3 years after Sikkim joined India as a new union state, the Indian National Dairy Development Board began organizing milk cooperatives in Sikkim, and in 1980 the Sikkim Co-operative Milk Producers' Union Limited was formed with operations at 5th Mile, Tadong and Jorethang. Over the next three decades, only limited efforts encouraged dairy farming within the three GPUs studied by Sharma et al. (2013). But by 2015, 16% of households in these three GPUs had taken up dairy farming. This represents a veritable explosion of new commercial activity: Of the dairying households, 65% started in 5 years since 2010, 15% started between 2005 and 2010, and only 20% had been practicing since before 2005. Dairy enterprises rely entirely on stall-feeding, setting up a flow of nutrients from forest (in the form of leaves gathered daily) to livestock and then to agricultural fields in the form of manure. Water, too, must in most cases be diverted from household supplies.

Meanwhile, the Sikkim Department of Animal Husbandry and Veterinary Services set up a Sikkim Poultry Development Corporation in 1995 to promote poultry activities. Growth was gradual until 2009, when a state Poultry Mission was launched with an objective "to achieve consumption level of poultry meat and table eggs at par with the developed countries of the world". Meat consumption is "to be increased to 18 kgs per capita and 43 table eggs per capita per annum." The Poultry Mission provided 1500 farmers with poultry subsidies of 2000–4000 mt/yr. "In Sikkim," claims the Department of Animal Husbandry and Veterinary Services website (http://www.sikkim-ahvs.gov.in/poultry_development.html), "livestock production is the ultimate answer to provide sustainable economic upliftment of the rural masses."

By 2015, around 20% of households surveyed in Sharma (2015) were raising poultry, but 75% of these households had started the enterprise just within the previous 5 years, 15% had started between 2005 and 2010, and a mere 10% had 10 years or more of experience. In addition, nearly 20% of surveyed households had initiated small commercial enterprises in floriculture over 5 years since 2010, with support from Sikkim Food Security & Agriculture Development. Goats and pigs are also kept.

Especially since 2010, each of these activities has been supported through campaigns by government agency. These campaigns have tended to target the larger and lower elevation wards, where as many as 40% of households have benefited (Table 7.2). In these wards, many households, observing the success of their neighbors, have also started small independent enterprises. However, a clear correlation exists between participation in government-livelihood schemes and the success of independent livelihood initiatives—both of which are highest in low-elevation wards.

As is evident from Table 7.2, these highly water-intensive, and commercially oriented, government extension activities are unevenly distributed across the study area. Low- and mid-elevation villages, which already had relatively secure access to water resources, benefited the most from the subsidies, support, and advice from government agencies. At the same time, government schemes for water distribution, such as subsidized piping and storage tanks also were able to reach the lower elevation villages first, but have scarcely touched the higher elevations. Currently, mean daily water consumption per capita in mid-elevation villages is twice that of high-elevation villages, whereas low-elevation villages consume at a rate over four times that of their high-elevation peers (Table 7.2). However, no clear trend occurs in mean domestic (personal) use per household—the large differences in total use per household are attributable almost entirely to livestock enterprises.

Thus, government agencies with different and sometimes conflicting mandates have to date not only failed to ameliorate the geohydrological deficits of marginalized high-elevation villages (e.g., through improved distribution and maintenance of water pipes and tanks), but have in effect exacerbated the extant inequalities through introduction of market-oriented schemes that preferentially benefit those who already enjoyed better market access.

TABLE 7.2 Estimated Water Consumption for a Range of Household (HH) Uses in Surveyed Wards at Three Relative Elevations

				Domestic Uses (L/d)								
Relative Elevation	Elevation (m a.s.l.)	Ward	Population (# HHs)	Drinking (L/d)	Hygiene and Other Domestic Use (L/d)	Total Domestic Use Per Ward (L/d)	Total Domestic Use Per HH (L/d/HH)	Mean Domestic Use Per HH (L/d/HH)	Total Livestock Use Per Ward (L/d)	Total Use Per Ward (L/d)	Total Use Per HH (L/d/HH)	Mean Total Use Per HH (L/d/HH)
High	1210–2050	Gupti	74	12	32	44	0.6	1.3	0	43	0.6	1.5
	1500–2000	Mungrang	58	32	82	114	2.0		30	144	2.5	
Mid	930–1610	Upper Ramabong	108	12	77	89	0.8	1.8	194	283	2.6	3.0
	600–1210	Beeling	68	35	135	170	2.5		19	188	2.8	
	840–1510	Manglabarey	86	26	142	168	2.0		152	320	3.7	
Low	160–1600	Upper Payong	93	30	122	152	1.6	1.7	649	802	8.6	6.8
	160–930	Melli Gumpa	112	30	133	163	1.5		500	663	5.9	
	500–1200	Melli Dara	120	23	203	226	1.9		472	697	5.8	

MGNREGS Linkage

According to the MGNREGA legislation of 2005, MGNREGS projects are specifically intended to benefit small farmers who are the vast majority throughout India as well as in the hill regions. Indeed it appears that projects focused on water management and building foot pathways and rural roads ("rural connectivity") have been very popular with beneficiaries and that the program as a whole has had measurable, poverty-alleviating impacts (McCartney and Roy, 2015; Esteves et al., 2013; Sridhar and Reddy, 2016). For instance, a survey in the state of Maharashtra (Narayanan et al., 2014) indicated that of 4800 surveyed participants, 90% felt that the assets created were useful—an extraordinarily positive assessment of any government program. In Sikkim, a 2010 assessment of MGNREGS activities (IRMA, 2010) described the Dhara Vikas-MGNREGS collaboration and emphasized the enthusiastic participation of women, for many of whom this was a first experience of paid work, and their anticipation of increased water availability. Almost every year since 2010, South Sikkim has been the recipient of awards for excellence in implementation of MGNREGS projects. In particular, Sikkim has been cited for the Dhara Vikas program, described as a model "convergence initiative" bringing together the goals and capacities of separate government agencies in a common purpose. In fact, it was partly the influence of the Dhara Vikas program in Sikkim that made it possible, after 2010, for MGNREGS in Sikkim to shift the main emphasis and budgeting from road building to water management.

One of the biggest challenges in Dhara Vikas has been in establishing and clarifying the incentives to villagers for building and maintaining the infiltration trenches. Convergence with MGNREGS helped overcome the construction problem, because village teams were able to prepare Dhara Vikas infiltration trenches while getting paid for it through MGNREGS. However, the recharge trenches also require ongoing maintenance, both to insure integrity of the pit sides and to clear out vegetation that tends to germinate in the moist, sheltered soil. *Cryptomeria japonica*, for instance, is a water-loving, non-native, invasive tree species that is common in the area and can quickly choke the trenches (Sharma, 2015). But no provision for MGNREGS has covered maintenance work, so no clear incentive existed to continue maintenance. Most importantly, assuming the trenches did improve water flow, the positive impacts would be felt far down the hillside. In the high-elevation villages where the water crisis is most severe, where the presumptive recharge areas lie, and where much of the MGNREGS work took place, no springs occur; while villagers at lower elevations did not and could not participate in the MGNREGS work, which took place too far from their home villages. Thus it appeared that due to the spatial separation of loci of investments and loci of benefits, Dhara Vikas may to some extent have ended up helping those who needed it less, while neglecting the neediest.

Investment Asymmetries

Villagers are acutely aware of the links between water availability and options for income generation. Sharma (2015) reported that of the households surveyed, 80% expressed willingness to pay for improved water services, in amounts ranging from 10 to 400 Rs/HH/month (mean 70 Rs/month). However, willingness to participate in voluntary collective-action groups to maintain the trenches constructed under Dhara Vikas was much lower and unequal across the region. As mentioned in South Sikkim Case Study: Contextual Complexity section, the situation displays certain elements of a classic "payments for ecosystem services" (PES) opportunity (Wunder et al., 2008; Ferraro, 2011; Rasul et al., 2011; but see critiques such as Muradian et al., 2013; Lakerveld et al., 2015). Arrangements could, in principle, be imagined whereby the communities lower down would pay those higher up to prepare and/or maintain the infiltration trenches. Implementing such a program would require (1) an unambiguously established, consistent relationship between trenches and increased spring discharge and (2) a clear legal framework into which such a contractual arrangement could be embedded. Neither of these requirements is close to being satisfied at present, though work is continuing to establish the former. Perhaps most importantly, the greatest need for water services is in fact among the high-elevation communities. Financial resources are also lowest in these same villages, but supplementing household incomes via a PES transaction would not solve their water deficit. The strongest positive quality-of-life impact for high-elevation villagers could probably be achieved by government investments in improving water distribution systems: community rainwater storage tanks, reliable piping, perhaps even (solar-powered?) pumps in some cases (Tambe et al., 2013).

In the bigger picture, no true "water deficit" actually exists in Sikkim, even in southern Sikkim. As a whole, the state receives some of the highest and most reliable annual precipitation in the country, although it is strongly and increasingly seasonal. (As points of comparison, the tropical desert state of Rajasthan annually receives between <100 mm and a maximum of 1600 mm, whereas the high desert around Leh in Ladakh receives only an average 100 mm annually.) Nevertheless, many households in South Sikkim suffer from deficits in water availability, especially in comparison with their neighbors. These deficits are highly correlated with deficits in other life opportunities, and they generally reflect broader social stratifications. To the extent that "drought-prone" pockets occur within Sikkim—seemingly related to rain shadow effects as well as to seasonality of precipitation—they must be regarded not so much as *supply* problems as the outcomes of chronically unresolved and perhaps increasing challenges in water storage and *distributional* inequities. These are the same inequities and inefficiencies that are seen in the surrounding urban conglomerations including Gangtok, Namchi and

Singtam in Sikkim, Darjeeling and Kalimpong just to the south in West Bengal state (Tiwari, 2012; Sharma et al., 2013).

CONCLUSION: DEFINING CLIMATE VULNERABILITY AND ADAPTATION

In conclusion, we reflect briefly on the nature of climate vulnerability in the rapidly changing social and economic contexts that today prevail all across the Himalayan forehills of India and Nepal. A good deal of the discussion about vulnerability, and the adaptive measures intended to counteract it, has focused on the effort to define ways in which the climate is likely to change in a given place, and then to imagine what kinds of impacts the change would have on people. Since the presumption is that one should be working with the range of climate scenarios most applicable to a particular location, this effort depends on good (realistic), regional-scale climate modeling. Unfortunately, it remains extremely difficult to convey to decision makers and the public the idea that "climate projections" are simply complex "if-then" models. It seems to be a natural human tendency to take scenarios literally—even trained scientists who work regularly with models and simulations often fall into the habit of speaking of them as though they were "predictions" (MacCracken, 2002; Bray and von Storch, 2009; Pielke and Wilby, 2012). For this psychological reason, as well as for practical reasons, it would seem better to discuss climate change adaptation as the need to prepare for great variation, unpredictability, and wide extremes, rather than attempting to define a new envelope of variation and to prepare for specific scenarios based within that envelope.

Regional downscaling of global circulation models (GCMs), performed with the purpose of clarifying likely future conditions for which adaptive strategies need to be developed, has made considerable technical advances in the last few years (e.g., Di Luca et al., 2015). However, these efforts are not uncontroversial. First occurs the inherent problem of error propagation from fuzzy large-scale models into more precise, but not necessarily more accurate, fine-scale ones. This is the professional challenge that the modeling community finds irresistible, although some members of that community already warn of diminishing pay-off from investments in ever-increasing detail and multiplication of input parameters that may build model detail without adding to model reliability (e.g., Racherla et al., 2012; Daron et al., 2015). Perhaps more important are a set of technical issues inherent to the process of relating high-resolution small areas to surrounding low-resolution modeled areas. For instance, regional dynamical downscaling is highly dependent on choice of lateral boundary conditions (i.e., climate conditions prevailing at the edge of the model domain), so the choice of domain boundary location can have extensive ramifications, although it is also essentially arbitrary from the meteorological perspective. Boundary conditions influence the regional model across this boundary, but in the model's universe, the influence is a one-way street: Modeled regional effects cannot be fed back into the GCM (Rockel et al., 2008; Pielke et al., 2012). Unrealistically, the regional domain thus becomes a partly closed system.

Statistical downscaling methods, computationally and conceptually simpler, may propagate GCM error even more directly into the regional models (Sunyer et al., 2012). Several innovative new downscaling approaches are being developed. Coordinated global and regional modeling (e.g., Scinocca et al., 2016) pursues a strategy of "strict physics compatibility," so that the global and regional model components share the same physical parameterizations from the ground up. This allows the boundary conditions to be more fully characterized in terms of atmospheric chemistry as well as temperature, moisture, and prevailing winds. This approach adds potential value to the GCM by linking regional scenarios more tightly to the dynamics of the surrounding context—in effect, model coordination potentially allows for better "top-down" modeling. The opposite approach is followed in the use of stochastic weather generators (e.g., Steinschneider and Brown, 2013; Whateley et al., 2016). These provide a large array of weather time series that simulates the natural variability of local weather patterns. Subsets of these climate sequences can then be filter selected, depending on the parameters of interest, for further investigation and impacts assessments. This could be described as an aid to "bottom-up" vulnerability assessment, because the generated scenarios can be treated independently of direct reference to surrounding conditions.

The goal of a bottom-up vulnerability assessment should be to estimate potential impacts, given a set of possible extremes, and to assess the degree of resilience to those impacts. This takes the emphasis and discussion off the climate scenario, and focuses attention on the set of achievable adaptive changes that can be made now or soon. It also has the potential effect of spurring greater communication and coordination among responsible government agencies, as well as civil society, because climate adaptation need not be defined as the responsibility of specialists or specialist cells within larger agencies—as it is presently in India—but rather of all environmentally interested agencies, from agriculture and rural development to infrastructure, as well as to forests and disaster risk management.

Climate vulnerability, in the South Sikkim case as in others, cannot usefully be separated from historical and current forms of vulnerability. In most cases, extant problems are likely to be exacerbated under climate change, and already marginalized groups further marginalized. In South Sikkim, the villages and the social groups most vulnerable to the loss of spring water are also the groups currently living under the most stressful conditions, and with the least degree of freedom for life choices and the development of capabilities (Sen, 1985, 1999). In complex mountain landscapes, and especially among social groupings that are relatively strictly stratified, we can assume that some degree of social filtering has taken place historically. The most difficult and inconvenient living sites tend to be occupied by powerless groups. The rapid pace of economic and infrastructural development and commercial market penetration currently underway across the southern face of the Himalaya is promoting some flexibility in the degree to which individuals are able to position themselves to take advantage of new opportunities, but very often this is taking place through cyclical (seasonal) or permanent migration to urban areas, or even out of the country.

If national, state, and district governments seriously want to counteract the drain of rural population and prepare for oncoming climate uncertainties, they will need to invest vigorously in expanding services to reduce the isolation of mountain populations. Water services are just one among the many services that are needed, irrespective of the pace and specific impacts of climate change. And targeted investments will be needed to promote coordination and cooperation among government agencies. The picture seen in South Sikkim is all too common: one set of agencies encouraging relatively comfortable citizens to build resource-consuming commercial enterprises in isolation from the halting and underfunded efforts of other agencies to keep up with the increasing demand for basic services. The mounting successes of the MGNREGA program show that agencies can attain challenging social goals in India, even at the largest of scales.

ACKNOWLEDGMENTS

Work on this chapter by RGS and YTW was partially funded by USAID-India (contract number 386-14-000024) and by the Barr Foundation (contract number 5808). Our thanks also to Dr. K.S. Bawa for important intellectual and financial support. Work on the paper by GS was partially funded by a grant to The Mountain Institute India from the National Bank for Agriculture and Rural Development (Head Office Mumbai, India). Thanks to Sikkim Government officials from the Rural Management and Development Department, Namchi; to Gram Panchayats of the three GPUs for providing us with information on Dhara Vikas; and Mr. P. Chettri for his assistance in field data generation.

REFERENCES

Agarwal, A., Agrawal, N.K., Nema, R.K., 2014. Life line: the Springs of Uttarakhand, India. International Journal of Innovative Research in Science, Engineering and Technology 3 (4), 11553–11560.

Andermann, C., Longuevergne, L., Bonnet, S., Crave, A., Davy, P., Gloaguen, R., 2012. Impact of transient groundwater storage on the discharge of Himalayan rivers. Nature Geoscience 5 (2), 127–132.

Annamalai, H., Sperber, K.R., 2016. South Asian summer monsoon variability in a changing climate. In: The Monsoons and Climate Change. Springer International Publishing, pp. 25–46.

Basistha, A., Arya, D.S., Goel, N.K., 2009. Analysis of historical changes in rainfall in the Indian Himalayas. International Journal of Climatology 29 (4), 555–572.

Bhutiyani, M.R., Kale, V.S., Pawar, N.J., 2007. Long-term trends in maximum, minimum and mean annual air temperatures across the Northwestern Himalaya during the twentieth century. Climatic Change 85, 159–177.

Bookhagen, B., Burbank, D.W., 2010. Toward a complete Himalayan hydrological budget: spatiotemporal distribution of snowmelt and rainfall and their impact on river discharge. Journal of Geophysical Research 115, 1–25.

Bray, D., von Storch, H., 2009. "Prediction" or "Projection"? The nomenclature of climate science. Science Communication 30 (4), 534–543.

Daron, J.D., Sutherland, K., Jack, C., Hewitson, B.C., 2015. The role of regional climate projections in managing complex socio-ecological systems. Regional Environmental Change 15 (1), 1–12.

Dash, S.K., Kulkarni, M.A., Mohanty, U.C., Prasad, K., 2009. Changes in the characteristics of rain events in India. Journal of Geophysical Research: Atmospheres 114 (D10).

Department of Animal Husbandry and Veterinary Services, Sikkim. http://www.sikkim-ahvs.gov.in/poultry_development.html.

Dhakal, D., Tiwari, A., Tambe, S., Sinha, U.K., Arrawatia, M.L., 2015. Isotope Studies to Identify the Origin and Recharge Area of Himalayan Springs as a Climate Change Adaptation Initiative: A Case Study from Sikkim, Eastern Himalaya. https://thewaternetwork.com/storage/NK-XwsU75vET0cTqmBL_vA/file/r053.pdf.

Di Luca, A., de Elía, R., Laprise, R., 2015. Challenges in the quest for added value of regional climate dynamical downscaling. Current Climate Change Reports 1 (1), 10–21.

Douglas, E.M., Niyogi, D., Frolking, S., Yeluripati, J.B., Pielke, R.A., Niyogi, N., Vörösmarty, C.J., Mohanty, U.C., 2006. Changes in moisture and energy fluxes due to agricultural land use and irrigation in the Indian Monsoon Belt. Geophysical Research Letters 33 (14).

Duncan, J.M.A., Dash, J., Atkinson, P.M., 2013. Analysing temporal trends in the Indian Summer Monsoon and its variability at a fine spatial resolution. Climatic Change 117 (1–2), 119–131.

Eriksen, S.H., Kelly, P.M., 2007. Developing credible vulnerability indicators for climate adaptation policy assessment. Mitigation and Adaptation Strategies for Global Change 12 (4), 495–524.

Esteves, T., et al., 2013. Environmental Benefits and Vulnerability Reduction Through Mahatma Gandhi NREGS: Synthesis Report, Ministry of Rural Development and GIZ, New Delhi.

Fatichi, S., Ivanov, V.Y., Paschalis, A., Peleg, N., Molnar, P., Rimkus, S., Kim, J., Burlando, P., Caporali, E., 2016. Uncertainty partition challenges the predictability of vital details of climate change. Earth's Future.

Ferraro, P.J., 2011. The future of payments for environmental services. Conservation Biology 25 (6), 1134–1138.

Ghosh, S., Das, D., Kao, S.C., Ganguly, A.R., 2012. Lack of uniform trends but increasing spatial variability in observed Indian rainfall extremes. Nature Climate Change 2 (2), 86–91.

Goswami, B.N., Venugopal, V., Sengupta, D., Madhusoodanan, M.S., Xavier, P.K., 2006. Increasing trend of extreme rain events over India in a warming environment. Science 314 (5804), 1442–1445.

Haigh, M., Rawat, J.S., Bisht, H.S., 1990. Hydrological impact of deforestation in the central Himalaya. Publ. No. 190. In: Proceedings of the Strbské Pleso Workshop Hydrology of Mountainous Areas (June 1988). IAHS.

Harris, C., Burgers, C., Miller, J., Rawoot, F., 2010. O-and H-isotope record of Cape Town rainfall from 1996 to 2008, and its application to recharge studies of Table Mountain groundwater, South Africa. South African Journal of Geology 113 (1), 33–56.

Hawkins, E., Sutton, R., 2009. The potential to narrow uncertainty in regional climate predictions. Bulletin of the American Meteorological Society 90 (8), 1095.

Hinkel, J., 2011. "Indicators of vulnerability and adaptive capacity": towards a clarification of the science–policy interface. Global Environmental Change 21 (1), 198–208.

Houze, R.A., 2012. Orographic effects on precipitating clouds. Reviews of Geophysics 50 (1).

IRMA [Institute of Rural Management, Anand], 2010. An Impact Assessment Study of the Usefulness and Sustainability of the Assets Created Under Mahatma Gandhi National Rural Employment Guarantee Act (MGNREGA) in Sikkim. IRMA, Anand, Gujarat.

Kerr, R.A., 2011. Vital details of global warming are eluding forecasters. Science 334 (6053), 173–174.

Kishtawal, C.M., Niyogi, D., Tewari, M., Pielke, R.A., Shepherd, J.M., 2010. Urbanization signature in the observed heavy rainfall climatology over India. International Journal of Climatology 30 (13), 1908–1916.

Klein, R.J., 2009. Identifying countries that are particularly vulnerable to the adverse effects of climate change: an academic or political challenge. Carbon and Climate Law Review 284.

Lakerveld, R.P., Lele, S., Crane, T.A., Fortuin, K.P.J., Springate-Baginski, O., 2015. The social distribution of provisioning forest ecosystem services: evidence and insights from Odisha, India. Ecosystem Services 14, 56–66.

MacCracken, M.C., 2002. Do the uncertainty ranges in the IPCC and US National assessments account adequately for possibly overlooked climatic influences? Climatic Change 52 (1), 13–23.

Mahamuni, K., Kulkarni, H., 2011. Hydrogeological Studies and Action Research for Spring Recharge and Development and Hill-Top Lake Restoration in Parts of Southern District, Sikkim State. Advanced Center for Water Resources Development and Management (ACWADAM) & Rural Management and Development Department (RMDD), Government of Sikkim. Submitted to GIZ India. ACWA/Hydro/2011/H-20.2011.

Mahmood, R., Pielke, R.A., Hubbard, K.G., Niyogi, D., Dirmeyer, P.A., McAlpine, C., Carleton, A.M., Hale, R., Gameda, S., Beltrán-Przekurat, A., Baker, B., 2014. Land cover changes and their biogeophysical effects on climate. International Journal of Climatology 34 (4), 929–953.

McCartney, M., Roy, I., 2015. A consensus unravels: NREGA and the paradox of rules-based welfare in India. European Journal of Development Research. http://dx.doi.org/10.1057/ejdr.2015.32.

Michener, R., Lajtha, K. (Eds.), 2008. Stable Isotopes in Ecology and Environmental Science. John Wiley & Sons.

Ministry of Law & Justice, September 7, 2005. The National Rural Employment Guarantee Act, 2005. The Gazette of India. http://nrega.nic.in/rajaswa.pdf.

Montgomery, D.R., Manga, M., 2003. Streamflow and water well responses to earthquakes. Science 300 (5628), 2047–2049.

Mooley, D.A., Parthasarathy, B., 1984. Fluctuations in all-India summer monsoon rainfall during 1871–1978. Climatic Change 6, 287–301.

Muradian, R., Arsel, M., Pellegrini, L., Adaman, F., Aguilar, B., Agarwal, B., Corbera, E., Ezzine de Blas, D., Farley, J., Froger, G., Garcia-Frapolli, E., 2013. Payments for ecosystem services and the fatal attraction of win-win solutions. Conservation Letters 6 (4), 274–279.

Myers, N., 1986. Environmental repercussions of deforestation in the Himalayas. Journal of World Forestry Resource Management 2, 63–72.

Narayanan, S., Ranaware, K., Das, U., Kulkarni, A., 2014. MGNREGA Works and Their Impacts: A Rapid Assessment in Maharashtra. WP-2014-042, Indira Gandhi Institute of Development Research (IGIDR), Mumbai. http://www.igidr.ac.in/pdf/publication/WP-2014-042.pdf.

Negi, G.C., Joshi, V., 2004. Rainfall and spring discharge patterns in two small drainage catchments in the Western Himalayan Mountains, India. Environmentalist 24 (1), 19–28.

NITI Aayog (Planning Commission), Government of India, UNDP, 2015. Dhara Vikas: Creating Water Security through Spring-shed Development in Sikkim. In: Social Sector Service Delivery: Good Practices Resource Book 2015. OneWorld Foundation India (Chapter 2.7).

Niyogi, D., Kishtawal, C.M., Tripathi, S., Govindaraju, R.S., 2010. Observational evidence that agricultural intensification and land use change may be reducing the Indian summer monsoon rainfall. Water Resources Research 46, W03533. http://dx.doi.org/10.1029/2008WR007082.

Pai, D.S., Sridhar, L., Badwaik, M.R., Rajeevan, M., 2015. Analysis of the daily rainfall events over India using a new long period (1901–2010) high resolution (0.25 × 0.25) gridded rainfall data set. Climate Dynamics 45 (3–4), 755–776.

Panthi, J., Dahal, P., Shrestha, M.L., Aryal, S., Krakauer, N.Y., Pradhanang, S.M., Lakhankar, T., Jha, A.K., Sharma, M., Karki, R., 2015. Spatial and temporal variability of rainfall in the Gandaki River Basin of Nepal Himalaya. Climate 3 (1), 210–226.

Parisi, S., Paternoster, M., Kohfahl, C., Pekdeger, A., Meyer, H., Hubberten, H.W., Spilotro, G., Mongelli, G., 2011. Groundwater recharge areas of a volcanic aquifer system inferred from hydraulic, hydrogeochemical and stable isotope data: Mount Vulture, southern Italy. Hydrogeology Journal 19 (1), 133–153.

Pielke Sr., R.A., et al., 2012. Dealing with complexity and extreme events using a bottom-up, resource-based vulnerability perspective. In: Sharma, A.S., et al. (Eds.), Complexity and Extreme Events in Geosciences, Geophys. Monogr. Ser. AGU, Washington, DC.

Pielke, R.A., Wilby, R.L., 2012. Regional climate downscaling: what's the point? Eos, Transactions American Geophysical Union 93 (5), 52–53.

Pielke, R.A., 2001. Influence of the spatial distribution of vegetation and soils on the prediction of cumulus convective rainfall. Reviews of Geophysics 39 (2), 151–177.

Prokop, P., Płoskonka, D., 2014. Natural and human impact on the land use and soil properties of the Sikkim Himalayas piedmont in India. Journal of Environmental Management 138, 15–23.

Racherla, P.N., Shindell, D.T., Faluvegi, G.S., 2012. The added value to global model projections of climate change by dynamical downscaling: a case study over the continental US using the GISS-ModelE2 and WRF models. Journal of Geophysical Research: Atmospheres 117 (D20).

Rahman, H., Karuppaiyan, R., Senapati, P.C., Ngachan, S.V., Kumar, A., 2012. An analysis of past three decade weather phenomenon in the mid-hills of Sikkim and strategies for mitigating possible impact of climate change on agriculture. In: Arrawatia, M.L., Tambe, S. (Eds.), Climate Change in Sikkim Patterns, Impacts and Initiatives. Information and Public Relations Department, Government of Sikkim. ISBN:978-81-920437-0-9, 424 pp.

Ramanathan, V., Carmichael, G., 2008. Global and regional climate changes due to black carbon. Nature Geoscience 1 (4), 221–227.

Rasul, G., Chettri, N., Sharma, E., 2011. Framework for Valuing Ecosystem Services in the Himalayas. ICIMOD, Kathmandu.

RMDD., 2014. Village Water Security Plan Lungchok-Kameray GPU, South Sikkim. Rural Management and Development Department, Government of Sikkim.

RMDD., 2015. Village Water Security Plan, Turuk Ramabong GPU, South Sikkim. Rural Management and Development Department, Government of Sikkim.

Rockel, B., Will, A., Hense, A., 2008. The regional climate model COSMO-CLM (CCLM). Meteorologische Zeitschrift 17 (4), 347–348.

Rosenfeld, D., Dai, J., Yu, X., Yao, Z., Xu, X., Yang, X., Du, C., 2007. Inverse relations between amounts of air pollution and orographic precipitation. Science 315 (5817), 1396–1398.

Sabin, T.P., Krishnan, R., Ghattas, J., Denvil, S., Dufresne, J.L., Hourdin, F., Pascal, T., 2013. High resolution simulation of the South Asian monsoon using a variable resolution global climate model. Climate Dynamics 41 (1), 173–194.

Scinocca, J.F., Kharin, V.V., Jiao, Y., Qian, M.W., Lazare, M., Solheim, L., Flato, G.M., Biner, S., Desgagne, M., Dugas, B., 2016. Coordinated global and regional climate modeling. Journal of Climate 29 (1), 17–35.

Sen, A., 1985. Commodities and Capabilities. North-Holland.

Sen, A., 1999. Development as Freedom. Oxford University Press.

Sharma, G., 2015. Spring Rejuvenation and Its Socio-Economic Impacts of Dhara-Vikas Initiative in Rain Shadow Areas of South Sikkim. The Mountain Institute-India, Gangtok. Submitted to National Bank for Agriculture and Rural Development, Mumbai. Ref. No. NB. DEAR./R&D/RM44/Sponsored Res.-2014-15.

Sharma, S., Kumar, K., Singh, K.K., September 2013. Water security in the mid-elevation Himalayan watershed, East district with focus in the State of Sikkim. Presented at Kohima workshop. Water Security, Water Planning and Management, 27 pp.

Shivanna, K., Tirumalesh, K., Noble, J., Joseph, T.B., Singh, G., Joshi, A.P., Khati, V.S., 2008. Isotope techniques to identify recharge areas of springs for rainwater harvesting in the mountainous region of Gaucher area, Chamoli district, Uttarakhand. Current Science 94 (8), 1003–1011.

Singh, A.K., Pande, R.K., 1989. Changes in spring activity: experiences of Kumaun Himalaya, India. Environmentalist 9 (1), 25–29.

Singh, D., Tsiang, M., Rajaratnam, B., Diffenbaugh, N.S., 2014. Observed changes in extreme wet and dry spells during the South Asian summer monsoon season. Nature Climate Change 4 (6), 456–461.

Soja, R., Starkel, L., 2007. Extreme rainfalls in Eastern Himalaya and southern slope of Meghalaya Plateau and their geomorphologic impacts. Geomorphology 84 (3), 170–180.

Spracklen, D.V., Arnold, S.R., Taylor, C.M., 2012. Observations of increased tropical rainfall preceded by air passage over forests. Nature 489 (7415), 282–285.

Sridhar, K.S., Reddy, A.V., 2016. India's Mahatma Gandhi National Rural Employment Guarantee Scheme (MGNREGS): wages, reservation wages and programme participation in Chitradurga district of Karnataka. Management and Labour Studies 40, 0258042X15627077.

Steinschneider, S., Brown, C., 2013. A semiparametric multivariate, multisite weather generator with low-frequency variability for use in climate risk assessments. Water Resources Research 49 (11), 7205–7220.

Sunyer, M.A., Madsen, H., Ang, P.H., 2012. A comparison of different regional climate models and statistical downscaling methods for extreme rainfall estimation under climate change. Atmospheric Research 103, 119–128.

Tambe, S., Kharel, G., Subba, S., Arrawatia, M.L., 2013. Rural Water Security in the Sikkim Himalaya: Status, Initiatives and Future Strategy. India Mountain Initiative, Summit in Kohima, Nagaland.

Tambe, S., Kharel, G., Arrawatia, M.L., Kulkarni, H., Mahamuni, K., Ganeriwala, A.K., 2012. Reviving dying springs: climate change adaptation experiments from the Sikkim Himalaya. Mountain Research and Development 32 (1), 62–72.

Tiwari, A., 2012. Water quality and quantity analysis in Sikkim, north eastern Himalaya. Current Science 103 (1), 41–45.

Tripathi, M., 2016. In the face of changing climate: groundwater development through artificial recharge in hard rock Terrain of Kumaun Lesser Himalaya. In: Geostatistical and Geospatial

Approaches for the Characterization of Natural Resources in the Environment. Springer International Publishing, pp. 937–947.

Valdiya, K.S., Bartarya, S.K., 1991. Hydrogeological studies of springs in the catchment of the Gaula river, Kumaun Lesser Himalaya, India. Mountain Research and Development 11 (3), 239–258.

Whateley, S., Steinschneider, S., Brown, C., 2016. Selecting stochastic climate realizations to efficiently explore a wide range of climate risk to water resource systems. Journal of Water Resources Planning and Management 142 (6), 06016002.

Wunder, S., Engel, S., Pagiola, S., 2008. Taking stock: a comparative analysis of payments for environmental services programs in developed and developing countries. Ecological Economics 65 (4), 834–852.

Chapter 8

Neotropical Mountains Beyond Water Supply: Environmental Services as a Trifecta of Sustainable Mountain Development

F.O. Sarmiento
University of Georgia, Neotropical Montology Collaboratory, Athens, Georgia, United States
E-mail: fsarmien@uga.edu

Chapter Outline

IS THERE AN ANDEAN VISION OF WATER?

In the "Andean vision of water," there are six criteria for the incorporation of water into the socioecological system: (1) water as a living being, (2) water as the basis of reciprocity and complementarity, (3) water as universal and community right, (4) water as an expression of flexibility and adaptability, (5) water as creator and transformer, and (6) water as social recreation (CONDESAN, 2003). The water collected as snow and ice from the snowpack volcanoes, and from the horizontal precipitation on the epiphytic gardens of the cloud forest have a special place in the mythology of Andean people, as they are integral to origin myths or have profound meaning to community and personal geographies. The *Kichwa* and *Aymara* share a mythical view of nature guided by the reciprocity principle or *Aynu*, that often

Mountain Ice and Water. http://dx.doi.org/10.1016/B978-0-444-63787-1.00008-1

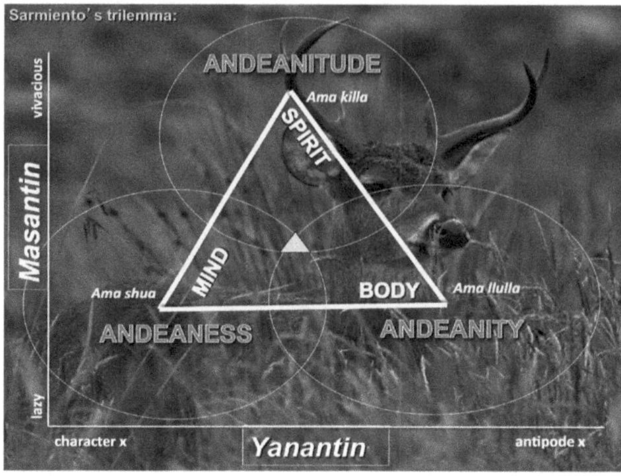

FIGURE 8.1 With the Andean deer or *tarugu* (*Odocoileus virginianus tropicalis*) in the background amidst the tillers of the bunchgrasses of the páramo, the Sarmiento's trilemma is depicted to understand the interaction of the trifecta that incorporates the dualism or *Yanantin* energized by the enthusiasm or *Masantin* in our new ethnoecological understanding of water management in the mountains.

fluctuates between two opposites, individualism or collectivism, of the character being reciprocated, such as water, as "complementary dualism" or *Yanantin* (Webb, 2012). This axis has an orthogonal variable of the intensity (or enthusiasm) displayed in acting on the character reciprocated or *Masintin*. This mythology of water, as a reciprocating good from "mother Earh" or *Pachamama*, clearly differentiates between the laziness of slow flows of calm waters of lakes or the vivaciousness of rapid flows, such as the one observed in the white waters of brooks and particularly in the waterfalls. Thus, the *Inka* precepts (*Ama llulla*, *Ama killa*, and *Ama shua*, to be neither lazy, lier, nor stealer) guide their ultimate appropriation of water and ice in a trifecta for sustainability, that achieves water conservation (material importance), community organizing and capacity building (mental agreement), and fulfillment of the reciprocity with the numen (spiritual significance) identified in the trilemma of Andean identity (Fig. 8.1).

Lately, interdisciplinary teams have placed the human dimension in the narrative of mountain water (e.g., Reinhardt, 1985; Rhoades, 1998; Orlove et al., 2008; Carey, 2010; Fonstad, 2013; Carey et al., 2014). In most of the Andean sites described therein land-use/land-cover change is evident; the forest cover of the Andean cloud forests has been replaced by a matrix of grass cover (Sarmiento, 2012) and the land-use shifted toward export-oriented agriculture. Only previously has attention shifted to the ancestral practices identified as rituals for the sacred geographies of the Andes, frequently involving water features including highland lakes and glaciated peaks (Allison, 2015; Borsdorf and Stadel, 2015).

The shrinking glaciers of tropical mountains have become icons of climate change due to their evident loss of glacial mass and diminished stream flows (Baraer et al., 2012). The alarming rate of significant retreat of many glacial tongues registered in many tropical snow-packed mountains—such as *Popocatepetl* and "Pico Orizabal" in Mexico (García-Romero et al., 2015; Bermúdez-Torres et al., 2015); "Pico Bolívar" and "Coromoto" in Venezuela (Balcázar et al., 2015); "Nevado del Ruiz" and "Sierra Nevada de Santa Marta" in Colombia (Rabatel et al., 2013); *Kutakachi*, *Kayampi*, *Antisana*, and *Chimburasu* in Ecuador (Rhoades et al., 2016; Frenierre and David, 2014); *Quelccaya*, *Waskaran*, *Salcantay*, and *Misti* in Perú (Fernández and Mark, 2016); *Intillimani* in Bolivia (Kaijser, 2013); and *Akonkija* in Argentina (Lázaro, 2016)—reinforce the message that crises related to water supply and delivery are imminent in the tropical Andes (Francou et al., 2004). Narratives associated with shrinking glaciers are now integral to climate change research and overall environmental policy (Taillant, 2015). Furthermore, extreme events, such as the disappearance of glaciers from glaciated peaks in the tropical Andes, have prompted policy shifts to protect the formerly glaciated mountains, such as the *Chacaltaya* in Bolivia (Hoffman and Requena, 2013; Raoul, 2015), generating public support to protect the threatened peaks.

Tropical glaciers are not, however, the only relevant part of the water cycle in the tropical Andes. Tropical Montane Cloud Forests (TMCFs) are now seen as one of the most important natural ecosystems in the Andean water cycle for their role in regulating horizontal precipitation. Since the first international workshop on cloud forest ecology held in Puerto Rico (Hamilton et al., 1993) and emphasizing the contribution of biodiversity conservation, to the second workshop on cloud forest taxonomy at the New York Botanical Gardens (Churchill et al., 1995) on systematics and applied botany, to the Convention of Biological Diversity report (Bubb et al., 2003), to the most late workshop on cloud forest environmental services (Bruijnzel et al., 2010), the importance of tropical mountain water has been highlighted only from the hydrological or botanical points of view (Martin and Bellingham, 2016). Lately, however, more attention is being paid to the linkages of culture and nature to inform perspectives of sociological and economic stresses on mountain communities due to global warming and climate change (Sarmiento et al., 2013). The indelible signature of the human impact can be detected with tools of ethnoecology, traditional knowledge, and emotional geography (Gade, 2016) (Fig. 8.2).

WEALTH OF THE CLOUD FORESTS

Spiritual dimensions map to relevant cultural ecosystem services, not only in the glaciated tropical peaks of the Americas, but also in TMCFs areas below the snowline. Glacier retreat in Ausangate, Peru, for instance, has prompted the prohibition for pilgrims of the ritualistic *Kuyllurit'i* pilgrimage (or "Qoyllur Riti") and associated yearly festivities of *Wayri Ch'unchu*, with more

FIGURE 8.2 A view of the Andean forest and páramo transitions from a sacred cove in the *Imbabura* volcano of Northern Ecuador, portraying the undeniable evidence that even the isolated hinterlands, the human imprint affect changes in the mountainscape. *Photo: César Cotacachi.*

than 10,000 pilgrims dancing and praying to the *Apu* (telluric mountain guardian), to cut ceremonial ice chunks to capture the essence of the glacial water therein and to bring them down to nearby towns to distribute among the wider population (Salas Carreño, 2010), as was the ancestral custom of this area, one of the oldest rituals of respect for mountains in the Andean world (Bolin, 2010).

The destruction of the livelihood associated with harvesting of glacial ice in the snow-pack volcanoes' nearby colonial cities has been captured in a documentary produced in Ecuador, about the icemen of Mt. *Chimburasu* or "Chimborazo" (Guayasamín and Guayasamín, 1980), who are no longer practicing that old trade. With the oblivion of the ancestral trade for which Baltazar Ushca was its last survivor (Tuaza, 2007), the harvesting of ice chunks is an extinct cultural practice in many tropical mountains or fast disappearing in others (Dunbar and Marcos, 2012). It can be asserted that because of the lack of the Andeanitude of water, the practice is no longer valued. Ever since dry ice was commercialized, and with the advent of refrigeration for the production of ice cubes, the delicacy of shaved, flavored ice from glacier blocks harvested from tropical mountains no longer calm the thirst of travelers and urbanites in the Andes (Lyons, 2005).

Much less ink and film has been spent on the cloud forests than on glaciers, despite their important role in the terrestrial water cycle. TMCF ecosystems remain shrouded by mist, continuously capturing the horizontal precipitation through aerial structures of epiphytes covered with mosses (Werner et al., 2005) all of them yielding pure water (Fig. 8.3). Most of them, isolated from current urban areas or the agricultural matrix of valleys and hills, remain shrouded in the higher reaches of headwaters and ridges of isolated watersheds

FIGURE 8.3 Capture of horizontal precipitation by branching pattern and the epiphytic garden of the Andean forests in monospecific groves of *Yawal* (*Polylepis reticulata* Hieronymus) with the works of channels to deliver water for irrigation of agricultural lowlands in the *Imbabura* province. *Photo: César Cotacachi.*

by a veil of mystery, superstition, or myth (Kappelle and Brown, 2001). TMCFs which are closer to urban settlements or penetrated by roads have already disappeared, converted to tropical highland grasslands (*Páramos* and *Punas*) or to pastures with introduced grass species in the tropical lowlands of the piedmont throughout the Andean crescent, both on the Pacific and Amazonian drainages (Borsdorf and Stadel, 2015). Erosion and other contaminants have lowered the purity of water, thus bringing issues of quality to the narrative of payment for environmental services.

Although global conservation organizations, such as the International Union for the Conservation of Nature and the UNESCO International Hydrology Program, promote engagement with tropical mountains and water issues at the global level, little has been done to bring the message to those who, while lacking the formal title of ownership, are the actual owners of the tropandean landscapes (Sarmiento et al., 2015). Instead of global agendas, local indigenous people are developing their own narratives by formulating new water acts (such as in Ecuador and Bolivia) or by promoting a friendlier approach to water usage, including the spiritual dimension (such as in Peru and Chile) (Thompson, 2016).

The appropriation of significant sites throughout the region by indigenous communities has reinforced a "communal" sense of ownership based on the cultural identity values posited in them. Whether it be waterfalls, such as in the lands of *Shuar* of southern Ecuador, the branches with funerary casts of the *Awaruna* of northern Peru, or the seepage coves in the lands of the *Kogi* in northeastern Colombia, mountain waters belong to the common good (Campo and Turbay, 2015). Most of these sites are already cataloged as protected areas such as national parks or ecological reserves within the national systems of conservation. Some of them, such as in the Colombian Massif, are governed by the indigenous people themselves within the "Resguardos" and other "ancestral" lands, such as the reserve in the Carchi province of Ecuador to integrate the *Cofan* territory (Sarmiento et al., 2015). Indigenous cultures living in tropical mountains have become part of the amenities that most tourists and mountaineers enjoy, making tourism an important sector that could also be affected by glacier shrinking and overall climate change (Barros et al., 2015).

However, market actors often contest the indigenous claim on the resources. In particular, multinational corporations are grabbing land for the expansion of mining operations following the acquisition of exploitation rights in the highlands. Most of these places have proven reserves of gold, copper, tin, and silver. Moreover, in many Montane sites and *Páramos*, mining activities have run continuously since colonial times. Important parcels have been expropriated by the government for extractive activities in lands that were denuded by gold mining activities, such as in Zaruma, Intag, Nambija, and other sites in the Zamora-Chinchipe province of Ecuador. New technologies of hydropressure washing the ore have dramatically affected the sloping terrain and contaminated not only the watercourse of the rivers, but also lakes and aquifers. The protests and political agency of indigenous *Kichwa* groups comprising the *Ecuarunari* collective in the highlands or of the Confederacy of Indigenous Nationalities of Amazonia, have demonstrated their commitment to preserving these lands from exploitative mining and highlighted the conservation and sustainability "vocation" of those tropical Andean landscapes (Ortiz, 2016).

BIOCULTURAL LANDSCAPES OF MIST AND MAGIC

Evidence is emerging to indicate a long history of human use of these TMCF environments, although they are now largely uninhabited (Price, 2006). Butte top constructions in otherwise flat terrain, for instance, remain unexplained in the longitudinal expanse of the mountains in the Americas (Scheiber and Nieves-Zedeño, 2015). Whether on the slopes of the active volcano Arenal or the awakened Irazú and Turrialba volcanoes of Costa Rica, the rounded stones, carved walls, and steles remind visitors of past cultural manipulations of the landscape in what has otherwise been considered pristine natural ecosystems. The National Institute for Biodiversity of the country has overemphasized the notion of "nature" in all its bioprospecting dissemination work (Mulligan, 1999) and is now no longer in existence (Fonseca, 2015). The branding of the whole of Costa Rica as a "natural paradise" fails to recognize the indigenous people who still occupy landscapes in the high frontier of the Talamanca range. We see the same trend in neighboring Panama, where one of the most important research centers for tropical science is headquartered. The emphasis of the Smithsonian Tropical Research Institute has historically been directed toward the discovery of flora and fauna of the lowland tropical rain forests and has only added the ethnoecology of the lowland peoples.

Most people identify the *Páramos* of the Sierra de Mérida in Venezuela as textbook examples of nature in the Andes; however, a close examination of the presence/absence of anthropic species in the hummock and the hidden evidence of burning and grazing makes them good examples of anthropogenic landscapes (Sarmiento and Frolich, 2002). In Colombia, within the *Tairona* National Park, hidden in the greenery of the cliffs, knolls, and cliffs remains "the lost city" of the *Tairona*, exhibiting a complex network of ditches and drainages to manage the hyper humid environment. Archaeological excavations toward and around these secluded TMCFs indicate an intricate network of trails, staircases, or round basements that appear to have been rather common centuries ago (Uribe and Osorio, 2001). Water capture for spiritual meaning or *aluna* was critical for indigenous groups living on the arid Guajira peninsula (Rodríguez-Navarro, 2000; Campo and Turbay, 2015) Only shamans or *mamas* can penetrate the veil of mystery in the elfin forest and the glaciated summit of the Sierra Nevada de Santa Marta to achieve a better sense of direction for the entire *Kogi* group. Coming down with ice chunks was a ritual to secure good omens (Ereira, 1993).

The use of sacred water from the mountain is readily observed in the *Imbakucha* watershed of Northern Ecuador, with the Otavalo ethnic group (*Atawallu runakuna*) actively engaged in maintaining their sacred sites to fuel their spirituality to the *Pachamama* and the telluric guardians or *Apukuna*, that surround the lake at the valley bottom (Sarmiento et al., 2008). The isolated higher reaches of the *Imbabura* volcano, the putative father of the "Otavaleños" retain bunchgrass páramo, small lake, and elfin forests dominating the summit, clogged with mosses, tank bromeliads, and spiny shrubs that capture the horizontal precipitation. The wisemen or the medicine-man or woman—*Yachag*—roam the terrain looking for specific plants or other paraphernalia for divination and cleansing, many of them converging on the town of *Illuman*, the putative capital of alternative, traditional medicine in the country. Once a year, the group of adolescent *Atawallu runakuna* converge on the sacred waterfall of *Piguchi* at the outlet of the *Imbakucha* lake—formerly known as Lago de San Pablo—to receive the ritualized initiation to adulthood during the summer solstice, in a generalized celebration of sun, harvest and awakening (*Inti Raymi*). The strength and cold temperature of the waterfall flow—*churru*—invigorates the young minds and bodies and directs them toward becoming better members of the group (Fig. 8.4). Further upstream, the wetlands along the valley bottom supports the cultivation of reeds—*tutura*—needed for the construction of a variety of household items, including boats, mats, baskets, fans, and so on (Burneo and Gunkel, 2003). This Andean practice of cultivating the shoreline has profound spiritual meaning, as workings of the tillers and other handcrafts with the reed are

FIGURE 8.4 The sacred waterfall of the *Atawallu runakuna*. Located following the brook at the outlet of the *Imbakucha* lake, the *Piguchi* waterfall remains as the archetype of cleansing rituals and divination, including the *churru* where the initiation ritual for young men during the summer solstice. *Photo: César Cotacachi.*

considered to bring the blessings of *Katekil*, the God of the calm waters, into the households of shoreline communities (Ninaquispe-Romero et al., 2012), or to those living directly on lake Titicaca, such as the *Uru* in Peru (Engelbert, 2001).

Even in the heart of one of the most impenetrable areas of the Andean crescent, between the Cotopaxi and the Tungurahua volcanoes in Ecuador, the *Llanganati* National Park registers a serendipitous discovery of monumental architecture with megalithic wall constructions showing capture and channeling of the horizontal precipitation for downstream use (Rostain et al., 2014) debunking once and for all the myth of the pristine Andes. Just as serendipitously, the discovery of the ruins of *Machu Picchu* in Peru by Hiram Bingham in 1911 has led, after a century of debate, to the formal recognition of the site as a National Sanctuary of Peru, the most widely recognized sacred site of the world, according to the internet survey for the new "marvels of the ancient world" (Reinhardt, 2007; Silverman, 2014). With the discovery of the walls at

"El Topo" headwaters in the *Llanganati* by a local explorer in 2014, must we wait another century to assimilate the full meaning of this sacred site in Ecuador (Sarmiento and Viteri, 2015)?

Many more sites throughout Peru and Bolivia affirm the presence of humans in the waterlogged but verdant TMCFs. An entire culture of Peru, called the *Chachapuya*,"the people of the clouds" has been described as masters of terracing, walling, zoomorphic decorating, and ceremonial burrowing in cliffs to preserve mummified ancestors (Guengerich, 2015; Schjellerup, 2015). Therein, shrouded with mist and fog, the livelihood of this extinct group, which predates the *Inka* presence in the warm hanging valleys of the cis-Andean slopes or *Yunga*, remains enigmatic (Epstein and Toyne, 2016). Clearer examples are found around other sacred sites in Peru, such as the "Gran Pajatén" in Río Abiseo, the *Chuki Kiraru*, the *Wakaywillka*, the *Urupampa*, and the sacred valley of the *Inka* in *Pisaq*, the *Puyupatamarka*, the *Wiñay Wayna*, the *Llaktapata*, the *Paltay* sites, and a plethora of other sites marked by water management (Reinhardt, 2007). In all these sites, special designs were employed to capture and use spring water for baths and semi enclosures, as is still the case in the brook of *Kundur Machay* near to the fortress of *Saksawaman* in the *Kushku* area with seepage coves as special sites for divination, cleansing ceremonies, or reservoirs.

Many stone constructions with royal design along the *Kapak Ñan*—or the Great Inca Road—take the designation of "royal baths" and are associated with *tampukuna* (rest areas) and *chullpakuna* (sepultures) buildings to symbolize the capture and usage of water for quick resting along the trip or to provide *tulpas* (ancestors' soul) respite for eternity. These structures are found throughout the mountain landscape from southern Colombia along the extent of the magnificent mountain road toward northern Chile and Argentina. In Ecuador, near the town of *Sicchu*, on the western slope of the Andes, the discovery of "baths of *Atawallpa*" brings the existence of major archaeological sites toward the trans-Andean domain, far away from the ruins of "Paredones" in the El Cajas National Park and the Cajas Massif Biosphere Reserve, once thought to be the western limit of *Inka* expansion toward the Pacific lowlands.

A final example of sacred sites driven by mountain water is given with the Argentinean settlement of *Killmi* or "Quilmes" near to the *Kallchaki* valleys. Archaeological studies in Argentina at the end of the XIX century brought news of a fascinating collection of ruined terraces or "andenes," walled constructions, reservoirs, and holes in the ground, and an expansive plaza or *kancha*, lost amidst the drier vegetation (Ambrosetti, 1897). At present, the ruins are now considered officially as the Sacred City of Quilmes (Sarmiento, 2013). Instead of having an open running water canal, such as the oldest canal in the Americas in the *Zaña* Valley of northern Peru, the Quilmes or *Diaguita* in Northwest Argentina kept paved underground reservoirs or *Puyus* and tendered irrigated terraces for cultivation, similar to those discovered for the *Nazca* and *Wari* cultures in the Lomas de *Antikipa*, Peru (Zaro, 2014).

THE ROLE OF THE SACRED IN CONSERVATION

In contrast to the value of provisioning services to thirsty capital cities, the growing agro industries, or even mountain mining, the greatest value of water in tropical Andean landscapes may instead be in the cultural ecosystem services embodied by the sacred geographies of the TMCFs integrating the triple reciprocity of the mental essence of place or Andeaness, biodiverse water resource base or Andeanity, and spiritual dimension or Andeanitude (Sarmiento, 2015). The hidden meanings of glacial and cloud water for the people of these areas create an overlooked but powerful impetus for conservation of cultural landscapes, an often intangible and incommensurable environmental service that could be the trifecta of sustainable mountain development in the Andes.

Throughout these places, the presence of water as an element of spiritual power has conferred a special value in the conservation toolbox. For example, in establishing prospective reservoirs for future consumption, the selection of lakes or coves to build the dams ought to avoid sacred sites. The discourse of environmental planners and politicians ought to include the nonmaterial heritage of water as well as the material importance for potable usage and for agricultural or industrial uses of water, particularly around protected areas or indigenous territories. Recognizing the community value of the spiritual dimension, locals could gather in habitual rituals, popularizing pilgrimages, or study trips to access the local knowledge on the TMCFs and the mountain waters (Sarmiento and Hitchner 2016).

The strong valuation of the spiritual dimension of water by local people is incorporated in some Andean countries, such as the National Sanctuary of Machu Picchu in Peru, the Cayapas-Mataje Reserve and the Sangay National Park in Ecuador, or the Sierra Nevada de Santa Marta in Colombia. However, many of other officially declared protected areas also have sacred sites within them. To manage them, Wild and McLeod (2008) recognized six principles as follows: (1) to identify and inventory those sites, (2) to integrate them into the planning process and management programs, (3) to promote stakeholder consent, participation, inclusion, and collaboration, (4) to encourage improved knowledge and understanding, (5) to protect the sacred good while providing access and use, and (6) to respect the rights of the custodians with an appropriate framework of national policy. Nevertheless, there are no specific guidelines for managing mountain sacred waters (Sarmiento et al., 2015).

Therefore, as an attempt to incorporate mountain waters in the planning of biocultural heritage conservation, the author applied trifecta framework to use the Sarmineto trilemma of Andean identity in Ecuador (see Table 8.1). He used *Andeanitude* (or the mountain's spiritual value as perceived by local people) as framework for a unique restructuring of the Ecuadorian national system to include six new categories, such as memory landscape, sacred site, spiritual park, religious reserve, cultural monument, and ancestral landscape;

TABLE 8.1 A Summary Framework to Integrate the Sarmiento's Trilemma of Andean Identity With the New Biocultural Heritage Landscapes to Be Protected in Ecuador. A Description of the Three Categories (Fundamental Ecuador, Constructed Ecuador, and Deep Ecuador) Explain the Application of the Trifecta for Sustainable Mountain Development in the Andes

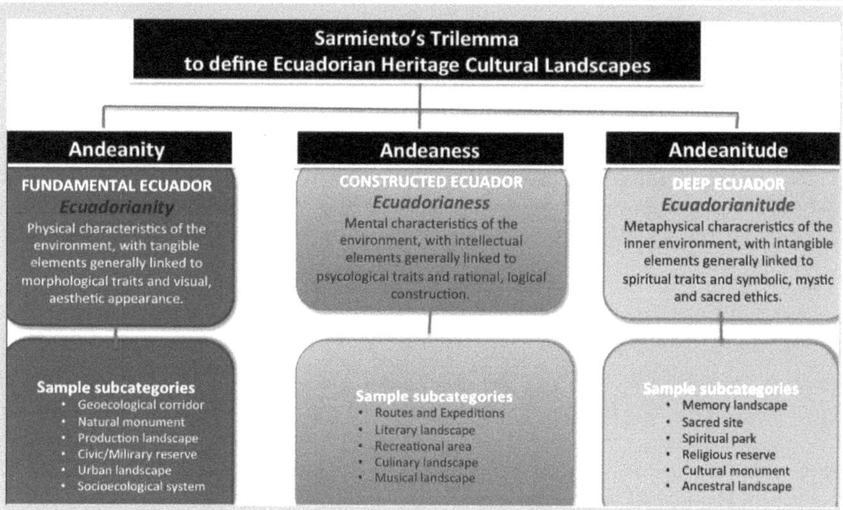

at least four of the six categories are amenable to incorporate the sacred mountain water (Sarmiento and Viteri, 2015). Five new categories of *Andeaness* (or the essential value as the mountain's social construct accruing to local people) were suggested, including routes and expeditions, literary landscape, recreational area, culinary landscape, and musical landscape; at least three of the five categories are amenable to include mountain waters. Finally, *Andeanity* (or the material content of biodiversity and physical attributes of the mountain) was listed to protect mountain waters in six new categories including geoecological corridor, natural monument, production landscape, civic military reserve, urban landscape or heritage city, and socioecological unit; all five categories are amenable to deal with protecting the water resource in the mountains.

CONCLUSION

The recovery of ancestral practices of respect toward water and the role of the TMCF belt as the actual catcher of the horizontal precipitation could contribute strongly to future sustainable mountain development. The anthropogenic *Páramos* and high Andean forests are as important as the tropical glaciers in providing environmental services of capture, sequestration, and delivery of water toward the lowlands. The *Páramos*, while misconstrued as natural ecosystems, represent a complex coupled nature—society hybrid of the orobiome. Andean communities include the sacred dimension in their construed mountainscape. To deal effectively with the water supply for thirsty cityscapes and farmscapes of the Andes, climate change adaptation and mitigation policies and programs should thus look beyond commodification to consider the sacred geographies of water: the wet scenic beauty, the protection of origin myths, the respect for glacial and lake water infused with the numen, and the veneration of sacred waters. Only by assuming the importance of the spiritual dimension, the necessity of community participation and the material importance of the TMCFs will make the water cycle be fully protected in the tropical Andes.

ACKNOWLEDGMENT

Much of the emphasis on Tropandean cloud forest conservation reflects the mentoring of Prof. Larry Hamilton. Funding for field work and graduate assistance came from the Neotropical Montology initiative of the Latin American and Caribbean Studies Institute's NRC Title VI (ED/NRC/P015A0046) grant and the Willson Center Research Cluster on Indigenous Foods and Fibers. I thank Sarah Hitchner for scholarly support on sacred natural sites, and Cesar Cotacachi for the use of his photographs of the sacred Imbakucha watershed.

REFERENCES

Allison, E.A., 2015. The spiritual significance of glaciers in an age of climate change. Wiley Interdisciplinary Reviews: Climate Change 6 (5), 493—508.

Ambrosetti, J.B., 1897. La antigua ciudad de Quilmes (Valle Calchaque). "La Buenos Aires" Imprenta, Papelería y Encuadernación.

Balcázar, W., Rondón, J., Rengifo, M., Ball, M.M., Melfo, A., Gómez, W., Yarzábal, L.A., 2015. Bioprospecting glacial ice for plant growth promoting bacteria. Microbiological Research 177, 1—7.

Baraer, M., Mark, B.G., McKenzie, J.M., Condom, T., Bury, J., Huh, K.I., Portocarrero, C., Gómez, J., Rathay, S., 2012. Glacier recession and water resources in Peru's Cordillera Blanca. Journal of Glaciology 58 (207), 134—150.

Bermúdez-Torres, K., Ferval, M., Legal, L., 2015. Lupinus species in Central Mexico in the Era of Climate Change: adaptation, migration, or extinction?. In: Climate Change Impacts on High-Altitude Ecosystems. Springer International Publishing, pp. 215—228.

Barros, A., Monz, C., Pickering, C., 2015. Is tourism damaging ecosystems in the Andes? Current knowledge and an agenda for future research. Ambio 44 (2), 82—98.

Bolin, I., 2010. Rituals of Respect: The Secret of Survival in the High Peruvian Andes. University of Texas Press.

Borsdorf, A., Stadel, Ch, 2015. The Andes: A Geographical Portrait. Springer.

Burneo, P.C., Gunkel, G., 2003. Ecology of a high Andean stream, Rio Itambi, Otavalo, Ecuador. Limnologica — Ecology and Management of Inland Waters 33 (1), 29—43.

Bruijnzel, L., Hamilton, L.S., Scatena, F.N. (Eds.), 2010. Montane Tropical Cloud Forests. Science for Conservation and Management. Cambridge University Press.

Bubb, P., Lysenko, I., Miles, L., Newton, A., 2003. Tropical montane cloud forests. CBD Technical Series No. 8, p. 70.

Campo, A.R.R., Turbay, S., 2015. The silence of the Kogi in front of tourists. Annals of Tourism Research 52, 44–59.

Carey, M., 2010. In the Shadow of Melting Glaciers: Climate Change and Andean Society. Oxford University Press.

Carey, M., Baraer, M., Mark, B.G., French, A., Bury, J., Young, K., McKenzie, J., 2014. Toward hydro-social modeling: merging human variables and the social sciences with climate-glacier runoff models. Journal of Hydrology 518 (Part A), 60–70.

Churchill, S., Balslev, H., Forero, E., Luteyn, J. (Eds.), 1995. Biodiversity and Conservation of Neotropical Montane Forests. The New York Botanical Garden. Bronx, NY.

CONDESAN, 2003. The andean vision of water. In: Proceedings of the Consultative Work for the World Vision of Water Meeting in The Hague. http://www.condesan.org/memoria/agua/AndeanVisionWater.pdf.

Dunbar, K.W., Marcos, K.D.M., 2012. Singing for shaved ice: glacial loss and raspadilla in the Peruvian Andes. In: Consumer Culture in Latin America. Palgrave Macmillan, US, pp. 195–205.

Engelbert, V., 2001. The end of the Uros. The World & I 16 (5), p. 156.

Epstein, L., Toyne, J.M., 2016. When Space is limited: a spatial exploration of pre-hispanic Chachapoya Mortuary and ritual microlandscape. In: Theoretical Approaches to Analysis and Interpretation of Commingled Human Remains. Springer International Publishing, pp. 97–124.

Ereira, A., 1993. The Elder Brothers: A Lost South American People and Their Wisdom.

Fernández, A., Mark, B.G., 2016. Modeling modern glacier response to climate changes along the Andes Cordillera: a multi-scale review. Journal of Advances in Modeling Earth Systems 8, 467–495.

Fonseca, P., 2015. A major center of biodiversity research crumbles. Scientific American – The Sciences Section. http://www.scientificamerican.com/article/a-major-center-of-biodiversity-research-crumbles/.

Fonstad, M., 2013. Geographies of water. Annals of the Association of American Geographers 103 (2), 251–252.

Francou, B., Vuille, M., Favier, V., caceres, B., 2004. New evidence for an ENSO impact on low-latitude glaciers: Antizana 15, Andes of Ecuador, 0°28'S. Journal of Geophysical Research 109, D18106.

Frenierre, L., David, J., 2014. Assessing the Hydrologic Implications of Glacier Recession and the Potential for Water Resources Vulnerability at Volcán Chimborazo, Ecuador (Doctoral dissertation). The Ohio State University.

Gade, D.W., 2016. Mysterious Ucumari: the Andean Bear in nature and culture. In: Spell of the Urubamba. Springer International Publishing, pp. 217–238.

García-Romero, A., Alanís-anaya, R.M., Muñoz-Jiménez, J., 2015. Environmental factors that affect primary plant succession trajectories on lahars (Popocatépetl Volcano, Mexico). Journal of Mountain Science 12 (5), 1254–1266.

Guayasamín, I., Guayasamín, G., 1980. Los hieleros del Chimborazo. Video Producción Ecuatoriana.

Guengerich, A., 2015. Settlement organization and architecture in late intermediate period Chachapoyas, Northeastern Peru. Latin American Antiquity 26 (3), 362–381.

Hamilton, L., Juvik, J., Scatena, F. (Eds.), 1993. Tropical Montane Cloud Forests. The East West Center, Honolulu.

Hoffman, D., Requena, C., 2013. Bolivia in a World Four Degrees Warmer: Sociopolitical Scenarios Under Climate Change for the Northern Altiplano in the Years 2030 and 2060. Instituto Boliviano de Montaña y Fundación PIEB, La Paz.

Kappelle, M., Brown, A. (Eds.), 2001. Bosques Nublados del Neotrópico. Instituto Nacional de Biodiversidad (INBIO), San José, Costa Rica.

Kaijser, A., 2013. White ponchos dripping away? Glacier narratives in Bolivian climate change discourse. Interpretive approaches to global climate governance: Deconstructing the greenhouse, p. 266.

Lázaro, R.P., 2016. La Nueva Cultura del Agua, el camino hacia una gestión sostenible. Causas e impactos de la crisis global del agua. Cuadernos de Trabajo Hegoa (68).

Lyons, B.J., 2005. Discipline and the arts of domination: rituals of respect in Chimborazo, Ecuador. Cultural Anthropology 20 (1), 97–127.

Martin, P.H., Bellingham, P.J. (Eds.), 2016. The Ecology of Tropical Montane Cloud Forests: A Global Synthesis. Journal of Tropical Ecology 32 (5), 345–354

Mulligan, S.P., 1999. For whose benefit? Limits to sharing in the bioprospecting 'regime'. Environmental Politics 8 (4), 35–65.

Ninaquispe-Romero, L., Weeks, S., Huelman, P., 2012. Totora: a sustainable insulation material for the andean parts of Peru. In: Pasive and Low Energy Architecture PLEA 2012.

Orlove, B.S., Wiegandt, E., Luckman, B.H., 2008. Darkening Peaks: Glacier Retreat, Science, and Society. Univ of California Press.

Ortiz, S., 2016. Marcha por el agua, la vida y la dignidad de los pueblos. Letras Verdes. Revista Latinoamericana de Estudios Socioambientales (19), 45–66.

Price, M.E., 2006. Global Change in Mountain Regions. Sapiens Publishing, Duncow, UK.

Rabatel, A., Francou, B., Soruco, A., Gomez, J., Cáceres, B., Ceballos, J.L., Basantes, R., Vuille, M., Sicart, J.E., Huggel, C., Scheel, M., 2013. Current state of glaciers in the tropical Andes: a multi-century perspective on glacier evolution and climate change. The Cryosphere 7 (1), 81–102.

Raoul, K., 2015. Can glacial retreat lead to migration? A critical discussion of the impact of glacier shrinkage upon population mobility in the Bolivian Andes. Population and Environment 36 (4), 480–496.

Reinhardt, J., 1985. Sacred mountains: an ethnoarchaeological study of high Andean ruins. Mountain Research and Development 5, 299–317.

Reinhardt, J., 2007. Machu Picchu: Exploring an Ancient Sacred Center, fourth ed. University of California, Los Angeles.

Rhoades, R.E., 1998. Participatory Watershed Research and Management: Where the Shadow Falls. IIED.

Rhoades, R.E., Zapata Ríos, X., Aragundy Ochoa, J., 2016. Mama Cotacachi: history, local perceptions, and social impacts of climate change and glacier retreat in the Ecuadorian Andes. In: Orlove, B., Wiegandt, E., Luckman, B.H. (Eds.), Darkening Peaks: Glacier Retreat, Science, and Society, pp. 216–225.

Rodríguez-Navarro, G.E., 2000. Indigenous knowledge as an innovative contribution to the sustainable development of the Sierra Nevada of Santa Marta, Colombia: the elder brothers, guardians of the "Heart of the World". AMBIO: A Journal of the Human Environment 29 (7), 455–458.

Rostain, S., de Saulieu, G., Lézy, E., 2014. El alto Pastaza precolombino en el Ecuador: del mito a la arqueología. In: Rostain, S. (Ed.), Amazonía. Memorias de las conferencias magistrales del 3 Encuentro Internacional de Arqueología Amazónica., Mccth/Senescyt/3eiaa, pp. 159–185. Quito.

Salas Carreño, G., 2010. Acerca de la antigua importancia de las comparsas de wayri chu'nchu y su contemporánea marginalidad en la peregrinación de Quyllurit'i. Anthropologica 28 (28).

Sarmiento, F.O., 2015. On the antlers of a trilemma: rediscovering andean sacred sites (Chapter 5). In: Rozzi, R., Pickett, S.T.A., Callicot, J.B., Chapin III, F.S.T., Power, M.E., Armesto, J.J. (Eds.), Earth Stewardship: Linking Ecology and Ethics in Theory and Practice. Springer, New York.

Sarmiento, F.O., 2013. Ruinas Reificadas: El revivir indígena, los paisajes culturales y la conservación de sitios sagrados. Revista Parques (1), 1–14.

Sarmiento, F., 2012. Contesting Páramo: Critical Biogeography of the Northern Andean Highlands. Kona Publishers, Higher Education Division, Charlotte, NC.

Sarmiento, F., Hitchner, S., 2016. Indigenous Revival and Sacred Sites Conservation in the Americas. Berhghan Books, Brooklyn, NY (in press).

Sarmiento, F.O., Bernbaum, E., Lennon, J., Brown, J., 2015. Managing cultural heritage. In: Worboys, G., et al. (Eds.), UCN's Global Guide for Protected Area Management. Earthscan, Sydney, Australia (E-Book).

Sarmiento, F., Viteri, X., 2015. Discursive heritage: sustaining andean cultural landscapes amidst environmental change. In: Taylor, K., Mitchell, N., StClair, A. (Eds.), Cultural Landscapes. Routledge.

Sarmiento, F.O., Russo, R., Gordon, B., 2013. Tropical mountains multifunctionality: dendritic appropriation of rurality or rhyzomic community resilience as food security Panacea. In: Pillarisetti, J.R., Lawrey, R., Ahmad, A. (Eds.), Multifunctional Agriculture, Ecology and Food Security: International Perspectives. Nova Science Publishers, New York, pp. 55–66.

Sarmiento, F.O., Cotacachi, C., Carter, L.E., 2008. Sacred Imbakucha: intangibles in the conservation of cultural landscapes in Ecuador. In: Mallarach, J.M. (Ed.), Cultural and Spiritual Values of Protected Landscapes, Protected Landscapes and Seascapes. IUCN and GTZ, vol. 2. Kaspareg Verlag, pp. 125–144.

Sarmiento, F.O., Frolich, L.M., 2002. Andean cloud forest tree lines: Naturalness, agriculture and the human dimension. Mountain Research and Development 22 (3), 278–287.

Scheiber, L.L., Nieves-Zedeño, M. (Eds.), 2015. Engineering Mountain Landscapes: An Anthropology of Social Investment. University of Utah Press.

Schjellerup, I., 2015. Over the Mountains, down into the Ceja de Selva Inka strategies and impacts in the Chachapoyas Region. The Inka Empire: A Multidisciplinary Approach 307.

Silverman, H., 2014. "New 7 Wonders of the World" Campaign. In: Encyclopedia of Global Archaeology. Springer, New York, pp. 5254–5255.

Taillant, J.D., 2015. Glaciers: The Politics of Ice. Oxford University Press.

Thompson, J.A., 2016. Intersectionality and water: how social relations intersect with ecological difference. Gender, Place and Culture 23 (4), 1–16.

Tuaza, L.A., 2007. Baltazar Ushca: el último hielero de Chimborazo. Revista ICONOS 173–176.

Uribe, M.V., Osorio, Á., 2001. Ciudad perdida un paisaje cultural en la Sierra Nevada de Santa Marta en Colombia. JANGWA PANA 1 (1).

Webb, H.S., 2012. Yanantin and Masintin in the Andean World: Complementary Dualism in Modern Peru. UNM Press.

Werner, F.A., Homeier, J., Gradstein, S.R., 2005. Diversity of vascular epiphytes on isolated remnant trees in the montane forest belt of southern Ecuador. Ecotropica 11, 21–40.

Wild, R., McLeod, C. (Eds.), 2008. Sacred Natural Sites: Guidelines for Protected Area Managers. IUCN, Gland, Switzerland.

Zaro, G., 2014. From terraces to trees: ancient and historical landscape changes in southern Peru. In: Thomas, H., Widgren, M. (Eds.), Landesque Capital: The Historical Ecology of Enduring Landscape Modifications. Left Coast Press, Walnut Creek, pp. 232–250.

Chapter 9

What Future for Mountain Glaciers? Insights and Implications From Long-Term Monitoring in the Austrian Alps

A. Fischer,*[,1] K. Helfricht,* H. Wiesenegger,[§] L. Hartl,* B. Seiser* and M. Stocker-Waldhuber*
*Institute for Interdisciplinary Mountain Research, Innsbruck, Austria; [§]Hydrographical Survey of the Federal Government of Salzburg, Salzburg, Austria
[1]Corresponding author: E-mail: andrea.fischer@oeaw.ac.at

Chapter Outline

Mountain Ice and Water. http://dx.doi.org/10.1016/B978-0-444-63787-1.00009-3

INTRODUCTION

This chapter presents the Austrian monitoring network in its historical context, together with results and challenges. The development of the Austrian monitoring network is a good example of historical contingency: Some mass balance time series are based on long-term scientific interest, while others were initiated as part of studies on waterpower. In length change monitoring, long-term series were often initiated and carried on by local people, and therefore personal commitment is part of today's program. Although not every site might be an optimum choice in the light of current global interests, existing long time series will be carried on if national activities and resources allow. A discussion of the background and development of the network should help to design a most effective monitoring network in other mountain regions. The high data density in the Austrian Alps also allows a discussion of the a priori representativeness of the monitoring sites. Based on the results of the monitoring network, basic outlines of future developments can be drawn.

Mountain glaciers are indicators of climate change, and are part of the local hydrological system and landscape. It is important to monitor changing glaciers to tackle changes in climate and local hydrology. Global climate change leads to a melt down of land ice and contributes to rising sea levels (Vaughan et al., 2013). The fate of disappearing mountain glaciers moves people, as the so-called eternal ice at the highest peaks is retreating at a historically unprecedented level (Zemp et al., 2015). Melting Alpine glaciers (Fig. 9.1) are not only a symbol of global warming, but also have implications for local hydrology and the economy in the populated mountain regions.

Glacier monitoring combines small-scale in situ measurements on hopefully representative glaciers and large-scale remote sensing methods. Haeberli et al. (2013a,b) trace the development of intentions and strategies of global glacier monitoring from early beginnings until today. Their modern integrated approach of global glacier monitoring links them into the framework of global climate observation strategies as Global Terrestrial/Climate Observing system and the concept of essential climate variables (Bojinski et al., 2014) They connect detailed local studies on individual glaciers with coverage of large glacier samples (Zemp et al., 2014). Mass balance and ice flow studies in the major climate zones improve process understanding and serve model calibration. In addition to that, a limited number of point measurements are combined with regional decadal mappings and long-term observations of glacier length changes to reflect regional patterns and dynamic responses. The World Glacier Monitoring Service is the international institution most engaged in the selection, compilation, and stimulation of national glacier monitoring activities to meet the needs of the international scientific communities. At national or local level, additional demands and restrictions shape the monitoring networks. Most local monitoring networks have not so much been designed with a detailed long-term national plan, but rather developed according to changing historical interests and needs.

FIGURE 9.1 Jamtal Ferner in Silvretta in 1929 (left, Archive of the Austrian Alpine Club) and 2010.

In past decades, human activities and infrastructure have edged closer to the glaciers. After the explorers conquered the highest summits, glacier tourism started in the mid- to late 19th century with glacier tours, horse-drawn sledges and mountain huts. Today, apart from mountaineering tourism, 8 glacier ski resorts are operating on 15 glaciers (Fischer et al., 2016), and glacier runoff is used in several hydropower reservoirs. Negative mass balances lead to a decrease in glacier area and altitude and necessitate ongoing adaptation of the infrastructure (Fischer et al., 2011).

In the Austrian Alps, 921 glaciers currently cover 415.47 km^2, that is 0.5% of the total Austrian territory (Fischer et al., 2015a). Glacier retreat affects local hydrology by changing seasonal and magnitude of glacier runoff, here future projections show great uncertainties (Huss et al., 2014). For the Austrian Alps, Kuhn and Batlogg (1998) present scenarios for 20 catchment areas in a warming climate. In essence, they found a sequence of three scenarios, which are likely to occur in future: (1) higher total runoff during increased ice melt, (2) an earlier runoff peak from seasonal snow with a second peak from ice melt, and (3) in the absence of significant glaciers, a runoff peak in summer, when precipitation becomes more important than snow and glacier melt.

Adaptation to future climate change, as exemplified by the above runoff scenarios, means tackling complex system changes based on the comparison of past system states, nowcasting and future scenarios. Capturing the system states and processes and designing robust models needs an accurate, homogenous, and reliable database. In the case of climate change, long time series are important to distinguish short-term weather extremes from long-term climate change effects, always bearing in mind that it will never be possible to prevent damages from small-scale weather extremes in Alpine environments, but it should be possible for future "climatic norm" conditions.

Climate and glaciers records start quite early for Scandinavia and the Alps. In Austria, records of air temperature start in 1760s, of precipitation in 1800s, and of sunshine duration in the 1880s (Auer et al., 2007). The early data thus cover the end of the Little Ice Age (LIA), when mountain glaciers advanced to a maximum state, as documented by the respective moraines. This makes these regions important sites for research on basic processes during glacier advance and retreat, although the main ice volume changes determining sea-level change in the next decades might take place in the Arctic (Vaughan et al., 2013).

The oldest picture of glaciers in the Austrian Alps dates from 1601, when the outburst of a lake dammed by the advancing Vernagt Ferner in the Ötztal Alps caused downstream devastation (Nicolussi, 2013). Even older written sources describe the glacier state at passes that were important for travel and trade. The first map showing a glacier (Groβferner) was published by Ygl in 1605 (Brunner, 2002). The Atlas Tyrolensis of Anich and Hueber (1774) shows basic glacier outlines, but no details (Fig. 9.2). In the early 19th century, the Napoleonic wars motivated the Austrian emperors to initiate mapping of the entire Austro-Hungarian Empire. The first map was compiled in 1801–05 and remained incomplete, while the second map (1816–21) covered the entire Austro-Hungarian territory. The maps show glacier outlines and a basic topography, but the representation of the specific glaciers varies in detail and accuracy. Mid-19th century maps of some glacier covered mountain groups were compiled from about 1880 onwards also displaying contour lines of altitude (Brunner, 2006). The Schlagintweit brothers (Schlagintweit and Schlagintweit, 1850) published the results of most fundamental scientific investigations in the Alps. Sonklar published early descriptions and detailed maps of the Ötztal Alps (Sonklar, 1857, 1860, 1877), of the Hohe Tauern, and the Zillertal Alps in 1877 (Kretschmer, 2002). Simony investigated Hallstätter Gletscher (Brunner, 2004) and the Ötztal group (Simony, 1863). Keil compiled maps of the ranges of Ankogel Group (1865), Glockner Group (1860), and Groβvenediger Group (1866; Müller, 1973). The Zillertal Group was mapped by Diener (1885). Richter (e.g., 1885, 1891, 1894) was amongt the most productive scientists compiling glacier data in the late 19th century. The third federal map (1869–87) displayed glacier area and elevation in much more detail and accuracy than the first and the second mappings.

Once Alpine tourism started, mountain clubs, scientific societies, and institutions set up climate and glacier monitoring programs. The glacier survey of the Austrian Alpine Club started with length monitoring in 1891. Mass balance programs started in 1952 and the first glacier inventory was compiled in 1969 (Patzelt, 1980; Groß, 1987).

Together with records of ice flow velocity and ice volume, the glaciers of the Austrian Alps (Fig. 9.3) today are quite well observed, allowing insights in short- to mid-term perspectives of current glacier shrinkage and its implications on local hydrology and economy. Time series of glacier inventories reveal the changes in glacier area and volume, allowing the calculation of geodetic mass balance. Today's monitoring network covers length changes, mass balance, area and volume changes, as well as glacier runoff. Each of

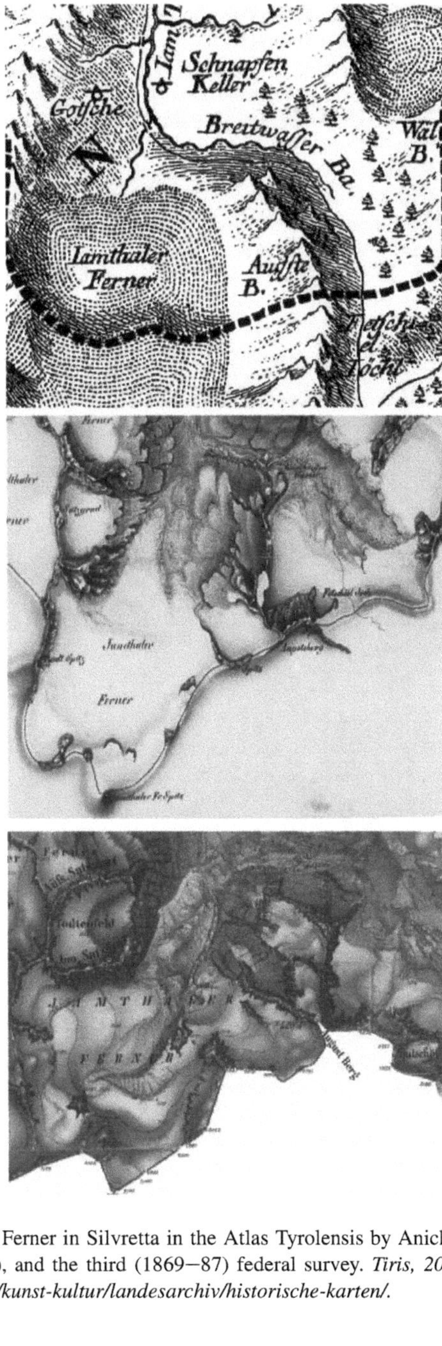

FIGURE 9.2 Jamtal Ferner in Silvretta in the Atlas Tyrolensis by Anich and Hueber (1774), in the second (1816–21), and the third (1869–87) federal survey. *Tiris, 2016. Historische Karten. https://www.tirol.gv.at/kunst-kultur/landesarchiv/historische-karten/.*

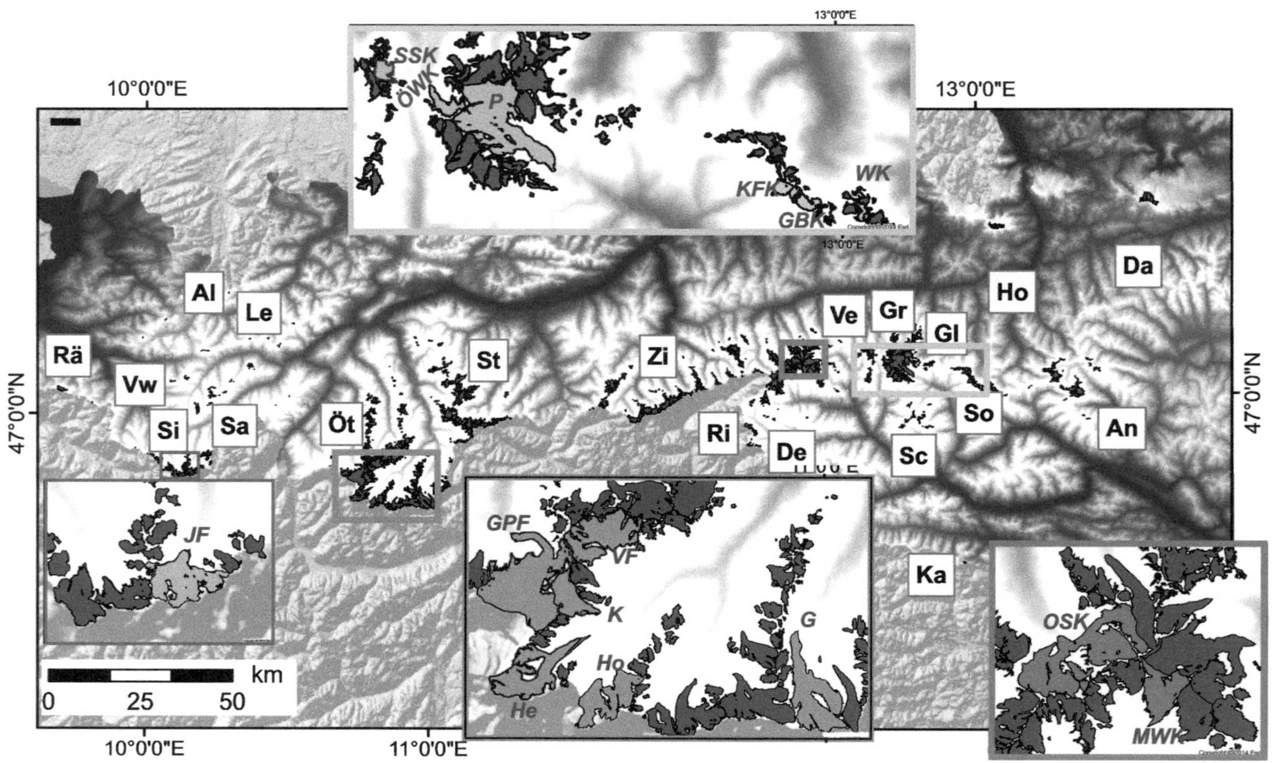

FIGURE 9.3 Austrian glaciers as recorded in the latest glacier inventory, with mountain ranges. *Al*, Allgäu Alps; *An*, Ankogel Group; *Da*, Dachstein; *De*, Defregger Alps; *Gl*, Glockner Group; *Gr*, Granatspitz Group; *Ho*, Hochkönig; *Ka*, Karnic Alps; *Le*, Lechtal Alps; *Öt*, Ötztal Alps; *Rä*, Rätikon; *Ri*, Rieserferner Group; *Sa*, Samnaun; *Sc*, Schober Group; *Si*, Silvretta; *So*, Sonnblick Group; *St*, Stubai Alps; *Ve*, Venediger Group; *Vw*, Verwall; *Zi*, Zillertal Alps. Insets: Specific regions with glacier names: *G*, Gurgler Ferner; *GBK*, Goldberg Kees; *GPF*, Gepatsch Ferner; *Ho*, Hochjoch Ferner; *JF*, Jamtal Ferner; *K*, Kesselwand Ferner; *KFK*, Kleines Fleiβ Kees; *MWK*, Mullwitz Kees; *OSK*, Obersulzbach Kees; *ÖWK*, Ödenwinkel Kees; *SSK*, Stubacher Sonnblick Kees; *WK*, Wurten Kees.

these parameters captures different aspects of glacial recession and their effect on hydrology in mountain regions. The number of observations varies for the specific parameters, as do effort and costs.

A BRIEF OVERVIEW OF METHODS

Glaciers react to climate change. With climate, mass balance, ice flow velocities, length, area, surface elevation, and surface properties change. Glacier mass balance, that is, the sum of mass gain (accumulation) and mass loss (ablation), is most directly related to the energy exchange with the atmosphere. Major drivers of mass balance of Alpine glaciers are radiation, air temperatures, and snowfall during summer, and accumulation of snow during winter. Positive or negative mass balances change glacier flow velocity, length, area, and volume. Length change is the result of mass balance (i.e., net mass loss or mass gain) and flow velocity changes. An increase in glacier length can occur not only when the glacier gains mass, but also when the ice flow velocity increases, as in a glacier surge.

At a single point of the glacier, thickness changes result from surface mass balance and flow dynamics (Cogley et al., 2011). The velocity vector points into the glacier in the accumulation area, and outwards in the ablation area, as submergence and emergence occur (Fig. 9.4).

All the above parameters respond on different time scales. The methods used to monitor them have changed over time, and the technique and effort necessary for monitoring specific parameters limit the feasibility of compiling of long-term records.

Length Change

The oldest method of glacier monitoring is measuring glacier length change, applied quite early by farmers or other local people when glaciers advanced. As advance and retreat of a glacier can be triggered not only by a large-scale climate signal, but also by internal dynamics of that glacier, glacier length changes do not provide an unambiguous climate signal. Therefore, only a monitoring network that includes a sufficient number of glaciers of various glacier types in different regions allows an interpretation that distinguishes large-scale climate signals from other factors.

Several survey points (e.g., rock outcrops or big boulders) are defined at the glacier tongues, and the distance between the glacier margin and the survey points in the direction of the ice flow is measured every summer. Fig. 9.5 shows the terminus of Jamtal Ferner with some of the survey points used for the length change surveys. The leading letter of the name of the survey point gives the position of the mark and the number, the year the mark was set up. When the distance to the glacier tongue becomes too large for a tape or laser distance measurement, or mass movements dislocate a mark, a new survey

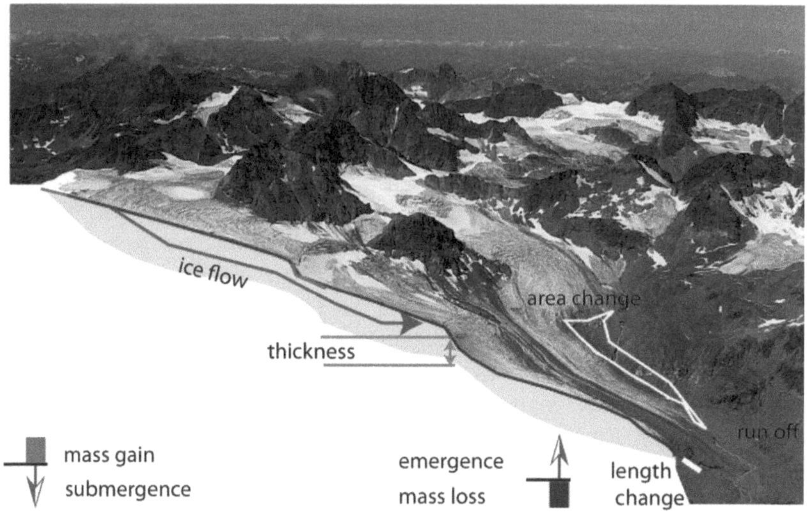

FIGURE 9.4 The balance of mass gain and loss dominantly governed by mass and energy exchange with atmosphere causes changes in flow velocity, length, area, and runoff changes. *Blue lines* (light gray in print versions) on the glacial surface indicate mass gain, and *red lines* (dark gray in print versions) on the glacial surface indicate mass loss. Ice flow transports mass from accumulation areas to ablation areas (*blue arrow* (light gray in print versions)). In areas with mass gain (*blue blocks* (light gray in print versions)), the vertical component of the glacier flow velocity (*arrow*) points into the glacier. Mass losses (*red block* (dark gray in print versions)) in lower lying areas are partially replaced by inflow of ice. The vertical component of the ice flow velocity points upwards there.

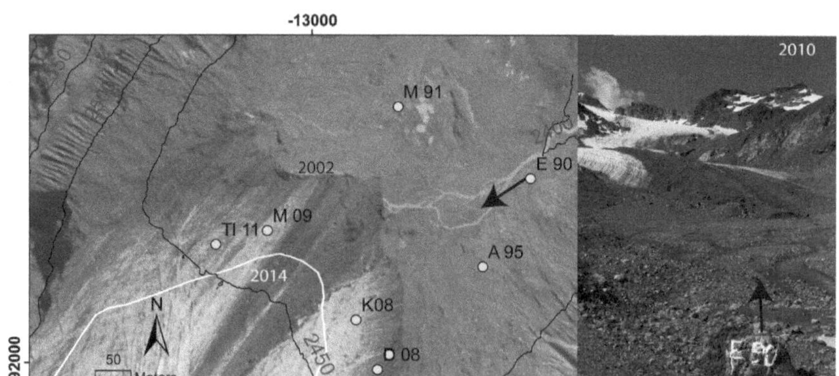

FIGURE 9.5 (Left) Orthophoto of the terminus of Jamtal Ferner, Silvretta (2002) with terminus position 2014 (*red line*(white in print versions)) and some of the survey points. (Right) The view from survey point E 90 (set up in 1990) toward the terminus in 2010.

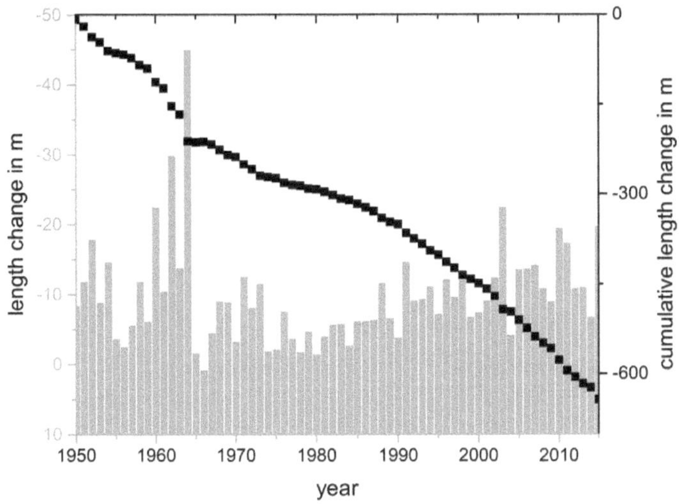

FIGURE 9.6 Length change measurements and cumulative length change 1950–2013 on Jamtal Ferner, Silvretta.

point is marked with a number and a measurement direction (preferably the direction of the ice flow). The mean annual glacier length change is calculated as the mean difference of these distances from survey point to glacier margin (Fig. 9.6). The annual changes show high interannual variability as a result of the specific geomorphological properties and processes at the terminus, for example collapse of thin ice layers or disconnection of parts of the glacier. The cumulative graph gives a better impression of the length change trend. If the reported glacier length change occurred over several years, recovering values for individual years is not straightforward and nor is the calculation of a regional annual mean value.

Mass Balance Measurements

Direct measurements of net glacier mass balance for the mass balance year (October 1 to September 30) uses ablation measurements at stakes in the ablation area and accumulation measurements in snow pits in the accumulation area (Fig. 9.7, Hoinkes, 1970; Cuffey and Paterson, 2010). Ablation stakes are set into holes drilled several meters into the ice and the length of the stake extending above the glacier surface is measured repeatedly, with the depth of ice melt calculated as the difference in length of exposed stake. Snow pits are dug into the snow and firn down to the dirt layer indicating the summer horizon of the previous year, something that in most years cannot be identified by snow probing or drilling. The density of snow and firn varies between 100 and 600 kg/m^2, so that the mass accumulated is the sum of the product of

FIGURE 9.7 Direct measurements of winter snow accumulation in a snow pit and drilling an ablation stake.

volume and density within the annual layer. Information on the spatial distribution of accumulation, for example gained by snow probing on ice or ground penetrating radar (GPR) measurements, can be used when density can be extrapolated from a snow pit.

The net mass balance *b* is calculated by manual extrapolation of the measured balance data at stakes and pits to the total glacier surface (Fig. 9.8). Specific net mass balance *b* is calculated by dividing the mass balance by the glacier area and given in millimeter water equivalent (mm w.e.). Some Austrian time series also include seasonal mass balances, that is winter mass balance from October 1 to April 30. Summer mass balance is calculated as the difference between net and winter mass balance. Usually specific mass balance for ablation/accumulation area or specific seasons are denoted with subscripts: ablation (a), accumulation (c), winter (w), and summer (s) mass balance. The equilibrium line altitude (ELA) historically developed from climatic snow lines but today it is defined as the spatial average of the zero mass balance contour line over a mass balance year.

Direct net mass balance measurements have high information content by providing hydrological and climatic interpretations of glacier changes plus a direct relation to the energy balance during the year, but they require great effort and can be done only for very few glaciers.

On Stubacher Sonnblick Kees, based on several years of direct measurements, a semidirect method has been applied, based on the observation of the snow line at the end of the season (Slupetzky, 2015). The method uses the direct mass balances during the season to develop a glacier-specific statistical relation between snow line and mass balance.

Geodetic mass balance is calculated as the product of mean density of glacier mass and glacier volume change. Differences between direct and geodetic mass balance arise from differences between measured and assumed densities (Huss, 2013), from basal and internal processes, and from

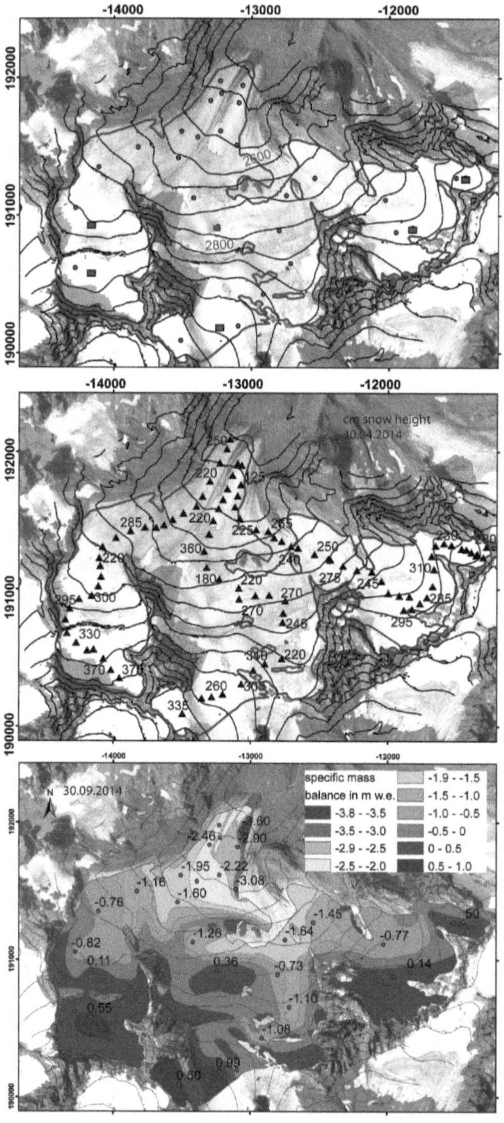

FIGURE 9.8 Network of stakes and pits on Jamtal Ferner (Silvretta), winter accumulation measurements and net mass balance map (Fischer and Markl, 2009).

measurement uncertainties in both volume change and direct mass balance (Fischer, 2011). Geodetic mass balances can be used as a control of direct mass balance time series, assuming that both results should be within the error bars of the measurements (Zemp et al., 2013).

As local volume change is the result of mass balance, ice flow, and density changes, it is not possible to calculate geodetic mass balance for a point or parts of a glacier. Geodetic mass balances have a lower temporal resolution than direct mass balance data but can be applied for a greater number of glaciers.

For long time series of direct mass balances, changes in the observing network as well as changes of the glacier itself influence the results. A homogenizing procedure should remove method-related biases in the time series (Fischer, 2010; Zemp et al., 2013). Changes in surface area, elevation, albedo, debris cover, firn extent, and all other geomorphological properties change the sensitivity of glacier mass balance to a climatic setting and its changes.

Glacier Inventories

The four Austrian glacier inventories (Fig. 9.9) are derived from airborne photography and light detection and ranging (LiDAR) data and contain information on the glacier extent and geomorphology including surface elevation (Fischer et al., 2015a). The surface elevation data can be used to calculate geodetic mass balance. The first (Patzelt, 1980) and second (Lambrecht and Kuhn, 2007; Kuhn et al., 2013) glacier inventories used airborne photogrammetry and mapped glacier outlines manually. Groß (1987) compiled a

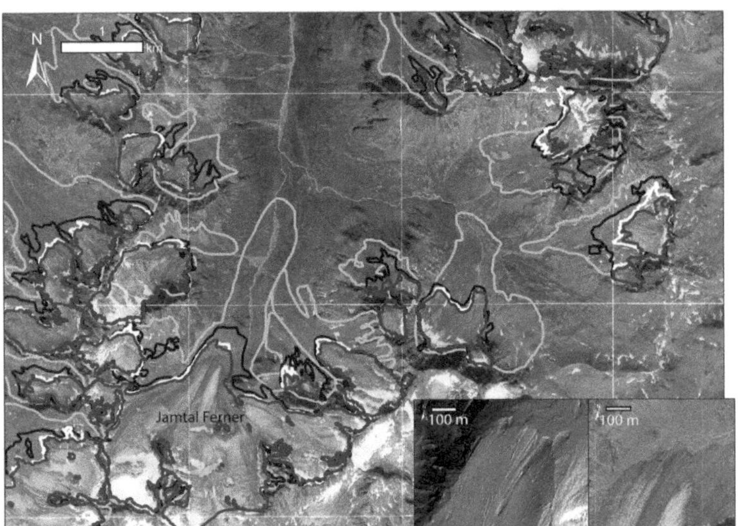

FIGURE 9.9 Glacier inventory data of Jamtal Ferner dating from LIA maximum (orange (light gray in print versions)), 1969 (blue (black in print versions)), 2002 (green (white in print versions)), and 2006 (red (gray in print versions)). The mapping is based on light detection and ranging hillshades (left inset) and orthophotos (right inset).

first analogue version of the LIA maximum outlines from field mappings and manual delineations of the moraine positions in the orthophotos of the first glacier inventory. The digitization of these data included the adaptation to the highly accurate LiDAR data: Previous mappings were based on low-resolution data, so that the glacier extent did not always fit the highly accurate moraine positions in the LiDAR data.

The majority of the glaciers in the third glacier inventory was mapped from LiDAR hillshades and elevation changes between the second and the third glacier inventories (Abermann et al., 2010). The delineation of glaciers is based on differences in the surface roughness of glaciers and periglacial areas and the maximum volume change at the glacier margin. Where no LiDAR data were available, RGB true color orthophotos with a nominal resolution of 20 × 20 cm were used for the delineation of glacier margins. Glacier names and ice divides were kept unchanged from the first to the third inventories to allow a glacier-by-glacier comparison.

Ice Volume

The first ice thickness measurements occurred on Hintereis Ferner in the Ötztal Alps from 1893 to 1922 by mechanic drilling (Blümcke and Hess, 1899), followed by seismic measurements from 1929 onwards (Span et al., 2005). Seismic and radar methods use the travel times of signals sent out and received at the glacier surface after being backscattered from the bedrock. From the two-way travel time, the distance of the bedrock and therefore ice thickness can be calculated if the signal velocity is known. Where boreholes to the bedrock exist, they allow calibration of the signal velocity of seismic and radar signals.

Today, GPR is widely used for measuring glacier thickness (Fig. 9.10). The small Alpine glaciers can be surveyed on the ground, which is not possible for larger glaciers or ice sheets, where airborne technologies must be applied. As Austria's glaciers are easily accessible (some of them even with cable cars), early attempts had been made to compile a database as the basis for numerical modeling of glacier thickness. Glacier thickness is an important parameter limiting future runoff. The loss of glacier area is determined by the melt rates and the diminishing volume. Even if future melt rates increase, glacier runoff will decrease when glaciers surface decreases, especially in lower parts of the glaciers which contribute most to the glacier runoff. Glacier thickness thus is an important predictor of future runoff. Ice thickness measurements on the world's glaciers are scarce, because no satellite-born methods are currently operational and ground based and airborne soundings require great effort, making numerical modeling of ice thickness even more important.

Fischer and Kuhn (2013) carried out surveys with a long-wave GPR in the Austrian Alps across a wide range of glacier types and sizes. They measured ice thickness measured at single points along profiles with a mean point

FIGURE 9.10 Ice thickness measurements with travel path of the signal from sender to receiver.

density of 36 points/km^2. The system used had a central frequency of 4.5–6 MHz, with penetration depths of several hundreds of meters even in ice with temperatures close to the pressure melting point and significant content of liquid water. On the one hand, the spatial resolution of the low-frequency data is lower than for higher frequency data, which can be considered as a major disadvantage. On the other hand, low frequencies have the advantage that the signal can pass even in firn areas with high liquid water content, which is a clear advantage over higher frequencies with higher absorption. The antennas were 30–50 m long, and the system was operated with 4 skiers during winter. A Fluke 105 B oscilloscope was used as receiver. The ice thickness h_p was then calculated from total signal travel times Δt from sender to receiver by assuming an isosceles triangle

$$h_p = \frac{1}{2}\sqrt{\left(\Delta t + \frac{1}{c_a}\right)^2 c_i^2 - a^2}$$

where c_a and c_i are the signal velocities in ice and air, respectively.

The locations of the measurements were recorded with handheld GPS at accuracies increasing from ~50 m in the 1990s to today's submeter range. From 2010, a Geophysical Survey Systems, Inc. (GSSI) system with multiple low frequency (MLF) antennas and central frequencies of 16–80 MHz was used for surveys, with Differential Global Positioning System (DGPS) records of the profile positions. Ice thickness was calculated using ReflexW Software.

The signal velocity within the glacier was assumed as 168 m/ns. The ice thickness distribution was interpolated manually (Fischer, 2009), accounting for lower ice thicknesses in crevassed or very steep zones that could not be surveyed.

Runoff

Direct measurements of glacier runoff require the installation of quite massive gauges, as the sediment and bedload transport as well as the seasonal and daily variabilities of the runoff are high. Changes in the terminus position and the relocations of the meltwater stream restrict the potential measurement sites. Good conditions are found in bedrock with a well-defined and not too steep channel bed, but often flat gravel covers the periglacial area.

A natural bedrock sill, a clearly defined hydraulic profile as well as an already existing footbridge of the Austrian Alpine Club, which could also be used for discharge measurements, was the main reasons for the Hydrological Service of the Regional Government of Salzburg to choose the particular location for the high-mountain gauging station Kees/Obersulzbach (Fig. 9.11). In 1988, when the operational series of records of the glacier-influenced discharge of Obersulzbach river started, the gauge, situated at approximately 2060 m above sea level (a.s.l.), was only 500 m away from the snout of Obersulzbach Kees. A simple battery-operated analogue bubbler gauge was used to record water level and hydrometric measurements, using propeller type current meters, allowed establishing stage–discharge relations at the gauge.

In the following years, and especially after the very hot summer of 2003, it became obvious that due to increased glacier melt, the cross-section profile of the gauge was insufficient to capture the total discharge of Obersulzbach river (Fig. 9.11). Specific discharges of up to 900 l/s km^2, far beyond the originally estimated values, were registered at the gauge. In 2009 a new gauge, Türkische Zeltstadt, was established further upstream at 2200 m a.s.l. at the outlet of the recent proglacial lake Obersulzbachsee (Fig. 9.12). The natural but stable

FIGURE 9.11 (Left) Runoff gauge Kees/Obersulzbach river during an unexpectedly high peak flow. (Right) Bubbler gauge.

FIGURE 9.12 Runoff gauge Türkische Zeltstadt with Lake Obersulzbachsee.

bedrock sole of the gauge transect provides the basis of a well-defined stage–discharge relation. Water level, water temperature, and electric conductivity are measured by means of redundant battery-powered data loggers.

The design of a gauge is not only a function of hydrological demands but also of current regulations. Given the gauge's prominent position in the Hohe Tauern National Park, no technical interventions (e.g., concrete or steel) were carried out to stabilize the discharge profile and the whole instrumentation of the gauging station had to be hidden in the newly built footbridge.

Accumulation of Seasonal Snow

Although accumulation is an important part of glacier mass balance, data are sparse. Monitoring of snow accumulation is done mainly in winter season lower down in the inhabited areas. Snow depth measurements are part of the monitoring network of the hydrological surveys of the federal states as well as the network of the National Weather Service. A few snow stakes are operated at high altitudes or close to the glaciers, as for instance in the vicinity of the observatory on Hoher Sonnblick (Schöner et al., 2009a). Apart from mass balance measurements, no regular surveys of water equivalent of the snow cover are operational.

LiDAR offers a new, promising method, which was tested at a site in the Ötztal Alps and hopefully will become an operational method for monitoring snow cover. In the last decade, LiDAR (Baltsavias, 1999; Wehr and Lohr, 1999) systems have been used to observe snow depths, using Terrestrial Laser Scanning (TLS) on individual slopes and Airborne Laser Scanning (ALS) over entire mountain catchments. Anttila et al. (2015) showed that the laser signal is backscattered from a depth of 0.5–1 cm for wet snow, and from the very surface for dry snow conditions. This low penetration depth allows quantification of snow depth and snow depth changes. Whereas TLS is the best choice to study wind drift, gravitational snow transport, and avalanches (e.g., Prokop et al., 2008; Deems et al., 2015), ALS is appropriate to survey the snow cover on gently inclined surfaces of less than 40 degrees and thus to observe the majority of the total snow volume (e.g., Bollmann et al., 2011; Jörg et al., 2012). Two surveys are needed to calculate surface elevation changes: a snow-free survey and snow-on survey (Deems et al., 2013). On glaciers, the emergence and submergence velocity have to be considered when interpreting surface elevation changes in terms of snow depths (Helfricht et al., 2014a).

Ice Flow Velocity

Flow velocity measurements have been carried out since the late 19th century. In most cases, they measured the displacement of stones located along a profile across the glacier in the ablation area. These measurements reveal only horizontal displacement rates, but no information on mass balance or vertical displacement rates. Stake measurements of ice flow velocity (Fig. 9.13) allow the calculation of local submergence and emergence but require the annual repositioning of the stakes. The calculation of ice flow velocities from satellite-born remote sensing using feature tracking algorithms or INSAR (Kargel et al., 2014) has not been applied operationally to Austria's glacier, as the topography is quite steep and the glaciers are small and with ice temperatures close to zero.

FIGURE 9.13 DGPS measurement of the annual displacement of an accumulation stake on Kesselwand Ferner, Ötztal Alps. The *arrow* indicates the direction of the horizontal ice flow.

GLACIER MONITORING NETWORK IN AUSTRIA

The Glacier Monitoring Network in Austria developed historically, fulfilling various needs and interests. To gain long-term data, adaptations and modernizations of the network can basically only add new data, but have to be based on often small samples of early measurements. While new techniques allow monitoring new variables, the long-term perspective emerges from continued measurement of variables that have already been monitored for years. The original aim of capturing the sources of natural disasters faded in the 20th century as no dangerous glacier-dammed lakes exist in the Austrian Alps today. Apart from basic hydrology, mass balance measurements are an important aspect today, as most glaciers are located in catchment areas of hydropower reservoirs. Glacier inventories in Austria have also legal implications, as regional policy, water regulations, and nature protection legislation include regulations for glaciers (Hautzenberg, 2013).

Length change records cannot be directly interpreted in terms of climate change or hydrology. Their value is the record length: climate science needs long time series (at least several decades) to distinguish short-term, more or less random, weather effects from the climate signal. The length change records are by far the longest records systematically measured, with the largest sample of glaciers (Figs. 9.14 and 9.15).

As the climatic setting, together with the topographic setting, governs the geomorphology and climate sensitivity of the glaciers (Abermann et al., 2011),

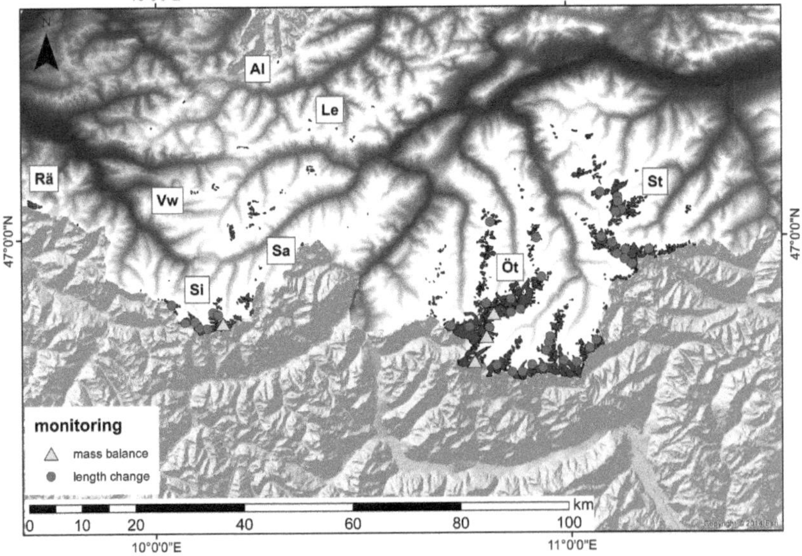

FIGURE 9.14 Overview map with glacier areas, length change, and mass balance monitoring network (west). *Al*, Allgäu Alps; *Le*, Lechtal Alps; *Öt*, Ötztal Alps; *Rä*, Rätikon; *Sa*, Samnaun; *Si*, Silvretta; *St*, Stubai Alps; *Vw*, Verwall.

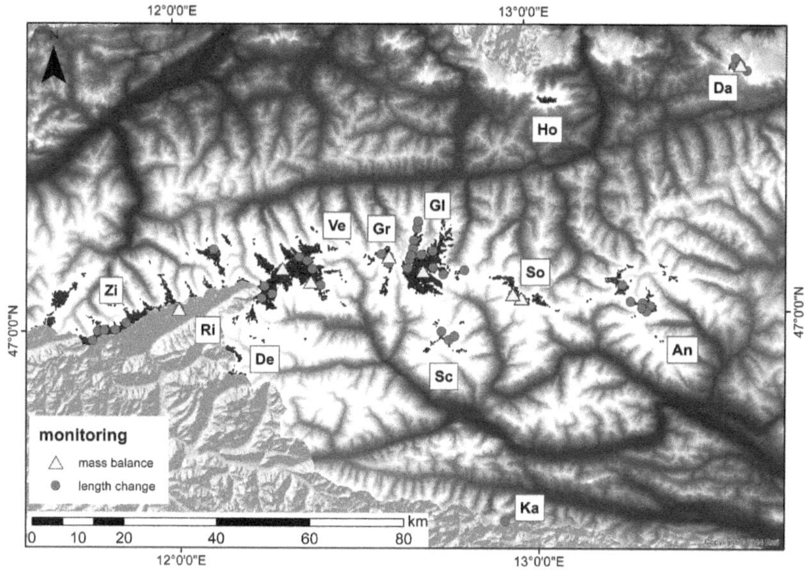

FIGURE 9.15 Overview map with glacier areas, length change, and mass balance monitoring network (east). *An*, Ankogel Group; *Da*, Dachstein; *De*, Defregger Alps; *Gl*, Glockner Group; *Gr*, Granatspitz Group; *Ho*, Hochkönig; *Ka*, Karnic Alps; *Ri*, Rieserferner Group; *Sc*, Schober Group; *So*, Sonnblick Group; *Ve*, Venediger Group; *Zi*, Zillertal Alps.

it is important to note that the Monitoring Network includes glaciers in all climatic settings, with high precipitation rates and low maximum elevations in the south (e.g., Carnic Alps, Schober Group) and north (Dachstein) as well as those with low precipitation and high maximum elevation (e.g., Ötztal Alps).

On an annual basis, length change monitoring is being done on about 100 glaciers of various sizes and glacier types, from valley glaciers to ice caps and cirque glaciers. The mass balance network maintained by various institutions so far includes 13 glaciers well distributed across different climatic and geomorphological settings.

So far four glacier inventories, dating from the LIA maximum until 2006, contain areas, surface elevations, and basic geomorphological information for all Austrian glaciers (Fischer et al., 2015a). Volume data based on GPR measurements are available for 64 glaciers (Fischer and Kuhn, 2013). Therefore, past volume changes calculated from the glacier inventories can be compared with present total volumes.

Length Changes: Simple Measurements, Complex Signal

Early reports on the fluctuation of glaciers in the Eastern Alps were made at the time of glacier advances dammed streams, thus creating lakes (Nicolussi, 2013), as was the case with Vernagt Ferner and Gurgler Ferner. These records were not made continuously and describe just a few glaciers. Regular monitoring was introduced on Hallstätter Glacier (Karls-Eisfeld) by Simony in 1840, by Seeland on Pasterze in 1879 (Lieb and Slupetzky, 2004), and on Vernagt Ferner in 1888 by Finsterwalder, Blümcke, Hess, and Kerschensteiner (Finsterwalder, 1897).

Two years later, Swiss glaciologist Forel established the first annual glacier report on all Alpine glaciers (Forel and Du Pasquier, 1896). In 1890, Forel reduced the scope of his reports to Swiss glaciers and Richter took over the Eastern Alps. As he thought that a long-term series on as many glaciers as possible was only feasible with the support of a central institution and the help of local observers, he published a call in the Journal of the Alpine Club to find local participants for the glacier surveys in 1891. The idea was to define a simple but accurate measuring principle to ensure that local observers could take over regular surveys without the need for complicated and expensive instruments, which were not readily available and could not be used without regular training. Observers reported not just figures but also changes in the geomorphology or specific events, like the emergence of new rock outcrops, accompanying their reports with sketches. Later, photography complemented the documents. In addition to length change in meters, the reports include a categorization in retreating, advancing, and stationary glaciers. Glaciers were classified as stationary when their length change was between -1.0 and $+1.0$ m/year.

In his report, Richter (1893) observed roughly 20 glaciers in todays' Austrian Alps, about 15 glaciers in South Tyrol and 1 glacier in Germany (Tables 9.1 and 9.2). Forel mentions additional glaciers in the Adamello Group (Mandron, Lobbia, and Presena) and in the Ortler Group (Vedretta de la Mare, Vedretta Rossa, Ebenewand Ferner, Zay Ferner, and Laaser Ferner). In the Ötztal Group, Forel's glacier list is quite similar to today's monitoring network, including Kesselwand Ferner, Taufkar Ferner, Rofenkar Ferner, Mitterkar Ferner, Marzell Ferner, Spiegel Ferner, Rotmoos Ferner, and Taschach Ferner, as well as Gaisberg Ferner. The Hochalmspitz Gletscher, monitored today, is also mentioned in Forel's report. Early data compilations have been done by Fritzsch (1898), Klebelsberg (1943), Patzelt (1970, 1977), and Kinzl (1979).

Today the monitoring network covers about 100 glaciers with annual length reports (Table 9.3). The glaciers are well distributed over Austria, allowing an annual interpretation of an average climate signal as well as regional variability (Fig. 9.16). Just a few years with positive accumulation anomalies can lead to glacier advances, negative accumulation anomalies can lead to strong retreats. These anomalies in accumulation result from atmospheric flow patterns leading to orographic precipitation on the northern or southern slope of the main Alpine ridge. For example, in the mass balance years 2012/2013 and 2013/2014, accumulation exceeded the long-term average in regions south of the main Alpine ridge (Fig. 9.17). Although the

TABLE 9.1 Glaciers Surveyed by Richter (1893) Within Today's Austrian Borders

Group	Glacier Name
Hochkönig	Übergossene Alm
Silvretta	Jamtal F.
	Vermunt F.
Zillertaler Alpen	Schwarzenstein K.
	Horn K.
Venediger Group	Obersulzbach K.
	Dorfer K.
	Schlaten K.
	Villtragen K.
	Frosnitz K.
Glockner Group	Pasterze
	Teischnitz K.
Stubai Alps	Grünau F.
	Fernau F.
	Daunkogel F.
	Sulztal F.
	Schwarzenberg F.
	Alpeiner F.
	Simming F.
Ötztal Alps	Vernagt F.
	Guslar F.
	Hochjoch F.
	Gepatsch F.

F., Ferner; *K.*, Kees.

glaciers generally showed extreme rates in the 21st century, these positive accumulation anomalies triggered short-term advances of small and fast reacting glaciers (Fischer et al., 2013a; Fig. 9.16).

The long-term trend of the total time series displays two periods of glacier advances. The classification in advancing, retreating, and stationary glaciers

TABLE 9.2 Glaciers Surveyed by Richter (1893) Outside Today's Austrian Borders

Group	Glacier
Cristallo	Cristallo G.
Southern Zillertal Alps	Glieder F.
Southern Stubai Alps	Feuerstein F.
	Übeltal F.
	Grübl F.
	Hangenden F.
Ortler group	Madatsch F.
	Ortler F.
	Trafoi F.
	Fürkele F.
	Zufall F.
	Sulden F.
	Schöntauf F.
	Rosim F.
Zugspitze	Schnee F.

F., Ferner; *G.*, Gletscher; *K.*, Kees.

(Fig. 9.18) shows that most glaciers have been retreating since the beginning of the observations. The first period of glacier advance was recorded between 1915 and 1926. With the exception of 1922 and 1924, more than 20% of the glaciers advanced during that period, with a maximum of 70% in 1920.

A second period of glacier advance started after the exceptional year of 1965, when the highest accumulation of the mass balance time series was recorded. As in the 1920s, the peak of 70% of advancing glaciers was reached in 1980, maximum states have occurred even later, for example for Kesselwand Ferner in the Ötztal Alps in 1985.

The number of surveyed glaciers reached a maximum in the 1980s. Prior to 1924, far fewer glaciers were being observed and some surveys were done biannually. Therefore, the average length change has been calculated for the years post-1960 (Fig. 9.19). The high percentage of advancing glaciers in the 1980s returns a positive mean length change. Mean length change is not necessarily positive during a shorter or less distinct period of glacier advance: Large glaciers have a longer reaction time and might continue to show high

TABLE 9.3 Length Change Monitoring Network in 2014 (Fischer et al., 2013)

Dachstein	**Glockner Group**	**Zillertal Alps**	**Ötztal Alps**
Schladminger G.	Pasterze	Wildgerlos K.	Gaiβberg F.
Hallstätter G.	Wasserfallwinkel K.	Schwarzenstein K.	Rotmoos F.
Schneeloch G.	Freiwand K.	Horn K.	Langtaler F.
Gr. Gosau G.	Brennkogl K.	Waxegg K.	Gurgler F.
	Bärenkopf K.	Furtschagl K.	Latsch F.
Ankogel Group	Karlinger K.	Schlegeis K.	Spiegel F.
	Schmiedinger K.		Firmisan F.
Winkel K.	Maurer K.	**Stubai Alps**	Diem F.
Westl. Tripp K.	Schwarzkarl K.	Grünau F.	Schalf F.
Hochalm K.	Kleineiser K.	Fernau F.	Mutmal F.
Groβelend K.	Unteres Riffl K.	Daunkogel F.	Marzell F.
Kälberspitz K.	Totenkopf K.	Alpeiner F.	Similaun F.
Kleinelend K.	Ödenwinkel K.	Verborgenberg F.	Niederjoch F.
		Berglas F.	Hochjoch F.
	Granatspitz G.	Bachfallen F.	Hintereis F.
Schobergruppe	Sonnblick K.	Schwarzenberg F.	Kesselwand F.
Horn K.	Landeck K.	Sulztal F.	Guslar F.
Göβnitz K.	Kalser Bärenkopf K.	Gaiβkar F.	Vernagt F.
Roter Knopf K.		Pfaffen F.	Mitterkar F.
	Venediger Group	Triebenkarlas F.	Rofenkar F.
	Untersulzbach K.		Rettenbach F.
Goldberg Group	Obersulzbach K.	**Silvretta Group**	Innerer Pirchlkar F.
	Krimmler K. I	Jamtal F.	Taschach F.
Kl. Fleiβ K.	Umbal K.	Totenfeld	Sexegerten F.
Ö. Wurten K.	Simony K.	Bieltal F. Mitte	Schweikert F.
Goldberg K.	Zettalunitz K	Vermunt G.	Gepatsch F.
	Frosnitz K.	Ochsental G.	Weiβsee F.
Carnic Alps	Schlaten K.	Schneeglocken G.	
Eiskar G.	Viltragen K.	M. Klostertaler G.	
		Litzner G.	

F., Ferner; *G.*, Gletscher; *K.*, Kees.

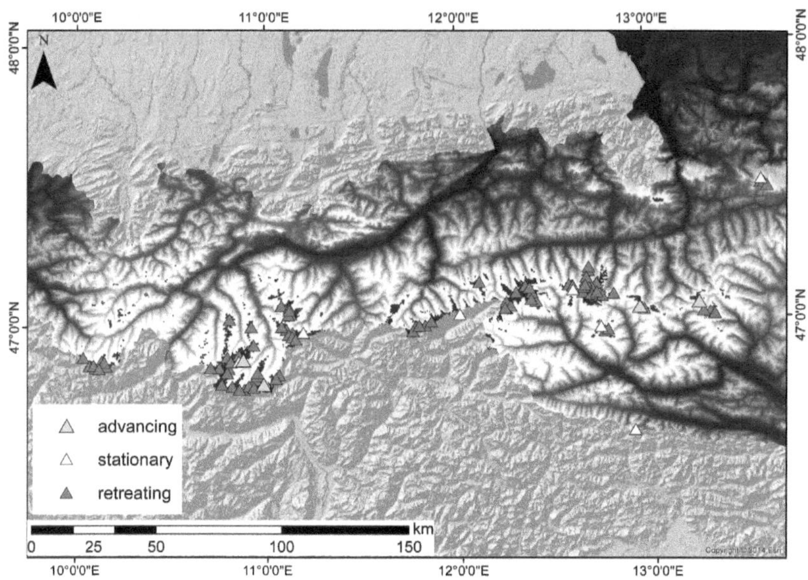

FIGURE 9.16 Results of glacier length monitoring in the year 2014.

retreat rates even when a high number of smaller glaciers are advancing. Mean retreat rates are also greatly influenced by the disintegration of the tongues of large glaciers. The selection of the surveyed glaciers has of course some impact on the sensitivity of the results to long-term climate change or short-term effects of atmospheric circulation.

Glaciers exhibit highly individual behavior, as demonstrated by Kuhn et al. (1985) for Hintereisferner and Kesselwandferner in the Ötztal Alps. They reflect long-term climate and short-term weather trends, as well as the local variability of snow accumulation and individual topographic effects. The example of four glaciers in the Silvretta Group in Fig. 9.20 illustrates some of the individual effects. Litzner Gletscher is the smallest glacier of the sample, Jamtal Ferner the largest. Ochsentaler Gletscher and Vermunt Gletscher had a common terminus during the LIA maximum. The absolute length change of Litzner Gletscher is the smallest, and of Ochsental Gletscher the largest of the sample.

The largest retreats occur at Jamtal Ferner in the late 1940s and 1950s. The maximum retreats of −73 m in 1944 and −44 m in 1964 have different reasons: In 1944 the glacier terminus was located on a rock ridge, so that the thin ice of the tongue melted quickly. In 1964 mass balances were extremely negative with nearly no snow left at the end of the ablation period, so that the retreat rates of all glaciers in the Silvretta were quite large.

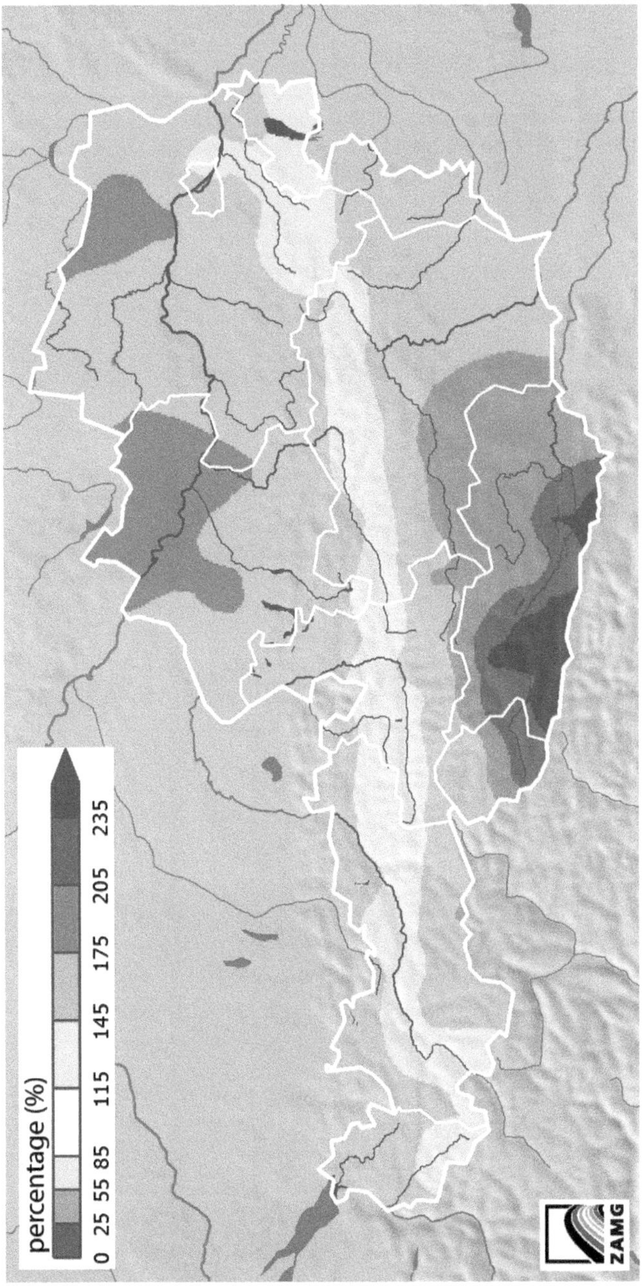

FIGURE 9.17 The anomaly of the precipitation sum during the accumulation season (October 1 to April 30) of the mass balance year 2013/14 in relation to the reference period 1981–2010. South of the main Alpine ridge, the accumulation was clearly above the long-term average.

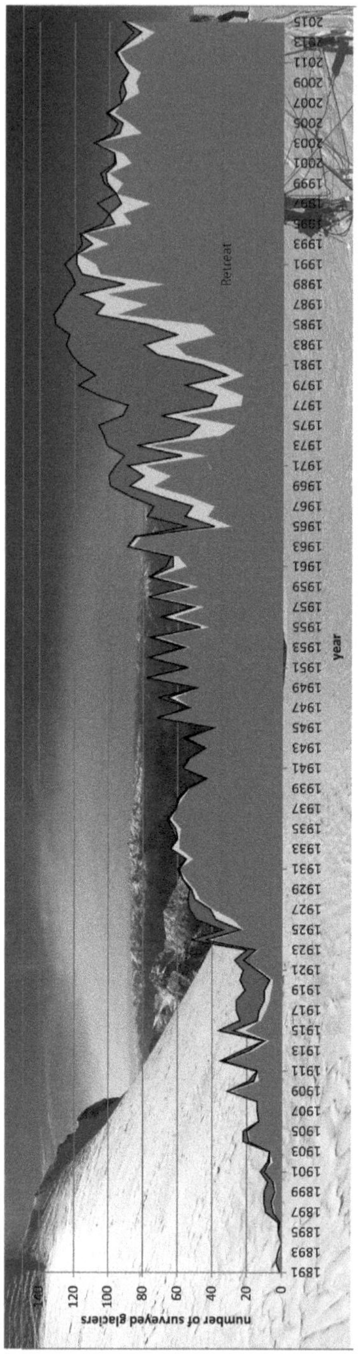

FIGURE 9.18 Number of surveyed glaciers with classification as advancing (blue (dark gray in print versions)), stationary (white), and retreating (red (gray in print versions)).

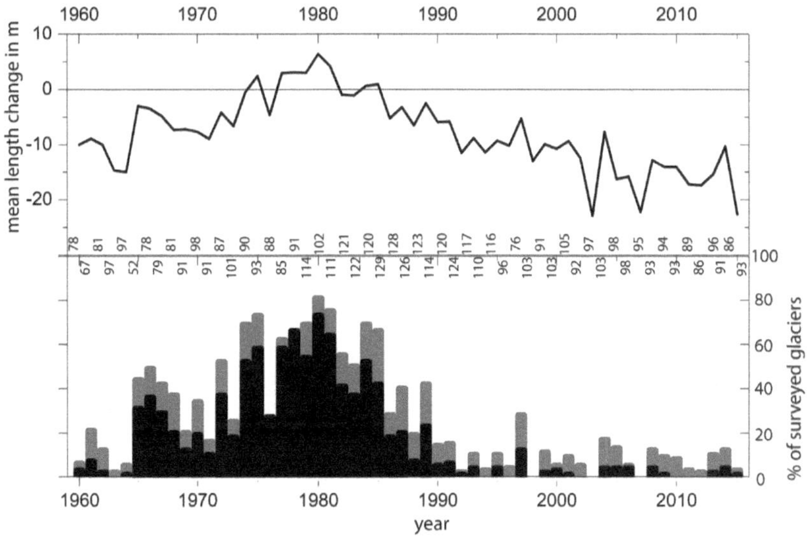

FIGURE 9.19 Mean length change, number of surveyed glaciers, and percentage of advancing (black), stationary (gray), and retreating glaciers (background color).

The length changes of Ochsentaler Gletscher are much larger than those of Vermunt Gletscher, as a result of the different topography of these glaciers: Ochsentaler Gletscher has a large and flat accumulation area and a steep and (Fig. 9.21) narrow glacier tongue. The steep tongue allows quick dynamic responses, both advance (when flow velocity increases) and retreat (by a quick meltdown of the thin ice of the tongue). The accumulation area of Ochsentaler Gletscher is higher than those of the other glaciers.

Litzner Gletscher showed its largest change rates prior to 1950s, and the latest and longest advance period of the four glaciers. The reason for the dampened and delayed behavior of Litzner Gletscher is not only its topography, but also a debris cover which has developed since the 1990s (Fig. 9.22). In 1983 the glacier tongue pushed a moraine and the tongue was partly covered by firn, so that submergence removed debris from the surface. In 1992 the glacier surface was concave and no accumulation occurred at the glacier tongue, so that emergence increased the thickness of the debris layer. In 2012, the debris layer was thick enough to reduce ablation and led to stationary behavior of the glacier terminus (which is also hard to define). Therefore, the measurements at Litzner Gletscher have been suspended. Note that the proglacial lake close to the tongue in 1983 and 1992 was almost completely filled with sediments in 2012.

The cumulative length change rates of four glaciers in the Venediger Group (Figs. 9.23 and 9.24) illustrate the general large-scale glacier retreat and individual governing factors. At Obersulzbach Kees, at the location of the

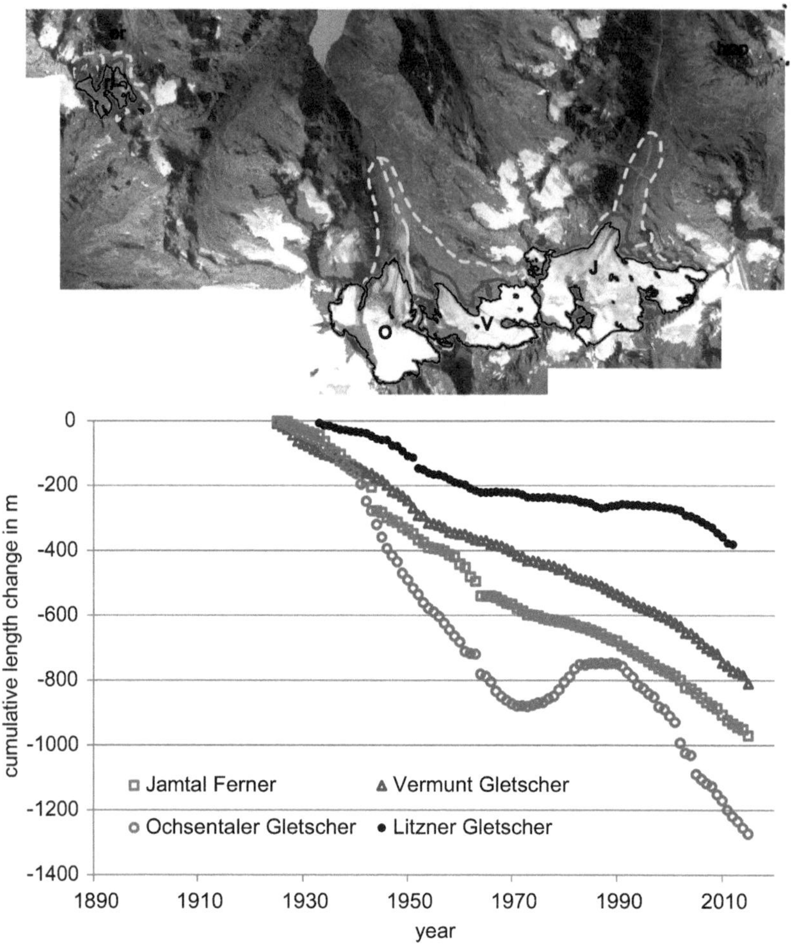

FIGURE 9.20 Length change of Jamtal Ferner (J), Vermunt Gletscher (V), Ochsentaler Gletscher (O), and Litzner Gletscher (L) in the Silvretta Group with outlines of LIA maximum (yellow (light gray in print versions)), 1969 (red (dark gray in print versions)), 1998 (blue (gray in print versions)), and 2006 (black).

former glacier tongue, a lake has developed. The new glacier tongues coming from four different firn basins meanwhile also retreated from this lake. The length change measurements are being continued on the western glacier tongue, but naturally the dynamic response of this smaller glacier branch differs from the former glacier. Obersulzbach Kees clearly advanced during the 1920s, but strongly retreated during the advance period in the 1980s, when Untersulzbach Kees and Schlaten Kees showed positive length changes. Viltragen Kees has a debris-covered tongue and presents the most steady behavior of these four glaciers.

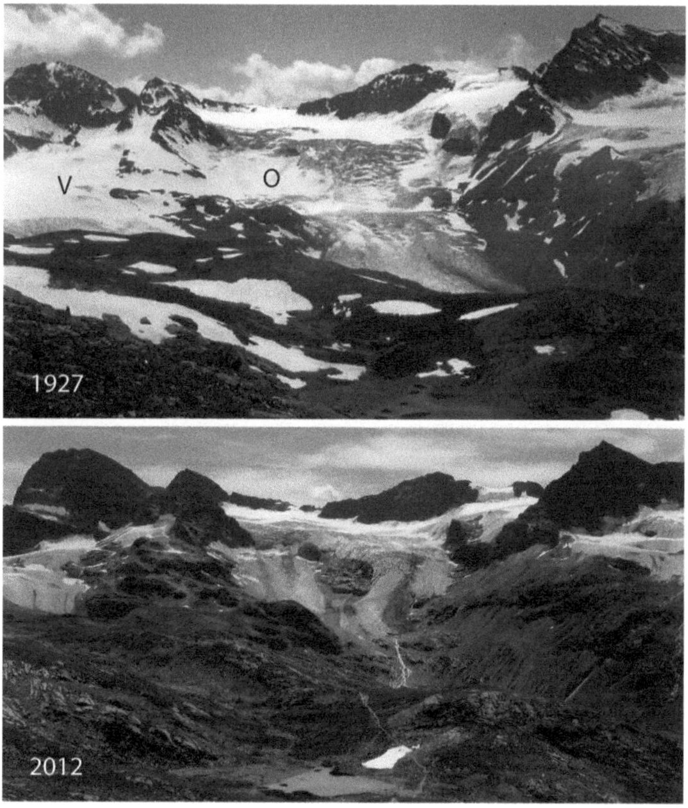

FIGURE 9.21 Vermunt Gletscher (V, left) and Ochsentaler Gletscher (O, right) in 1927 (photography by H. Kinzl) and 2012 (photography by G. Groβ).

Apart from individual glacier behavior and the importance of selecting representative glaciers, the continuity of individual series despite temporal difficulties, such as proglacial lakes and steep or debris-covered glacier tongues, is a key point of long-term monitoring. Although the current network still includes most of the initial glaciers, some of the time series have short gaps, for instance when a glacier advances too fast and covers survey points, when mud flows destroy survey points, or in situations during retreat periods, when the glacier tongue ends and an inaccessible rock cliff. Where parts of the glacier tongues become disconnected from the main glacier during retreat, the time series is continued at the new glacier margin from new survey points. Thus, the sum of the length changes is not equal to the total glacier retreat as derived from inventories. This strategy results from the aim of monitoring climate- and not topography-driven glacier retreat and removes sudden jumps in glacier length, which would otherwise bias the annual means of the full glacier sample.

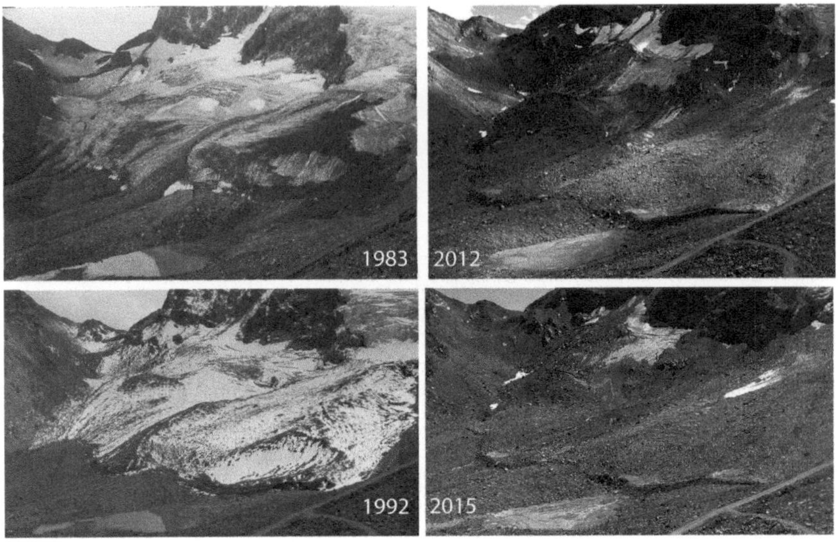

FIGURE 9.22 Changes in debris cover on the tongue of Litzner Gletscher in the Silvretta Group. *Photography by G. Groß.*

Mass Balance Measurements

In Austria, the first annual measurements of ablation were carried out on the tongue of Hintereis Ferner 1894/1895 (Hess, 1904) using the stakes of the deep drills on Hintereis Ferner in the Ötztal Alps. At that time, annual monitoring programs focused more on length change and on measuring horizontal displacement rates of stone lines. Regular annual mass balance monitoring was initiated on Hintereis Ferner (Kuhn et al., 1999) and Kesselwand Ferner in the mass balance year 1952/1953. These monitoring programs included not only measuring ablation with ablation stakes, but also measuring accumulation in snow pits. Basic scientific studies, including measurements of energy balance plus radiative balance, contributed to the development of today's direct mass balance methods (Hoinkes, 1970), the degree day method (Hoinkes and Steinacker, 1975), and remote-sensing methods (Rott, 1977). Another mass balance time series was initiated on Vernagt Ferner (with runoff gauge) in 1963/1964 (Mayer et al., 2013). All three glaciers are located in the inner Ötztal Valley, where the ice-dammed Lake Rofen triggered early glaciological studies (Richter, 1892; Nicolussi, 2013). Further mass balance series have been initiated in Hohe Tauern on Stubacher Sonnblick Kees (Slupetzky, 2015; Slupetzky and Ehgartner, 2014), later on Kleinfleiss Kees, Goldberg Kees, and Wurten Kees (Schöner et al., 2009b; Fischer et al., 2014a). In the west of Austria, Jamtal Ferner, Vermunt Gletscher, and Ochsental Gletscher (Silvretta Group) have been measured since 1990. The surveys on Vermunt Glacier and Ochsental Glacier were

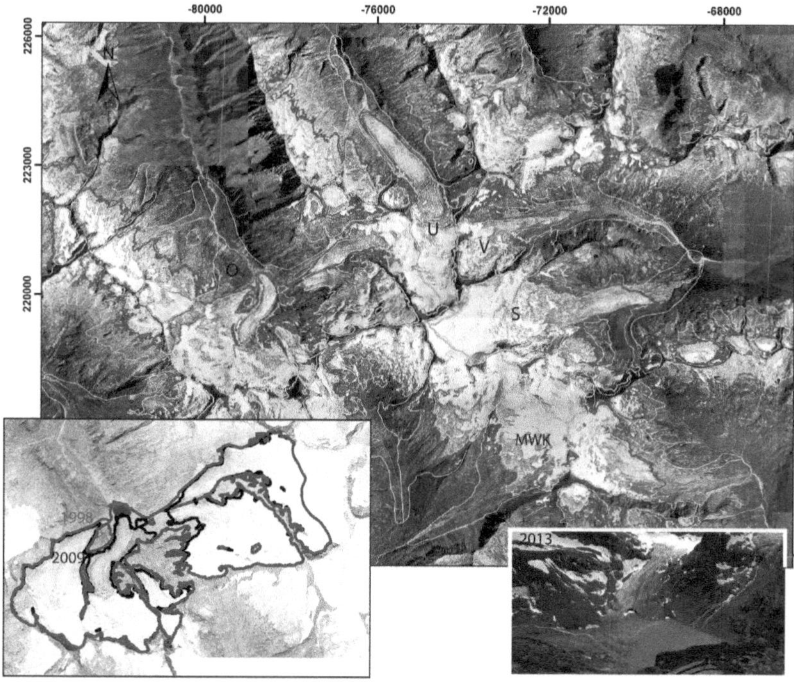

FIGURE 9.23 Obersulzbach Kees (O), Untersulzbach Kees (U), Viltragen K. (V), Schlaten Kees (S), and Mullwitzkees (MWK) in the Venediger Group with glacier inventory margins at the LIA maximum (yellow (light gray in print versions)), 1969 (red (dark gray in print versions)), 1998 (blue (black in print versions)), and 2006 (green (gray in print versions)). The tongue of Obersulzbach Kees (with a famous crevasse zone named "Türkische Zeltstadt") disintegrated, and a lake formed (insets).

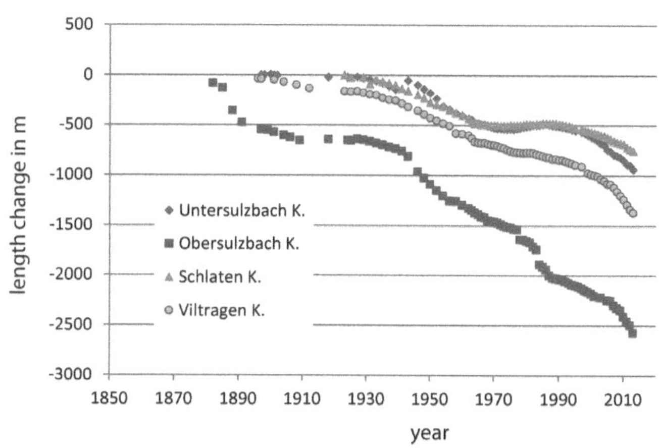

FIGURE 9.24 Cumulative length change of Untersulzbach Kees, Obersulzbach Kees, Schlaten Kees, and Viltragen Kees.

terminated in 1998/1999. Between 1965 and 1975, mass balance measurements were carried out at the Übergossene Alm (Goldberger and Hoffmann, 2013), a glacier with early length change observations, nearly disappeared today. The glacier is located in the Northern Limestone Alps, which are often subject to heavy precipitation events and show high accumulation. To include again a glacier in this climate zone in the monitoring network, mass balance at Hallstätter Gletscher started in 2006, a glacier with an even longer history of scientific research, initiated by Simony in the 1840s (Stocker-Waldhuber et al., 2015). In Hohe Tauern National Park, one glacier south of the main Alpine ridge (Mullwitz Kees) and north of the main Alpine ridge (Venediger Kees) were added to the monitoring program to find the gradients along the ridge and investigate differences in the reaction of the larger north-facing and the smaller and higher south-facing glaciers. For Austria's largest glaciers, where a time series of profile measurements has been carried out for several decades (Lieb and Slupetzky, 2011), detailed mass balance measurements started in 2004.

The cumulative mass balance time series of 11 glaciers (Fig. 9.25) reveals a mass loss since the individual beginning of the measurements. The neighboring glaciers Hintereis Ferner and Kesselwand Ferner are exposed to basically the same climate signal, but respond differently. The mass balance of Kesselwand Ferner is less negative than the one of Hintereis Ferner, with positive mass balances on Kesselwand Ferner for several decades, when Hintereis Ferner lost mass almost continuously.

In contrast to the length change, the two periods of glacier advances in the 1920s and 1980s (Fig. 9.18) are not so evident from the mass balance measurements: The first was not captured, as the mass balance measurements started after the length change records. During the second period of advance, the length change records document advances for a great number of glaciers. The glaciers measured for mass balance in the same period, however, returned much less positive responses: Only Kesselwandferner and Vernagtferner advanced significantly. The reason for the absence of advances in a period of positive mass balances is the larger size of the mass balance glaciers, which causes a delayed and dampened advance. The small sample of mass balance glaciers therefore might not capture short-term regional trends; the hydrological and climatological interpretation of mass balance data is more straightforward than for length change data.

In addition to the long-term complete direct mass balance series, shorter measurements were carried out for instance on Rotmoos Ferner or Sulztal Ferner. Only ablation was measured at a few stakes for several years on Gepatsch Ferner, Schlaten Kees, Taschach Ferner, Schaufel Ferner, Rettenbach Ferner, Tiefenbach Ferner, Mittelberg Ferner, Daun Ferner, Gaiβkar Ferner, Weissee Ferner, and Ödenwinkel Kees.

Of the Austrian direct mass balance time series, a homogenization was done for 60 years of measurements on Hintereis Ferner (Fischer, 2010). Direct

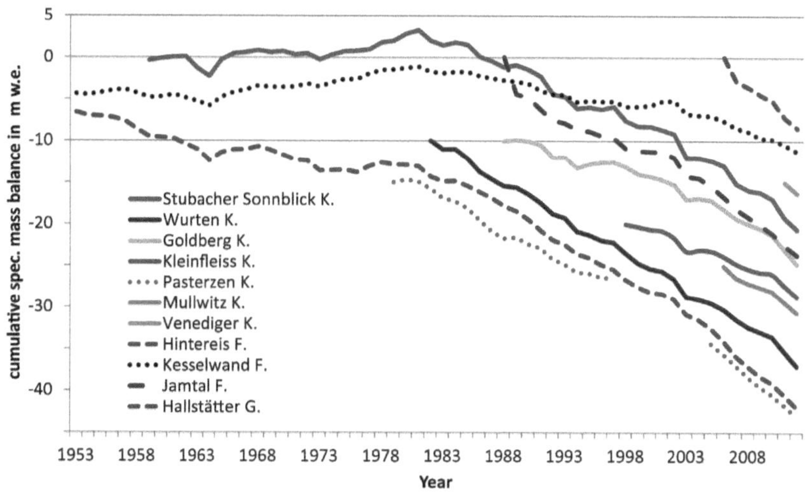

FIGURE 9.25 Selected mass balance time series in the Austrian Alps. *Adapted from Fischer, A., Stocker-Waldhuber, M., Seiser, B., Hynek, B., Slupetzky, H., 2014a. Glaciological monitoring in the Hohe Tauern National Park. Ecomont 6 (1), 5–62 and complemented by Fischer, A., Markl, G., Kuhn, M., 2013. Glacier Mass Balances and Elevation Zones of Hintereisferner, Ötztal Alps, Austria, 1952/1953 to 2010/2011. Institut für Interdisziplinäre Gebirgsforschung der Österreichischen Akademie der Wissenschaften, Innsbruck. http://dx.doi.org/10.1594/PANGAEA.818898; Fischer, A., Markl, G., Schneider, H., Abermann, J., Kuhn, M., 2014b. Glacier Mass Balances and Elevation Zones of Kesselwandferner, Ötztal Alps, Austria, 1952/1953 to 2012/2013. Institut für Interdisziplinäre Gebirgsforschung der Österreichischen Akademie der Wissenschaften, Innsbruck. http://dx.doi.org/10.1594/PANGAEA.818757; Fischer, A., Markl, G., Kuhn, M., 2014c. Glacier Mass Balances and Elevation Zones of Jamtalferner, Silvretta, Austria, 1988/1989 to 2013/2014. Institut für Interdisziplinäre Gebirgsforschung der Österreichischen Akademie der Wissenschaften, Innsbruck. http://dx.doi.org/10.1594/PANGAEA.818772; Fischer, A., Stocker-Waldhuber, M., Reingruber, K., Helfricht, K., 2015b. Glacier Mass Balances and Elevation Zones of Hallstätter Gletscher, Dachstein, Austria, 2006/2007 to 2013/2014. Institut für Interdisziplinäre Gebirgsforschung der Österreichischen Akademie der Wissenschaften, Innsbruck. http://dx.doi.org/10.1594/PANGAEA.806609.*

and geodetic balances were compared for Goldberg Kees, Kleinfleiss Kees, and Wurten Kees (Hynek et al., 2012).

Glacier Inventories: Area and Volume Changes

The areas of all Austrian glaciers have been compiled for the LIA maximum (GI LIA), 1969 (GI 1), 1998 (GI 2), and 2004–11 (GI 3; Fischer et al., 2015a). The LIA maximum covers time spans from several years to decades around 1850s, because the time when a glacier reached its maximum state after the LIA varies from glacier to glacier. The glacier area was derived from moraine mappings of Groß (1987), adapted to the LiDAR data. The moraine positions of the glacier tongues of larger glaciers can be mapped with smaller

TABLE 9.4 Area Reduction of Austrian Glaciers Between LIA and Today (Fischer et al., 2015a) in Absolute Numbers and Percentage of LIA Area

	Size Classes (km^2)						Total Area (km^2)	Area (%)	Loss (%)
	<0.1	0.1–0.5	0.5–1	1–5	5–10	>10			
GI LIA	Small glaciers potentially missing						941.13	100	
GI 1	11.30	96.03	79.08	220.30	84.73	73.43	564.88	60	−40
GI 2	18.83	80.01	65.89	192.97	65.89	47.07	470.67	50	−50
GI 3	20.77	70.63	49.86	170.34	70.63	33.24	415.47	44	−56

uncertainties, the glacier margins in the former accumulation area show greater uncertainties. Some glaciers which have melted away since the LIA maximum are missing in the inventory.

The inventories show a decrease of glacier area by 66% between GI LIA and GI 3 (Table 9.4). The number and area of small glaciers increases between GI 1 and GI 3. Of the five glaciers with an area of more than 10 km^2, only two are left in this size class in GI 3. The disintegration of larger glaciers into several smaller ones has implications on flow dynamics and thus on further glacier retreat: The inflow of ice toward the glacier tongues is reduced; assuming the same melt rate, less of the melted ice is replaced and the glacier retreat rate increases. Small glaciers often are located at high altitudes and in favorable settings, that is in shaded and avalanche-fed areas. The altitude with the main glacier area rose between GI 3 and GI 2 (Fig. 9.26).

Lambrecht and Kuhn (2007) calculated the mean thickness loss between GI 1 and GI 2 to be 0.3 m/annum. They find a maximum volume change in the elevation band from 2800 to 2850 m, which is about 100 m lower than the altitude zone with the largest glacier area. Regional studies (Abermann et al., 2009) indicate that absolute volume change between GI 2 and GI 3 is similar to the one between GI 1 and GI 2, although the period is much shorter. For the Ötztal Alps they calculated a mean annual thickness change of −0.3 m/year for GI 1 to GI 2 and −0.9 m/year for GI 2 to GI 3. They found the greatest changes for glaciers between 0.5 and 1 km^2. Thickness loss of individual glaciers differs greatly, with the largest losses at the glacier tongues. On Jamtal Ferner up to 24 m of ice thickness was lost within 10 years at the glacier tongue, with about half of the glacier area losing less than 8 m (Fig. 9.27). In the upper parts of the glacier, major thickness losses occur at steep and exposed parts of the glacier, that is crevasse zones and ice crests.

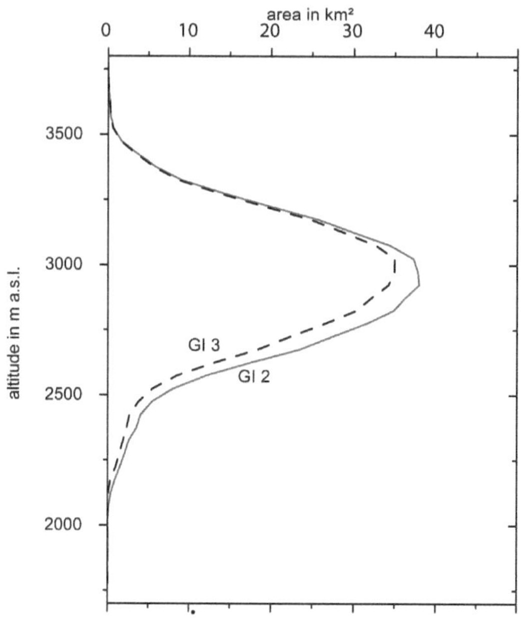

FIGURE 9.26 Altitudinal distribution of glacier area in GI 2 and GI 3.

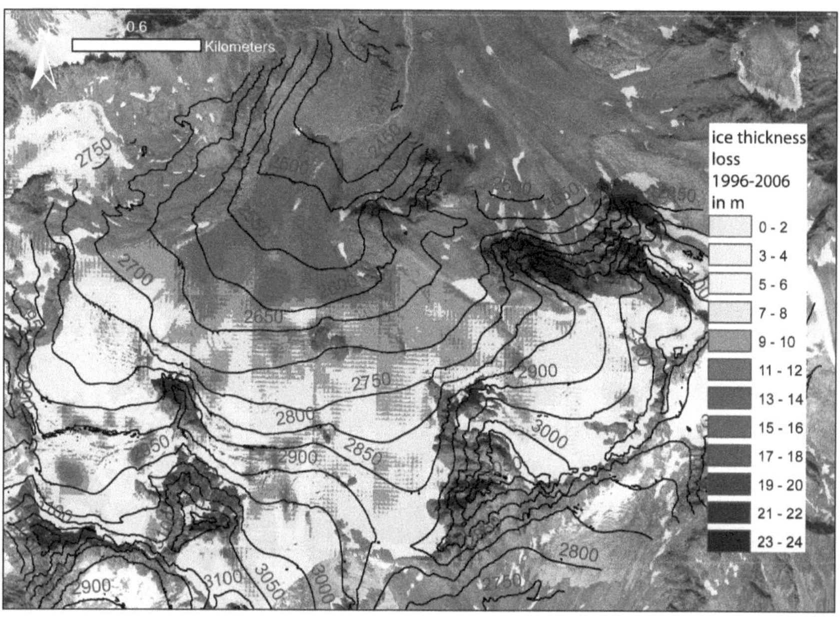

FIGURE 9.27 Thickness loss between GI 2 (1996) and GI 3 (2006) for Jamtal Ferner, Silvretta.

Ice Volume

The first glacier depths were measured between 1893 and 1922 on Hintereis Ferner in the Ötztal Alps by 12 mechanical drillings (Hess, 1904). From 1929 onwards, Mothes, Brockamp, Thyssen, Förtsch, and Vidal applied seismic methods for ice thickness measurements on Austrian glaciers (Span et al., 2005). Radio-echo sounding was applied in 1982 by Wächter on Hintereis Ferner. In the 1960s and 1970s, Aric and Brückl surveyed eight Austrian glaciers with seismic sounding (Span et al., 2005; Fischer et al., 2007).

From 1995 to 2010, 64 Austrian glaciers with areas from 0.001 to 18.4 km^2 were surveyed with low-frequency GPR (Fischer and Kuhn, 2013). These glaciers extend over a total of 223.3 ± 3.6 km^2, that is 47% of the total glacier area. The glacier volume is 11.9 ± 1.1 km^2, the mean thickness 50 ± 3 m. The maximum ice thickness ranges from 26 to 311 m, with the mean maximum being more than twice the mean ice thickness (Fig. 9.28).

The distribution of ice thickness within the glaciers varies significantly. Typically, maximum ice thicknesses at the glacier tongues are found where the glacier bed is overdeepened as a result of the ice flow history, that is where

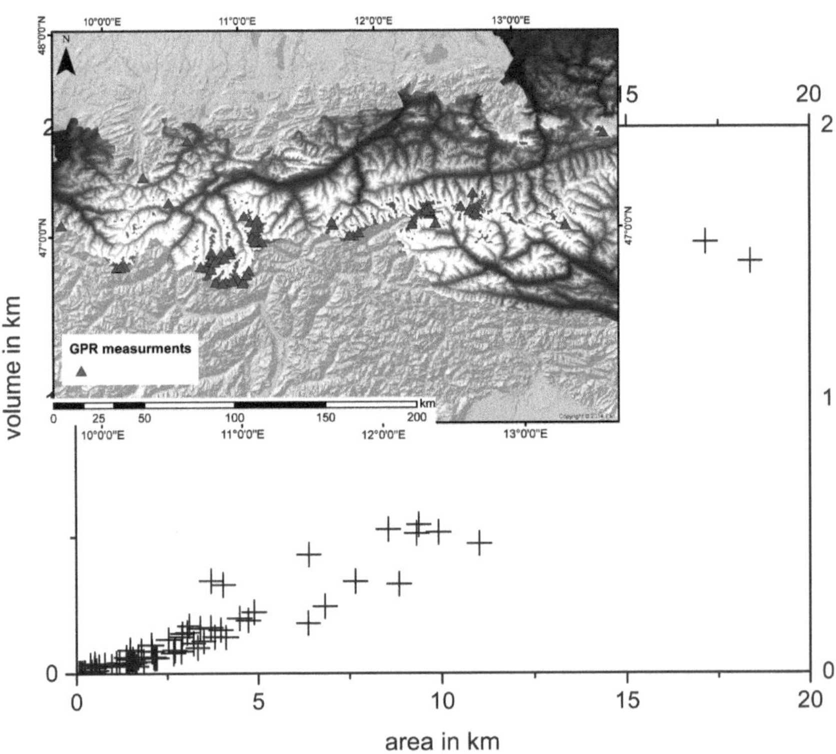

FIGURE 9.28 Volume and area of the GPR-surveyed Austrian glaciers (inset: Map of surveyed glaciers).

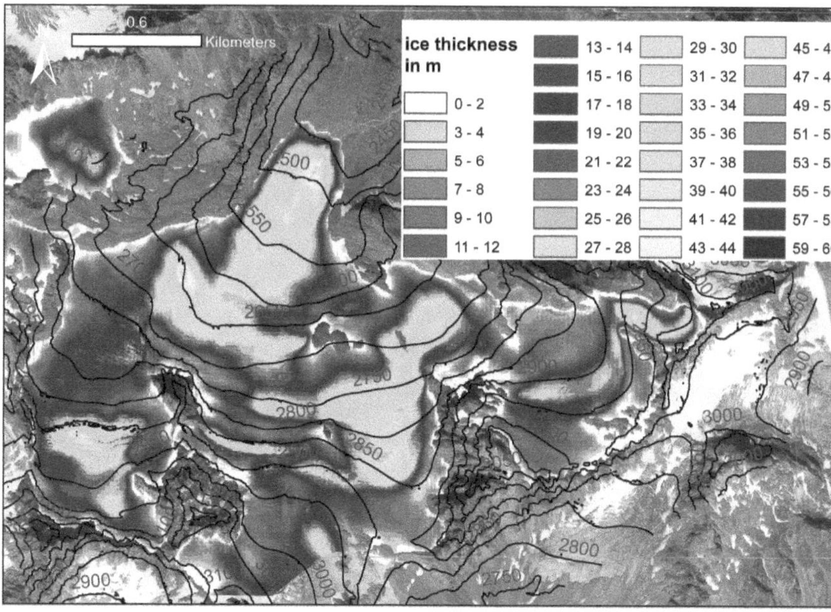

FIGURE 9.29 Ice thickness of Jamtal Ferner in 2015.

current or historical glacier tributaries joined and high flow velocities led to bed erosion (Fig. 9.29). The maximum ice thicknesses on Jamtal Ferner have been measured at the glacier tongue, where ice from three (until 1988 four) tributary glaciers joined.

Runoff

Continuous measurements of glacier runoff are rare, with the longest time series at Vernagt Ferner (Escher-Vetter and Reinwarth, 2013) and shorter ones at Obersulzbach Kees (since 1988) and Stubacher Sonnblick Kees (since 2002). Manual runoff observations had been part of the early mass balance monitoring on Hintereis Ferner and Kesselwand Ferner (Finsterwalder, 1891; Hoinkes and Lang, 1961). From 1897 onwards, an automatic gauge in Vent has recorded the runoff from Hintereis Ferner, Kesselwand Ferner, Vernagt Ferner, and Hochjoch Ferner as well as a number of smaller glaciers (Gattermayr, 2013). Only a few of the historical gauges are still operating today, although long time series are necessary to identify trends and variability on a catchment scale. Measured parameters are water level and temperature, discharge, suspended load, and bedload transport. The seasonal runoff from the glacier-covered catchment area of Rofenache in Vent (Fig. 9.30, data provided by the Hydrographical Service of the Federal Government of Tyrol) shows a clear peak in July and August as a result of ice melt. Minimum runoffs are low, as

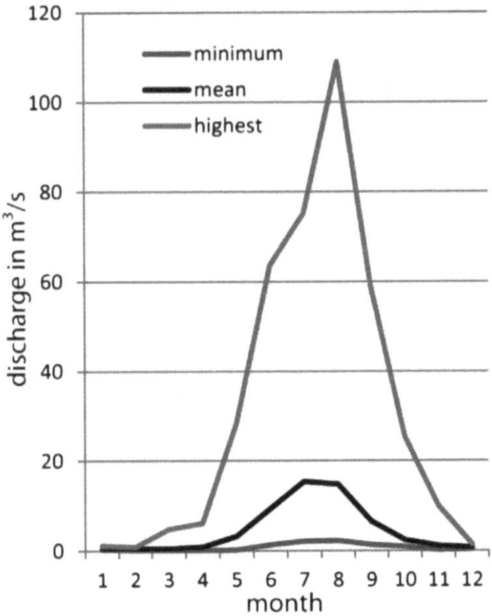

FIGURE 9.30 Mean, minimum, and maximum discharge calculated for the period 1971–2012 for the runoff gauge in Vent (98.1 km^2, 38% of the area is covered by glaciers).

they occur when ice melt is suspended during summer cold fronts. Maximum runoffs stem from heavy precipitation events occurring at the same time as the maximum glacier runoff, which reaches the gauge in late afternoon. Annual transport of suspended sediment has been measured to vary between 150,000 and 240,000 tons. The annual transported bedload sediment is even higher (340,000 to 720,000 tons, Gattermayr, 2013).

The variance of total runoff is higher the smaller the basin and the higher the fraction of the ice-covered area within the basin. The total catchment area of the gauge Türkische Zeltstadt at Obersulzbach Kees is 17.8 km^2, approximately 55% of it covered by glaciers, Obersulzbach Kees, and its tributaries (status 2009). A high seasonal discharge fluctuation (from 50 l/s in winter to 18,000 l/s in summer), very low electric conductivity (3–10 μS/cm), and daily water-level variations of 20–30 cm during summer times at Lake Obersulzbachsee are the main characteristics of the gauge Türkische Zeltstadt, situated approximately 800 m upstream of the gauge Kees (Fig. 9.31). The year-to-year variability of the monthly mean discharge is high, and even higher is the variability of the daily discharges.

Snow Accumulation on Glaciers

Conventional methods provide a few point measurements but no information on the distribution of snow cover in high Alpine catchments. In the upper

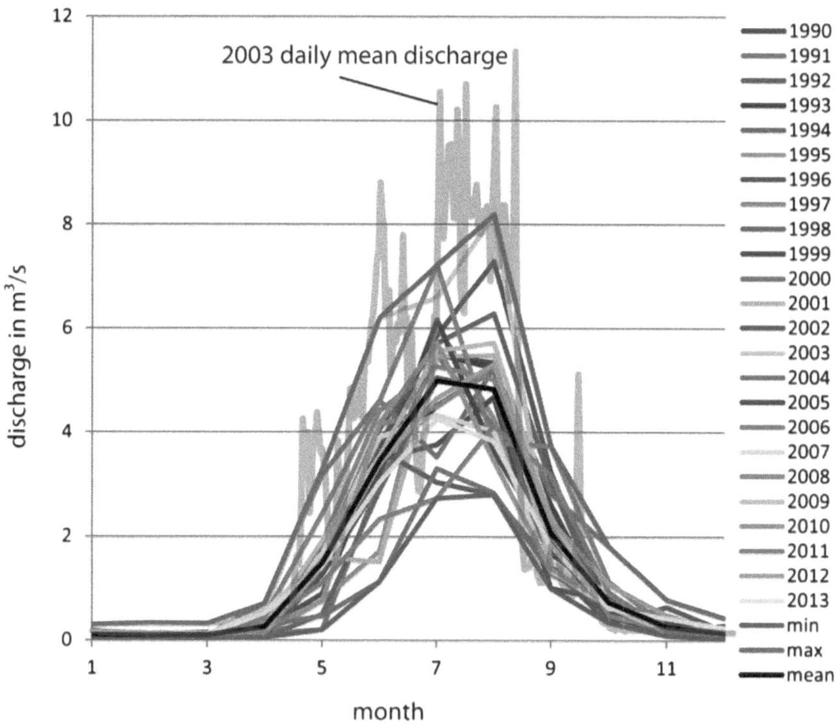

FIGURE 9.31 Monthly mean discharge at the gauge Kees for the years 1990–2013 with mean, minimum, and maximum during this period. Superimposed (purple (gray in print versions)) is the daily mean discharge of the extreme year of 2003.

Rofental valley (46.81N, 10.79E, Tyrol, Austria), the snow accumulation of a 36 km^2 large, partly glacierized, catchment was analyzed on the basis of multiple ALS surveys (Helfricht et al., 2014b). The influence of glacier flow at glaciers like Hintereis Ferner and Vernagt Ferner is small, so that seasonal surface elevation changes can be interpreted as snow depths in this region. However, actual snow depths are significantly underestimated by LiDAR--derived surface elevation changes in present firn areas. This is caused by the summation of submerging ice flow and densification of the firn layer (Helfricht et al., 2014a).

ALS surveys of the seasonal snow depth distributions at the end of five accumulation seasons allow extensive statistical analysis of the persistence of snow accumulation patterns. The snow-covered area at the end of the accumulation season varied by 16% and the mean snow depth of the total area ranged by a factor of 2 between the 5 years.

Fig. 9.32A shows the mean snow depth based on the ALS data. Exposed mountain crests, steep slopes, and lower altitude south-facing areas remain

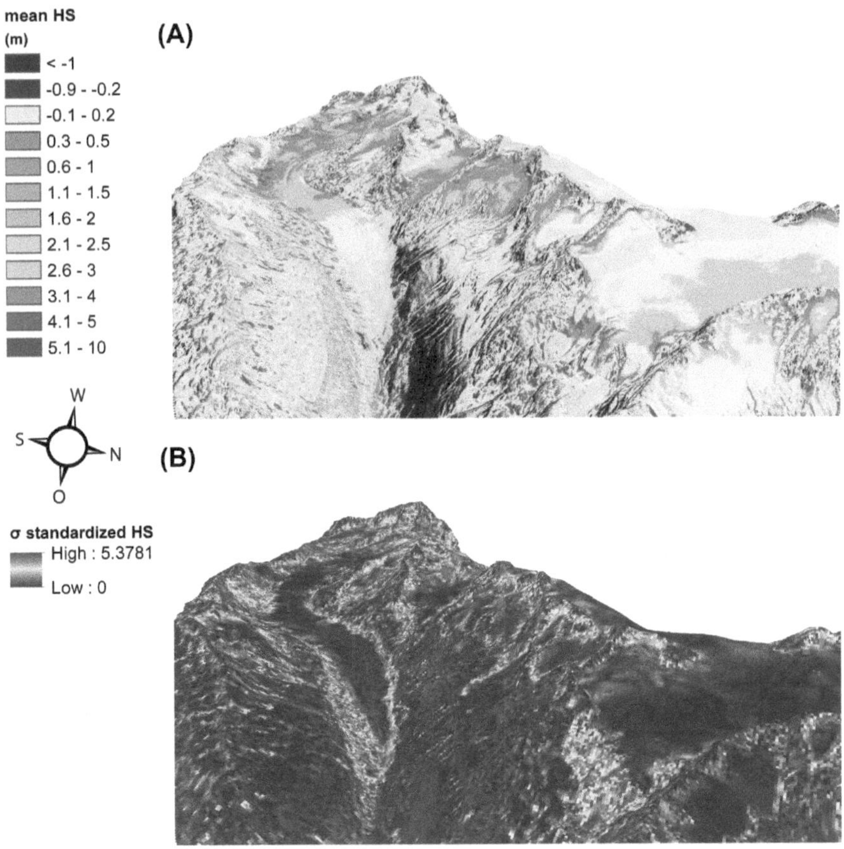

FIGURE 9.32 (A) Mean snow depth (HS) at the end of five accumulation seasons in the Upper Rofental valley (46.81N, 10.79E, Tyrol, Austria) calculated from ALS surveys. Note that surface elevation changes of ±0.2 m are transparent on the underlying shaded relief. (B) The 5-year-standard deviation (σ) of seasonal standardized HS as a measure of interannual variability of accumulation patterns.

almost snow free up to the end of accumulation season, whereas high snow depths can be expected in wind sheltered areas, in surface depressions, at the foot of steep slopes or rock walls, and more generally on glacier surfaces. Helfricht et al. (2014b) found that standardized snow depths are suitable for analyzing the interannual persistence of snow depth patterns. Fig. 9.32B shows the multiannual standard deviation of standardized snow depths. High values are caused by high interannual variability, and vice versa. Snow depth distribution patterns on glaciers and in less steeply inclined terrain present high interannual persistence. Extreme snow depths along rather narrow bands along the glacier margins showed little persistence between years. Helfricht et al. (2014b) also found that snow cover underwent a change from a homogeneous distribution on the former glacier surface to a more heterogeneous snow cover in the recently deglaciated terrain.

Knowledge on interannual persistence of accumulation patterns supports conventional winter mass balance measurements and analysis. Additionally, the LiDAR-derived snow depth distribution, in combination with assumptions on snow density to calculate Snow Water Equivalent, yield a basis for calibration and validation of hydrological models (e.g., Schöber et al., 2014; Helfricht et al., 2015).

In general, LiDAR is a costly surveying technique. New lightweight systems allow using laser scanners on Unmanned Aerial Vehicles, which will fill the gap between the TLS and ALS surveys with respect to flexibility in survey planning and extensive spatial covering.

Ice Flow Velocity

Ice flow measurements were carried out in the late-19th century on Hochjoch Ferner and Hintereis Ferner. Today annual ice flow velocity is measured along profiles with the help of stone lines on Hintereis Ferner (Span et al., 1997), Kälberspitz Kees (e.g., Fischer, 2016), and Pasterze (Lieb and Slupetzky, 2004). Stone line measurements allow the calculation of horizontal displacement rates only, and these showed a decrease during the glacier retreats in the 21st century.

Stake measurements allow the calculation of horizontal and vertical displacement rates. The longest time series of stake measurements of ice flow velocity has been measured on Kesselwand Ferner in the Ötztal Alps. The time series was initiated by H. Schneider in 1965 (Schneider, 1970; Span and Kuhn, 2003).

Kesselwand Ferner has had a period of positive mass balances from 1955 to 1984. During that time, L6 was an accumulation stake, often close to the ELA (Fig. 9.33). L2 still is an accumulation stake in most of the years, and L8 and L10 are ablation stakes. The highest ice flow velocities along the glacier surface are recorded at stakes L8 and L10, the lowest at stake L2. All stakes have shown decreasing flow velocities once the mass balance turned negative after 1985 (In that year, Kesselwand Ferner reached its maximum length for the 1980s glacier advance period.). Flow velocity at stake L10, where the glacier is now completely gone, is down to zero as the inflow of ice toward the terminus has become drastically reduced and the glacier tongue and the main glacier have nearly lost connection (Fig. 9.34).

During glacier advance, emergence velocity on Kesselwand Ferner was highest at stake L10. After 1985 emergence velocity decreased to values lower than 1 m for all stakes, while submergence velocity at the highest stake L2 remains high (Fig. 9.35).

Emergence/submergence velocity is nearly as high as local ablation/accumulation, so that most of the glacier surface was in an almost steady state

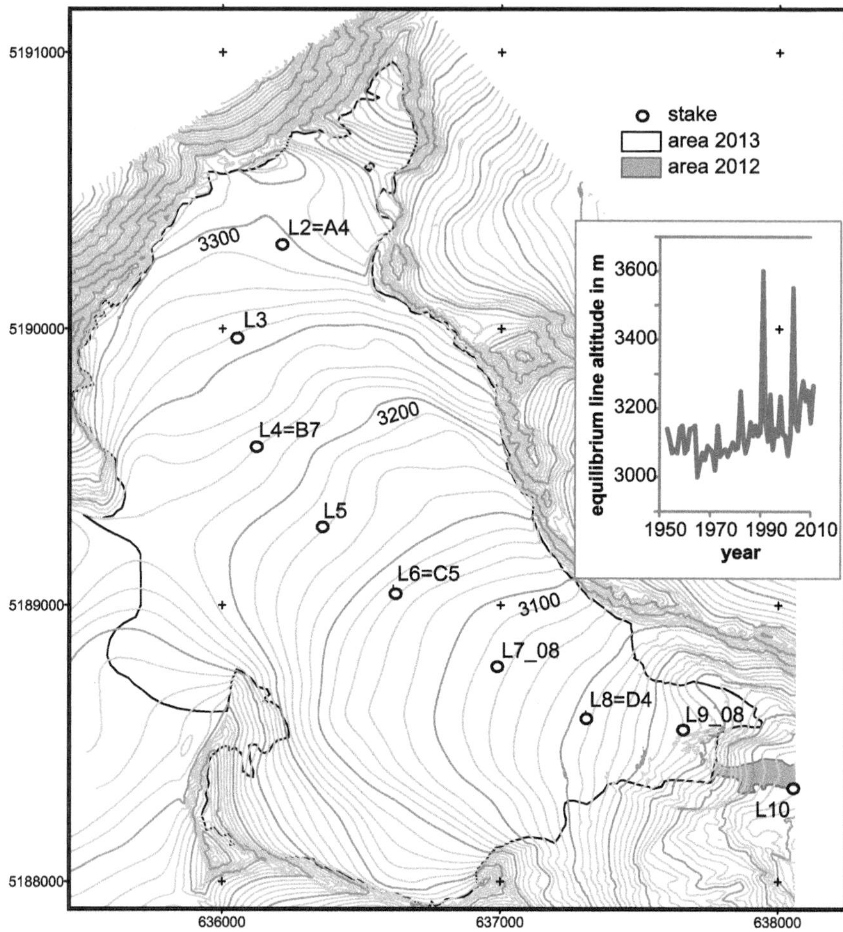

FIGURE 9.33 Map of stake network on Kesselwand Ferner (Ötztal Alps) in 2013. Inset: Equilibrium line altitude from direct mass balance measurements.

prior to 1985. The local thickness changes had been very small for most of the stakes prior to 1985 and positive for the stakes at the advancing glacier tongue. After 1985, slightly negative thickness changes occurred at the lower stakes as immediate reaction to the negative mass balances. The elevation of the upper stakes L2 and L6 remains nearly unchanged (Fig. 9.36).

The response of ice flow velocity to changes in glacier mass balance is quite quick, and the interpretation of time series of horizontal and vertical ice flow velocities at several locations along altitudinal transects allows estimating the rough mass balance state of a glacier.

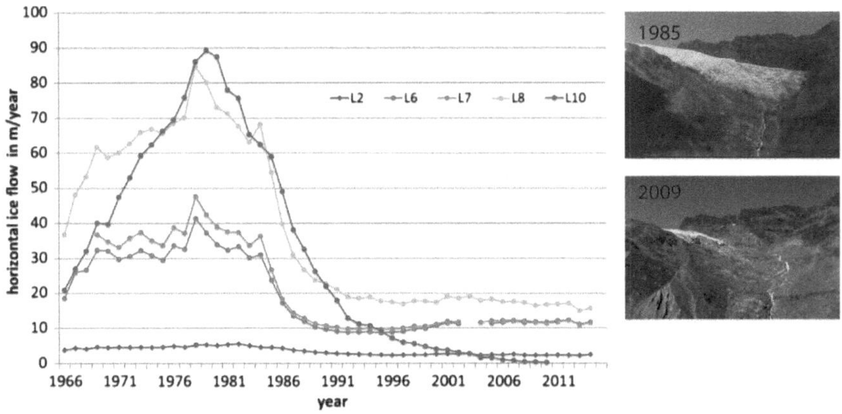

FIGURE 9.34 Horizontal ice flow on Kesselwand Ferner at stakes L2, L6, L7, L8, L9 and the lowest stake L10. Inset: glacier terminus at the final stage of advance in 1985 and in 2009.

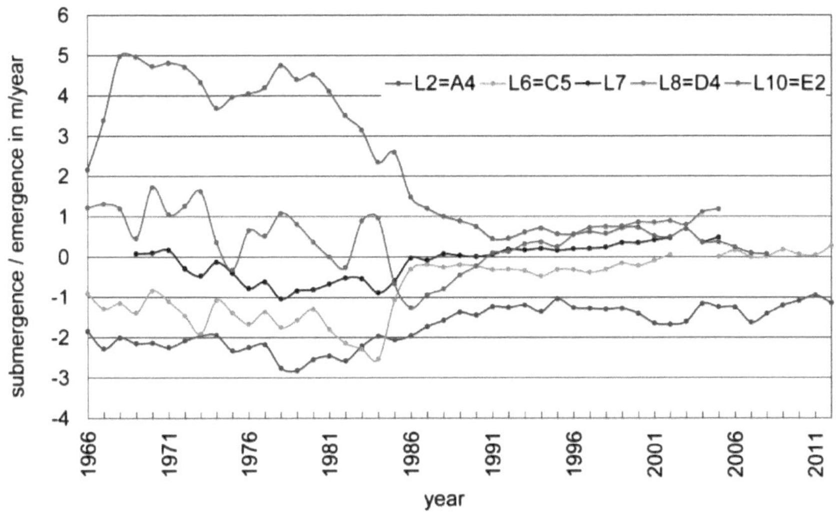

FIGURE 9.35 Emergence (negative values: submergence) velocity on Kesselwand Ferner at stakes L2, L6, L7, L8, L9 and the lowest stake L10.

ARE THE SURVEYED GLACIERS REPRESENTATIVE OF ALL AUSTRIAN GLACIERS?

The various monitoring parameters, that is length change, mass balance, glacier areas and elevation in inventories, ice volumes, runoff, and seasonal snow cover, have historically developed to fulfill different needs. With respect to technical and operational limitations, the sample sizes surveyed for the

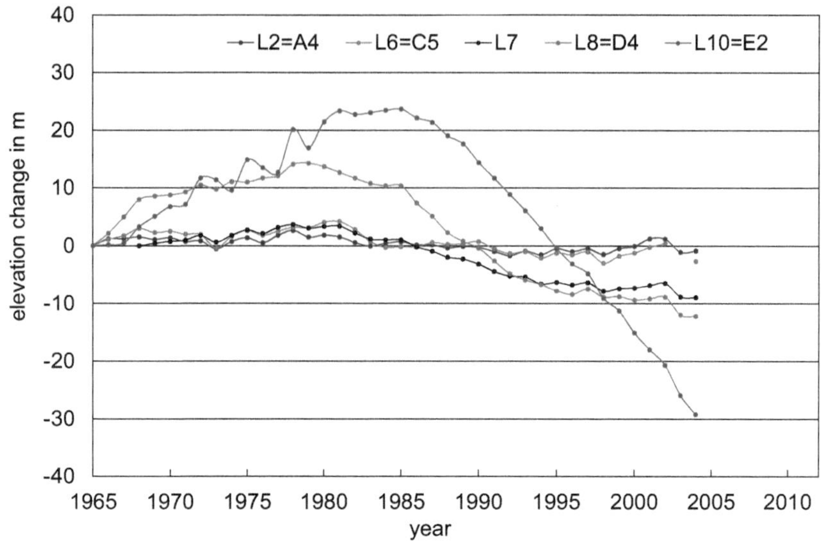

FIGURE 9.36 Elevation change on Kesselwand Ferner at the stakes L2, L6, L7, L8, L9 and the lowest stake L10 (Abermann et al., 2007).

specific parameters differ. Where glacier inventories basically cover all glaciers they do so with a low temporal resolution. Runoff, ice flow, and mass balance are measured on just a handful of sites (Fig. 9.37). Runoff and mass balance data are needed especially in hydrology, ice flow, and mass balance for

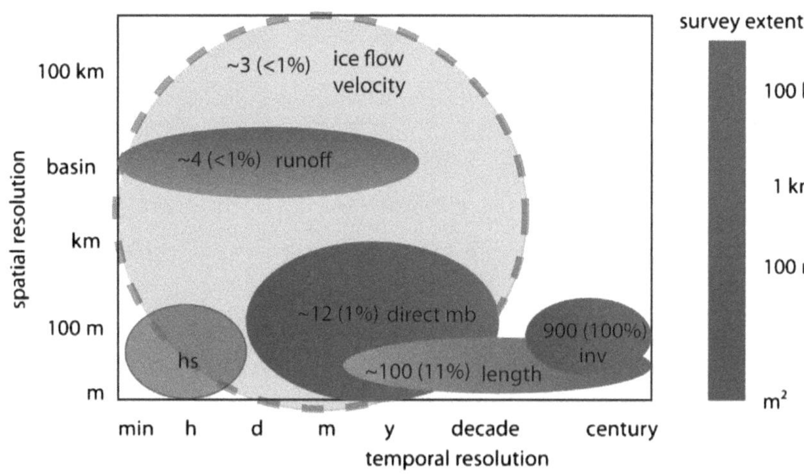

FIGURE 9.37 Spatial and temporal resolution of the cryospheric monitoring parameters with the potential covered area. The numbers represent the Austrian glaciers for which that parameter is surveyed and their percentage of all Austrian glaciers. *hs*, height of snow; *inv*, inventory (glacier area and surface elevation); *mb*, mass balance.

capturing climate change. It is not possible to achieve high spatial coverage with these parameters, unless satellite technologies for measuring ice flow and snow cover become operational for mountain glaciers. The length change time series is the longest time series with a high temporal and spatial resolution and a large sample size, and was initiated with special attention to tracking climatic change. Do we need all these "measuring the same but different" time series, or should we just decide which ones to keep and focus all resources on those series?

One part of the answer is legal: For all infrastructure projects in glacier-covered environments, the effects of the measures on glaciers must be investigated to decide if the project should be approved. Therefore, a scientific glaciological database for all glaciers is necessary for studying local effects. Usually, these studies rely on the glacier inventories and mass balance time series. The glacier inventories provide volume changes and glacier areas for all glaciers, and the mass balance time series are used for converting volume changes to mass changes and for hydrological interpretations.

The second part of the answer concerns the time horizon: New monitoring strategies should not truncate the length of time series. Therefore, length change time series are valuable as they provide the longest time series.

The third part of the answer concerns the global aspect: National and local monitoring strategies should fulfill the needs of global monitoring. In the case of the Austrian Alps, the variety of measured parameters, together with the length of the time series, allow comparative studies on the information content of the various parameters for specific research questions, for example for estimating the hydrological effects of glacier retreat. Regional effects on hydrology result from the reaction of all the glaciers in the catchment area. Even if the reaction of an individual glacier has a high variability, the variability of the basin glacier runoff might be small, so that measurements on a basin scale might complement and smooth measurements taken on individual glaciers. Most sophisticated is the relationship between regional-scale data and local measurements in the interpretation of global past climate based on local moraines or documents. Here, past and very local states of a *specific* glacier are used to find indications of a *general and larger scale* climatic change. Current global glacier monitoring strategies are therefore based on different monitoring levels (Zemp et al., 2014).

A key question is the representability of time series with a low number of measurements for the range of changes of all glaciers in the region. One consequence of the answer to this question would allow the selection of appropriate glaciers for a monitoring strategy. As in most cases, sites evolve historically independent of strategies, and the a priori selection of glaciers with respect to their long-term future properties might be difficult, so this consequence might remain a theoretical one. The practical consequence remains as limits of interpretation of large-scale data for small-scale changes (e.g., glacier inventory data for local mass balance) or the interpretation of small-scale data

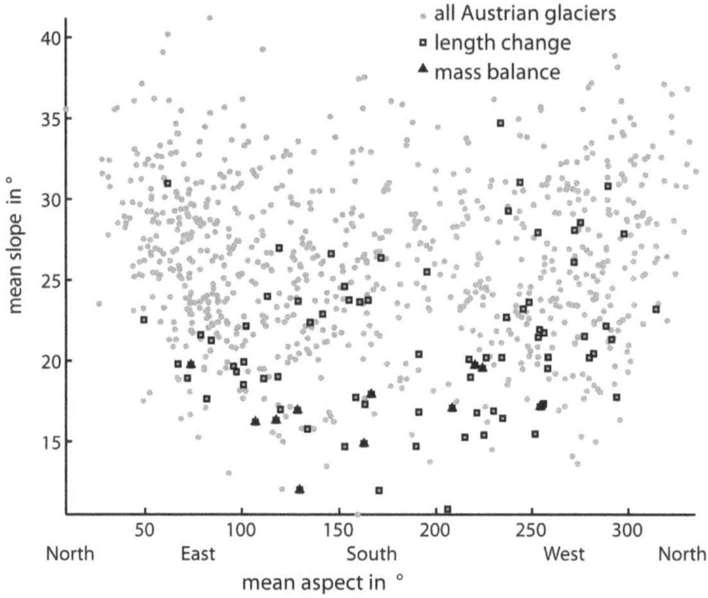

FIGURE 9.38 Mean slope and aspect for all Austrian glaciers (GI 2) and the glaciers with length change and mass balance monitoring programs.

for large-scale changes (e.g., local mass balance time series for regional hydrological changes).

As a first approach, basic topographic properties of all glaciers are compared to the ones of glaciers in the length change and mass balance monitoring network. The total sample shows mean slopes of up to 40 degrees, with a mean slope of 35 degrees for the steepest length change glaciers and slopes of 20 degrees for the steepest mass balance glaciers. The actually steepest glaciers are not represented in the monitoring network. Unlike slope, most other mean aspects are covered by the mass balance and length change subsamples (Fig. 9.38).

Austria has only three glaciers larger than 10 km^2 and seven glaciers with areas between 5 and 10 km^2 in the second glacier inventory GI 2 (Fischer et al., 2015a). Most of these large glaciers are part of length and mass balance monitoring, as well as some of the many small glaciers, so that the area distribution of the monitored glaciers in the total glacier sample is well balanced. The monitored glaciers also present a good distribution with respect to area changes (Fig. 9.39).

Mass balance and length change monitoring is carried out on glaciers which are clearly larger than the average of all Austrian glaciers. Large mountain glaciers have glacier tongues reaching far down into the valley, where air temperatures and ice melt are higher than at higher elevations.

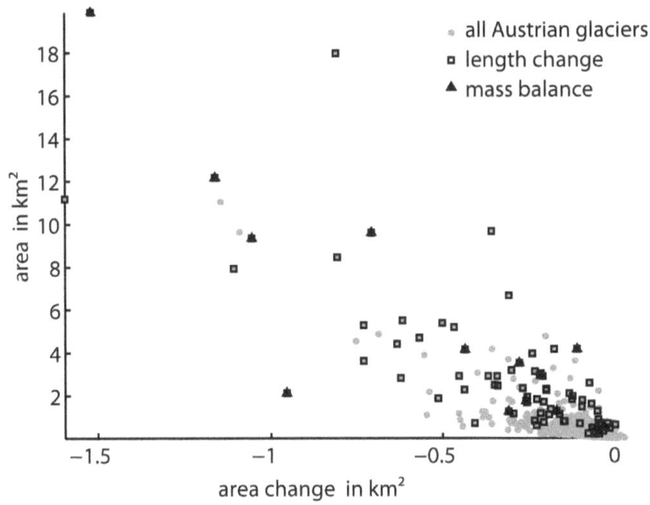

FIGURE 9.39 Area (GI 2, 1998) and area changes (between GI 1, 1969 and GI 2, 1998) for all Austrian glaciers and the glaciers of the length change and mass balance monitoring programs.

Therefore, these larger glaciers tend to return more negative mass balances than the smaller glaciers located at higher elevations. The comparison of the altitudinal reach of the glaciers in the specific monitoring network thus reveals some information on climate sensitivities of the glaciers. Fig. 9.40 shows the mean and extreme upper- and lower-elevation limit of the glaciers of each monitoring network for all three glacier inventories. Between GI 1 (dating from 1969) and the latest glacier inventory, the mean and extreme lower limits of the glacier have risen. As the largest retreat rates are observed on the largest glaciers, the mean minimum elevation rose most for mass balance glaciers between GI 1 and GI 2 (43 m) and for length change glaciers between GI 2 and GI 3. The smallest samples, which are the mass balance glaciers, cover the smallest elevation band. On average, mass balance glaciers reach further down into the valley than length change glaciers.

The representativity of subsamples of glaciers is changing with glacier retreat or advance on timescales of decades. The a priori selection of an optimum small sample which will be representative for all glaciers for the next decades might be not possible, so that large enough samples should be conceived to cover all types of glaciers.

FUTURE CHALLENGES FOR GLACIER MONITORING IN AUSTRIA

Current glacier retreat is a challenge for glacier monitoring in three ways: First, the rapid glacier retreat with disintegration of large glaciers into smaller

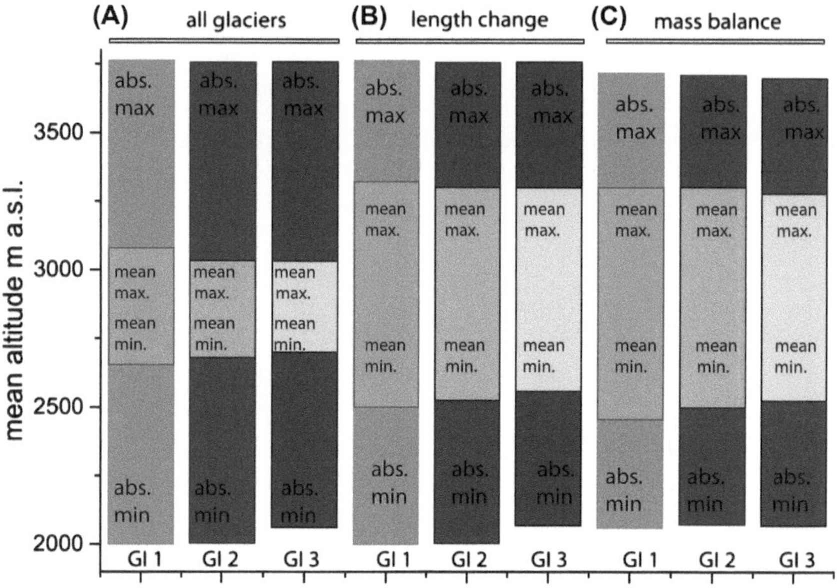

FIGURE 9.40 Mean and extreme uppermost and lowermost altitudes of (A) all Austrian glaciers, (B) length change glaciers, and (C) mass balance glaciers in the three glacier inventories GI1, GI2, and GI 3.

ones, the formation of periglacial lakes, and the increased debris cover makes it difficult to keep up length change and mass balance measurements. The glacier area subject to mass balance measurements and the dynamics of surveyed glacier tongues is changing quite quickly. The localization of the glacier terminus for length changes is complicated by debris cover and lakes, as often ice is buried under lake sediments. Increasingly, area changes do not necessarily happen at the glacier tongues but also at the highest elevations.

Second, major parts of the monitoring systematics were developed for glaciers closer to balance as they appear to be now. On some of the Austrian mass balance glaciers the ELA in the last decade was above summits for most years, even though the glaciers still had an accumulation area. This is possible when the mean mass balance at all altitudinal zones is negative in the presence of accumulation, for example, when a steep east-facing slope with extremely high ablation rates borders on a shady trough with extreme accumulation rates. Then, per definition, the ELA is above summit, as no elevation zones with positive mean mass balances exist. Therefore, the systematics should be adapted to capture and quantify even such extreme events.

Third, all the above-mentioned processes need a thorough and consistent documentation, which has ideally to be developed and coordinated in real time at national and international level. As the example of Litzner Glacier shows, increased debris cover changes the glacier's response to climate change. The

FIGURE 9.41 Changes on Jamtal Ferner between 1950 and 2015 from the documentation of the Austrian Alpine Club: from lateral rock walls covered by ice in the 1950s, rock falls transport debris to the glacier surface (*yellow line and arrow* (light gray in print versions)), the upper end of the supraglacial moraine raises (*red arrows* (dark gray in print versions)), and rock outcrops appear (*blue lines* (gray in print versions)). Firn cover reduces, steep slopes loose firn, and ice cover (*black arrows* (black in print versions)).

increase of debris cover is also related to rock falls from formerly ice-covered lateral rock walls and changes in the flow regime of glaciers (Fig. 9.41).

The response of glaciers to climate change generally is not only governed by the climate, but also by glacier changes (Fischer, 2010). A reduction of glacier area in zones with high ablation rates leads to reduction in glacier

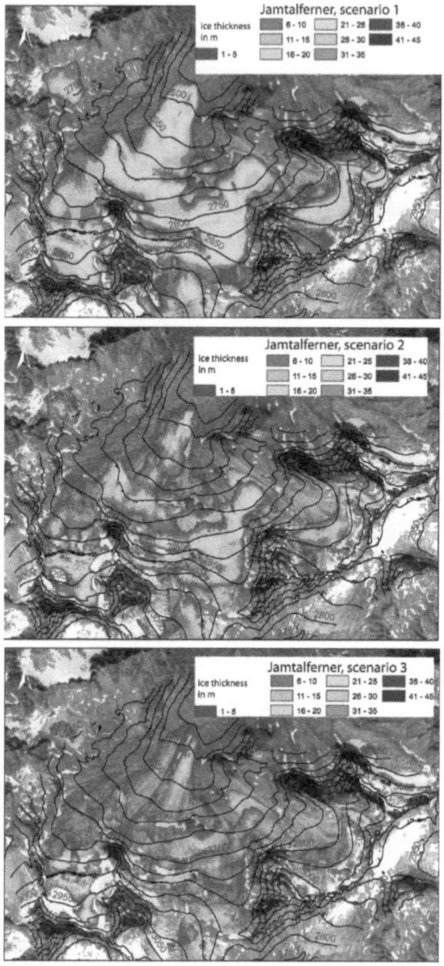

FIGURE 9.42 Scenarios of ice thickness and glacier area for Jamtal Ferner in Silvretta, based on volume change 1996–2006.

runoff and a less negative specific mass balance (Marzeion and Kaser, 2014). Regional and local scenarios of future climate still include great uncertainties, especially for the precipitation regime (Christensen et al., 2007), which is characterized by steep gradients in high-mountain regions (Whitemann, 2000). A rough but robust future scenario can be developed by extrapolation of past volume changes into the future. This method neglects the effect of still rising temperatures, but includes past changes in the dynamic regimes. In any case, there is great uncertainty about future glacier scenarios (Huss et al., 2014), so that we do not propose a time frame for potential future changes here. The

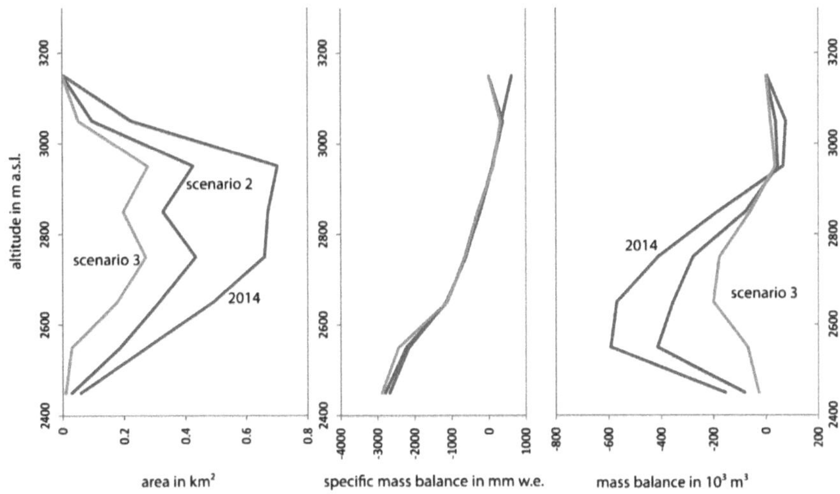

FIGURE 9.43 Implications of area reduction on mass balance as measured in the year 2014.

extrapolation of one-fold (scenario 1), two-fold (scenario 2), and three-fold (scenario 3) past volume changes, calculated from the glacier inventory data 1997 and 2006 into the future for Jamtal Ferner in Silvretta, returns a glacier area reduction to 60% for scenario 1, 33% for scenario 2, and a nearly complete loss of ice in scenario 3 (Figs. 9.42 and 9.43).

Assuming that the mass balance year 2014, with a specific mass balance of −846 mm w.e., occurred on a glacier area like that in scenarios 1 and 2, the specific mass balance is reduced to −621 mm w.e. and −480 mm w.e., respectively. The glacier runoff, which had been 6,190,000 m^3 in 2014, would go down to 1126 m^3 in scenario 1 and 486 m^3 in scenario 2. The significant reduction in glacier runoff effects the total runoff only in small basins with high proportions of glacier cover, but not so much in larger basins with lower glacier cover (Kaser et al., 2010). In any case, the midterm glacier retreat as presented in these scenarios is challenging for capturing the speeding up disintegration processes to develop local adaption and mitigation strategies, including, for example, flood protection downstream.

ACKNOWLEDGMENTS

This chapter presents a monitoring network to which numerous individuals and institutions have contributed for more than 100 years. It is not possible to thank all contributors personally, although such monitoring would not be possible without the people committed to it. We want to name at least a few of the institutions engaged in monitoring: The Austrian Alpine Club, with the help of 20 observers, maintains the glacier survey of the glacier length measurements. The hydrological services of Tyrol, Salzburg, Upper Austria, and Hohe

Tauern National Park contribute to the measurements and operate runoff gauges. Measurements are also carried out by the ZAMG, the Bavarian and Austrian Academies of Sciences, the universities of Innsbruck, Graz and Salzburg, and by the Glacier and Climate Association. We are also grateful to Brigitte Scott for the English editing, which improved the final text.

REFERENCES

Abermann, J., Fischer, A., Lambrecht, A., Geist, T., 2010. On the potential of very high-resolution repeat DEMs in glacial and periglacial environments. The Cryosphere 4, 53–65. http://dx.doi.org/10.5194/tc-4-53-2010.

Abermann, J., Kuhn, M., Fischer, A., 2011. Climatic controls of glacier distribution and changes in Austria. Annals of Glaciology 52 (59), 83–90.

Abermann, J., Lambrecht, A., Fischer, A., Kuhn, M., 2009. Quantifying changes and trends in glacier area and volume in the Austrian Ötztal Alps (1969–1997–2006). The Cryosphere 3, 205–215. http://dx.doi.org/10.5194/tc-3-205-2009.

Abermann, J., Schneider, H., Lambrecht, A., 2007. Analysis of surface elevation changes on Kesselwand glacier. Comparison of different methods. Zeitschrift für Gletscherkunde und Glazialgeologie 41, 147–167.

Anttila, K., Hakala, T., Kaasalainen, S., Kaartinen, H., Nevalainen, O., Krooks, A., Kukko, A., Jaakkola, A., 2015. Calibrating laser scanner data from snow surfaces: correction of intensity effects. Cold Regions Science and Technology 121, 52–59. http://dx.doi.org/10.1016/j.coldregions.2015.10.005.

Auer, I., Böhm, R., Jurkovic, A., Lipa, W., Orlik, A., Potzmann, R., Schöner, W., Ungersböck, M., Matulla, C., Briffa, K., Jones, P.D., Efthymiadis, D., Brunetti, M., Nanni, T., Maugeri, M., Mercalli, L., Mestre, O., Moisselin, J.-M., Begert, M., Müller-Westermeier, G., Kveton, V., Bochnicek, O., Stastny, P., Lapin, M., Szalai, S., Szentimrey, T., Cegnar, T., Dolinar, M., Gajic-Capka, M., Zaninovic, K., Majstorovic, Z., Nieplova, E., 2007. HISTALP – historical instrumental climatological surface time series of the greater Alpine region 1760–2003. International Journal of Climatology 27, 17–46.

Baltsavias, E., 1999. Airborne laser scanning: basic relations and formulas. ISPRS Journal of Photogrammetry and Remote Sensing 54, 199–214. http://dx.doi.org/10.1016/S0924-2716(99)00015-5.

Blümcke, A., Hess, H., 1899. Untersuchungen am Hintereisferner. Wissenschaftliche Ergänzungshefte zur Zeitschrift des Deutschen und Österreichischen Alpenvereins 1 (2), 1–87.

Bojinski, S., Verstraete, M., Peterson, T.C., Richter, C., Simmons, A., Zemp, M., 2014. The concept of essential climate variables in support of climate research, applications, and policy. Bulletin of the American Meteorological Society 95 (9), 1431–1443. http://dx.doi.org/10.1175/BAMS-D-13-00047.1.

Bollmann, E., Sailer, R., Briese, C., Stötter, J., Fritzmann, P., 2011. Potential of airborne laser scanning for geomorphologic feature and process detection and quantifications in high alpine mountains. Zeitschrift für Geomorphologie 55 (2), 83–104.

Brunner, K. (Ed.), 2004. Das Karls-Eisfeld. Forschungsarbeiten am Hallstätter Gletscher. Wissenschaftliche Alpenvereinshefte, vol. 38.

Brunner, K., 2002. Regionalkarten von Tirol des Matthias Burgklechner und ihre Vorläufer. Mitteilungen der Österreichischen Geographischen Gesellschaft 144, 237–254.

Brunner, K., 2006. Karten dokumentieren den Rückzug der Gletscher seit 1850. Kartographie als Kommunikationsmedium. Wiener Schriften zur Geographie und Kartographie 17, 191–200.

Christensen, J.H., Hewitson, B., Busuioc, A., Chen, A., Gao, X., Held, I., Jones, R., Kolli, R.K., Kwon, W.-T., Laprise, R., Magaña Rueda, V., Mearns, L., Menéndez, C.G., Räisänen, J., Rinke, A., Sarr, A., Whetton, P., 2007. Regional climate projections. In: Solomon, S., Qin, D., Manning, M., Chen, Z., Marquis, M., Averyt, K.B., Tignor, M., Miller, H.L. (Eds.), Climate Change 2007: The Physical Science Basis. Contribution of Working Group I to the Fourth Assessment Report of the Intergovernmental Panel on Climate Change. Cambridge University Press, Cambridge, United Kingdom and New York, NY, USA.

Cogley, J.G., Hock, R., Rasmussen, L.A., Arendt, A.A., Bauder, A., Braithwaite, R.J., Jansson, P., Kaser, G., Möller, M., Nicholson, L., Zemp, M., 2011. Glossary of Glacier Mass Balance and Related Terms. IHP-VII Technical Documents in Hydrology 86, IACS Contribution 2. UNESCO-IHP, Paris.

Cuffey, K.M., Paterson, W.S.B., 2010. The Physics of Glaciers. Butterworth-Heinemann/Elsevier, Burlington, MA, 693 pp.

Deems, J.S., Gadomski, P.J., Vellone, D., Evanczyk, R., LeWinter, A.L., Birkeland, K.W., Finnegan, D.C., December 2015. Mapping starting zone snow depth with a ground-based LiDAR to assist avalanche control and forecasting. Cold Regions Science and Technology 120, 197–204. http://www.sciencedirect.com/science/article/pii/S0165232X15001883.

Deems, J., Paintner, T., Finnegan, D., 2013. LiDAR measurement of snow depth: a review. Journal of Glaciology 59, 467–479. http://dx.doi.org/10.3189/2013JoG12J154.

Diener, C., 1885. Studien an den Gletschern des Schwarzensteingrundes. Zeitschrift des Deutschen und Oesterreichischen Alpenvereins 16, 66–78.

Escher-Vetter, H., Reinwarth, O., 2013. Meteorologische und hydrologische Registrierungen an der Pegelstation Vernagtbach – Charakteristika und Trends ausgewählter Parameter. Zeitschrift für Gletscherkunde und Glazialgeologie 45/46, 117–128.

Finsterwalder, S., 1891. Die Pegelstation Vent. Mitteilungen des Deutschen und Österreichischen Alpenvereins 7 (17), 65–66.

Finsterwalder, S., 1897. Der Vernagtferner. Seine Geschichte und seine Vermessung in den Jahren 1888 und 1889. Wissenschaftliche Ergänzungshefte zur Zeitschrift des Deutschen und Österreichischen Alpenvereins 1 (1), 112 pp.

Fischer, A., 2016. Gletscherbericht 2014/2015. Sammelbericht über die Gletschermessungen des Österreichischen Alpenvereins im Jahre 2015. Bergauf 02/2016 77 (141), 6–13.

Fischer, A., 2009. Calculation of glacier volume from sparse ice thickness data, applied to Schaufelferner, Austria. Journal of Glaciology 55 (191), 453–460.

Fischer, A., 2010. Glaciers and climate change: interpretation of 50 years of direct mass balance of Hintereisferner. Global and Planetary Change 71 (1–2), 13–26. http://dx.doi.org/10.1016/j.gloplacha.2009.11.014.

Fischer, A., 2011. Comparison of direct and geodetic mass balances on a multi-annual time scale. The Cryosphere 5, 107–124. http://dx.doi.org/10.5194/tc-5-107-2011.

Fischer, A., Helfricht, K., Stocker-Waldhuber, M., 2016. Local reduction of decadal glacier thickness loss through mass balance management in ski resorts. The Cryosphere Discuss. http://dx.doi.org/10.5194/tc-2016-61. http://www.the-cryosphere-discuss.net/tc-2016-61/.

Fischer, A., Kuhn, M., 2013. GPR measurements of 64 Austrian glaciers as a basis for a regional glacier volume inventory. Annals of Glaciology 54 (64), 179–188.

Fischer, A., Markl, G., 2009. Mass balance measurements on Hintereisferner, Kesselwandferner and Jamtalferner 2003 to 2006: database and results. Zeitschrift für Gletscherkunde und Glazialgeologie 42 (1), 47–83.

Fischer, A., Markl, G., Kuhn, M., 2013. Glacier Mass Balances and Elevation Zones of Hintereisferner, Ötztal Alps, Austria, 1952/1953 to 2010/2011. Institut für Interdisziplinäre Gebirgsforschung der Österreichischen Akademie der Wissenschaften, Innsbruck. http://dx.doi.org/10.1594/PANGAEA.818898.

Fischer, A., Markl, G., Kuhn, M., 2014c. Glacier Mass Balances and Elevation Zones of Jamtalferner, Silvretta, Austria, 1988/1989 to 2013/2014. Institut für Interdisziplinäre Gebirgsforschung der Österreichischen Akademie der Wissenschaften, Innsbruck. http://dx.doi.org/10.1594/PANGAEA.818772.

Fischer, A., Markl, G., Schneider, H., Abermann, J., Kuhn, M., 2014b. Glacier Mass Balances and Elevation Zones of Kesselwandferner, Ötztal Alps, Austria, 1952/1953 to 2012/2013. Institut für Interdisziplinäre Gebirgsforschung der Österreichischen Akademie der Wissenschaften, Innsbruck. http://dx.doi.org/10.1594/PANGAEA.818757.

Fischer, A., Olefs, M., Abermann, J., 2011. Glaciers, snow and ski tourism in Austria's changing climate. Annals of Glaciology 52/58, 89–96.

Fischer, A., Patzelt, G., Kinzl, H., 2013a. Length Changes of Austrian Glaciers 1969–2013. Institut für Interdisziplinäre Gebirgsforschung der Österreichischen Akademie der Wissenschaften, Innsbruck. http://dx.doi.org/10.1594/PANGAEA.82182.

Fischer, A., Seiser, B., Stocker-Waldhuber, M., Mitterer, C., Abermann, J., 2015a. Tracing glacier changes in Austria from the Little Ice Age to the present using a LiDAR-based high-resolution glacier inventory in Austria. The Cryosphere 9 (2), 753–766. http://dx.doi.org/10.5194/tc-9-753-2015.

Fischer, A., Span, N., Kuhn, M., Massimo, M., Butschek, M., 2007. Radarmessungen der Eisdicke Österreichischer Gletscher. Band II: Messungen 1999 bis 2006. Österreichische Beiträge zu Meteorologie und Geophysik 39, 142 pp.

Fischer, A., Stocker-Waldhuber, M., Reingruber, K., Helfricht, K., 2015b. Glacier Mass Balances and Elevation Zones of Hallstätter Gletscher, Dachstein, Austria, 2006/2007 to 2013/2014. Institut für Interdisziplinäre Gebirgsforschung der Österreichischen Akademie der Wissenschaften, Innsbruck. http://dx.doi.org/10.1594/PANGAEA.806609.

Fischer, A., Stocker-Waldhuber, M., Seiser, B., Hynek, B., Slupetzky, H., 2014a. Glaciological monitoring in the Hohe Tauern National Park. Ecomont 6 (1), 5–62.

Forel, F.A., Du Pasquier, L., 1896. Les variations périodiques des glaciers. Ier Rapport, 1895. Extrait des Archives des Sciences physiques et naturelles 101/4 (2), 129–147.

Fritzsch, M., 1898. Verzeichnis der bis zum Sommer 1896 in den Ostalpen gesetzten Gletschermarken. Vienna.

Gattermayr, W., 2013. Das hydrographischer Regime der Ötztaler Ache. In: Koch, E.-M., Erschbamer, B. (Eds.), Wetter und Klima im Wandel. Innsbruck University Press, Innsbruck, pp. 121–155.

Goldberger, J., Hoffmann, F., 2013. Glacier mass balance Übergossene Alm, Austria, pp. 1965–1975. http://dx.doi.org/10.1594/PANGAEA.821517.

Groß, G., 1987. Der Flächenverlust der Gletscher in Österreich 1850-1920-1969. Zeitschrift für Gletscherkunde und Glazialgeologie 23 (2), 131–141.

Haeberli, W., Huggel, C., Paul, F., Zemp, M., 2013a. Glacial responses to climate change. In: Shroder, J.F., James, L., Harden, C.P., Clague, J.J. (Eds.), Treatise on Geomorphology, 13. Academic Press, San Diego, pp. 152–175.

Haeberli, W., Paul, F., Zemp, M., 2013b. Vanishing glaciers in the European Alps. In: Crutzen, P.J., Bengtsson, L., Ramanathan, V. (Eds.), Fate of Mountain Glaciers in the Anthropocene, Proceedings of the Working Group 2-4 April 2011, Pontifical Academy of Sciences, Scripta Varia, vol. 118, pp. 1–9.

Hautzenberg, M., 2013. Schutz und Nutzung der Gletscher im alpinen Rechtsraum. In: Fachbeiträge des Österreichischen Alpenvereins, vol. 38. Alpine Raumordnung, Innsbruck, 188 pp.
Helfricht, K., Hanzer, F., Strasser, U., Kuhn, M., 2015. Airborne scanning: a basis for hydrological studies in glacierized catchments. Zeitschrift für Gletscherkunde und Glazialgeologie 47/48, 121–134.
Helfricht, K., Kuhn, M., Keuschnig, M., Heilig, A., 2014a. LiDAR snow cover studies on glaciers in the Ötztal Alps (Austria): comparison with snow depths calculated from GPR measurements. The Cryosphere 8 (1), 41–57. http://dx.doi.org/10.5194/tc-8-41-2014.
Helfricht, K., Schöber, J., Schneider, K., Sailer, R., Kuhn, M., 2014b. Interannual persistence of the seasonal snow cover in a glacierized catchment. Journal of Glaciology 60 (223), 889–904. http://dx.doi.org/10.3189/2014JoG13J197.
Hess, H., 1904. Die Gletscher. Friedrich Vieweg und Sohn, Braunschweig, 426 pp.
Hoinkes, H., 1970. Methoden und Möglichkeiten von Massenhaushaltsstudien auf Gletschern. Zeitschrift für Gletscherkunde und Glazialgeologie 6, 37–90.
Hoinkes, H., Lang, H., 1961. Über Niederschlag und Abfluß im Gebiet des Hintereisferners, 1957/58 und 1958/59. Wetter und Leben Sonderheft IX 102–106.
Hoinkes, H., Steinacker, R., 1975. Zur Parametrisierung der Beziehung Klima-Gletscher. Rivista Italiana di Geofisica 1, 97–104.
Huss, M., 2013. Density assumptions for converting geodetic glacier volume change to mass change. The Cryosphere 7, 877–887. http://dx.doi.org/10.5194/tc-7-877-2013.
Huss, M., Zemp, M., Joerg, P.C., Salzmann, N., 2014. High uncertainty in 21st century runoff projections from glacierized basins. Journal of Hydrology 510, 35–48.
Hynek, B., Unger, R., Schöner, W., Binder, D., Weyss, G., 2012. Comparison of direct and geodetic mass balances of three small glaciers in the Eastern Alps. Geophysical Research Abstracts 14. EGU2012–9025.
Joerg, P., Morsdorf, F., Zemp, M., 2012. Uncertainty assessment of multi-temporal airborne laser scanning data: a case study on an Alpine glacier. Remote Sensing of Environment 127, 118–129. http://dx.doi.org/10.1016/j.rse.2012.08.012.
Kargel, J.S., Leonard, G.J., Bishop, M.P., Kaab, A., Raup, B. (Eds.), 2014. Global Land Ice Measurements from Space. Springer-Praxis.
Kaser, G., Großhauser, M., Marzeion, B., 2010. Contribution potential of glaciers to water availability in different climate regimes. Proceedings of the National Academy of Sciences 107, 20223–20227. http://dx.doi.org/10.1073/pnas.1008162107.
Kinzl, H., 1979. Die Gletscher der Österreichischen Alpen 1977/78. Zeitschrift für Gletscherkunde und Glazialgeologie 15 (1), 105–113.
Klebelsberg, R.V., 1943. Die Alpengletscher in den letzten 30 Jahren (1911–1941). Petermanns Mitteilungen 89, 23–32.
Kretschmer, I., 2002. Carl Sonklars Atlas der Ötztaler Gebirgsgruppe – ein früher Gletscheratlas der Ostalpen. Cartographica Helvetica 25, 11–20.
Kuhn, M., Batlogg, N., 1998. Glacier runoff in alpine headwaters in a changing climate. IAHS Publication 79, 79–88.
Kuhn, M., Dreiseitl, E., Hofinger, S., Markl, G., Span, N., Kaser, G., 1999. Measurements and models of the mass balance of Hintereisferner. Geografiska Annaler A, Physical Geography 81 (4), 659–670.
Kuhn, M., Lambrecht, A., Abermann, J., 2013. Austrian Glacier Inventory 1998. http://dx.doi.org/10.1594/PANGAEA.809196.

Kuhn, M., Markl, G., Kaser, G., Nickus, U., Obleitner, F., Schneider, H., 1985. Fluctuations of climate and mass balance: different responses of two adjacent glaciers. Zeitschrift für Gletscherkunde und Glazialgeologie 21, 409–461.

Lambrecht, A., Kuhn, M., 2007. Glacier changes in the Austrian Alps during the last three decades, derived from the new Austrian glacier inventory. Annals of Glaciology 46, 177–184.

Lieb, G.K., Slupetzky, H., 2004. Gletscherweg Pasterze. In: Naturkundlicher Führer zum Nationalpark Hohe Tauern, vol. 2. Innsbruck, 122 pp.

Lieb, G.K., Slupetzky, H., 2011. Die Pasterze. Anton Pustet, Salzburg, 160 pp.

Marzeion, B., Kaser, G., 2014. Peak Water: An Unsustainable Increase in Water Availability From Melting Glaciers. Mountains and Climate Change: A Global Concern. Centre for Development and Environment (CDE), Swiss Agency for Development and Cooperation (SDC) and Geographica Bernensia, Bern, pp. 47–50.

Mayer, C., Escher-Vetter, H., Weber, M., 2013. 46 Jahre glaziologische Massenbilanz des Vernagtferners. Zeitschrift für Gletscherkunde und Glazialgeologie 45/46, 219–234.

Müller, G., 1973. Franz Keil. Ein Alpenforscher und Pionier der plastischen Gebirgsdarstellung (1822–1876). Mitteilungen der Gesellschaft für Salzburger Landeskunde 116, 287–310.

Nicolussi, K., 2013. Die historischen Vorstöße und Hochstände des Vernagtferners 1600–1850 AD. Zeitschrift für Gletscherkunde und Glazialgeologie 45/46 (2011/12).

Patzelt, G., 1970. Die Längenmessungen an den Gletschern der österreichischen Ostalpen 1890–1969. Zeitschrift für Gletscherkunde und Glazialgeologie 6 (1–2), 151–159.

Patzelt, G., 1977. Statistik der Längenmessungen an den österreichischen Gletschern 1960 bis 1975. Zeitschrift für Gletscherkunde und Glazialgeologie 12 (1), 91–94.

Patzelt, G., 1980. The Austrian Glacier inventory: status and first results. IAHS Publication 126, 181–183.

Prokop, A., Schirmer, M., Lehning, M., Stocker, M., 2008. A comparison of measurement methods: terrestrial laser scanning, tachymetry and snow probing, for the determination of spatial snow depth distribution on slopes. Annals of Glaciology 49, 210–216.

Richter, E., 1885. Beobachtungen an den Gletschern der Ostalpen. II. Die Gletscher der Ötztaler Gruppe im Jahre 1883. Zeitschrift des Deutschen und Österreichischen Alpenvereins 16, 54–65.

Richter, E., 1891. Geschichte der Schwankungen der Gletscher der Ostalpen. Zeitschrift des Deutschen und Oesterreichischen Alpenvereins 22, 1–74.

Richter, E., 1892. Urkunden über die Ausbrüche des Vernagt- und Gurglergletschers im 17. und 18. Jahrhundert Forschungen zur deutschen Landes- und Volkskunde 6, 345–440.

Richter, E., 1893. Bericht über die Schwankungen der Gletscher der Ostalpen 1888–1892. Zeitschrift des Deutschen und Oesterreichischen Alpenvereins 24, 473–485.

Richter, E., 1894. Die wissenschaftliche Erforschung der Ostalpen seit der Gründung des Österreichischen und Deutschen Alpenvereins. Zeitschrift des Deutschen und Oesterreichischen Alpenvereins 25, 1–94.

Rott, H., 1977. Analyse der Schneeflächen auf Gletschern der Tiroler Zentralalpen aus Landsat-Bildern. Zeitschrift für Gletscherkunde und Glazialgeologie 12 (1), 1–28.

Schlagintweit, K., Schlagintweit, A., 1850. Untersuchungen über die physikalische Geographie der Alpen in ihren Beziehungen zu den Phänomenen der Gletscher, zur Geologie. Meteorologie und Pflanzengeographie, Leipzig, 600 pp.

Schneider, H., 1970. Die Grundlagen der Vermessungen am Kesselwandferner(Ötztaler Alpen) und die Bewegung dieses Gletschers in den Haushaltsjahren 1965/66, 1966/67 und 1967/68 (Dissertation). University of Innsbruck, Innsbruck, 186 pp.

Schöber, J., Schneider, K., Helfricht, K., Schattan, P., Achleitner, S., Schöberl, F., 2014. Snow cover characteristics in a glacierized catchment in the Tyrolean Alps – improved spatially distributed modeling by usage of LiDAR data. Journal of Hydrology 519, 3492–3510. http://dx.doi.org/10.1016/j.jhydrol.2013.12.054.

Schöner, W., Auer, I., Böhm, R., 2009a. Long term trend of snow depth at Sonnblick (Austrian Alps) and its relation to climate change. Hydrological Processes 22, 1052–1063. http://dx.doi.org/10.1002/hyp.7209.

Schöner, W., Hynek, B., Böhm, R., Binder, D., Koboltschnig, G., Holzmann, H., 2009b. Glazialhydrologisches Monitoring im Bereich des Rauriser Sonnblicks (Hohe Tauern, Österreich). Internationale Forschungsgesellschaft Interpraevent Klagenfurt Austria Österreichische Wasser- und Abfallwirtschaft 61 (1), 1–8. http://dx.doi.org/10.1007/s00506-008-0059-z.

Simony, F., 1863. Beitrag zur Kunde der Ötztaler Alpen. Mitteilungen des Deutschen und Oesterreichischen Alpenvereins 1, 1–24.

Slupetzky, H., 2015. Die Massenbilanzreihe vom Stubacher Sonnblickkees 1946 bis 2014 und die semidirekte Berechnung des Massenhaushalts von Gletschern. Zeitschrift für Gletscherkunde 47/48, 167–200.

Slupetzky, H., Ehgartner, G., 2014. Glacier Mass Balances of Stubacher Sonnblickkees, Hohe Tauern Range, Eastern Alps, Austria, 1958/1959 to 2012/2013. http://dx.doi.org/10.1594/PANGAEA.829950.

Sonklar, K.V., 1857. Das Ötztaler Eisgebiet. Mitteilungen der Geographischen Gesellschaft Wien, pp. 1–15.

Sonklar, K.V., 1860. Die Ötztaler Gebirgsgruppe, mit besonderer Rücksicht auf Orographie und Gletscherkunde. Gotha, 292 pp.

Sonklar, K.V., 1877. Studie über den Gurgler Gletscher in der Ötztaler Gebirgsgruppe. Zeitschrift des Deutschen und Oesterreichischen Alpenvereins 8, 1–14.

Span, N., Kuhn, M., Schneider, H., 1997. 100 years of ice dynamics of Hintereisferner 1894–1994. Annals of Glaciology 24, 297–302.

Span, N., Fischer, A., Kuhn, M., Massimo, M., Butschek, M., 2005. Radarmessungen der Eisdicke österreichischer Gletscher, Band I: Messungen 1995 bis 1998. Österreichische Beiträge zur Meteorologie und Geophysik 33, 146 pp.

Span, N., Kuhn, M., 2003. Simulating annual glacier flow with a linear reservoir model. Journal of Geophysical Research 108 (D10). ACL 8-1–8-9.

Stocker-Waldhuber, M., Helfricht, K., Hartl, L., Fischer, A., 2015. Glacier surface mass balance 2006–2014 on Mullwitzkees and Hallstätter Gletscher, Austria. Zeitschrift für Gletscherkunde und Glazialgeologie 47, 101–119.

Tiris, 2016. Historische Karten. https://www.tirol.gv.at/kunst-kultur/landesarchiv/historische-karten/.

Vaughan, D.G., Comiso, J.C., Allison, I., Carrasco, J., Kaser, G., Kwok, R., Mote, P., Murray, T., Paul, F., Ren, J., Rignot, E., Solomina, O., Steffen, K., Zhang, T., 2013. Observations: cryosphere. In: Stocker, T.F., Qin, D., Plattner, G.-K., Tignor, M., Allen, S.K., Boschung, J., Nauels, A., Xia, Y., Bex, V., Midgley, P.M. (Eds.), Climate Change 2013: the Physical science basis. Contribution of Working Group I to the Fifth Assessment Report of the Intergovernmental Panel on Climate Change. Cambridge University Press, Cambridge, United Kingdom and New York, NY, USA.

Wehr, A., Lohr, U., 1999. Airborne laser scanning – an introduction and overview. ISPRS Journal of Photogrammetry and Remote Sensing 54, 68–82. http://dx.doi.org/10.1016/S0924-2716(99)00011-8.

Whiteman, C.D., 2000. Mountain Meteorology: Fundamentals and Applications. Oxford University Press, New York, 376 pp.

Zemp, M., Armstrong, R., Gärtner-Roer, I., Haeberli, W., Hoelzle, M., Kääb, A., Kargel, J.S., Singh Khalsa, S.J., Leonard, G.J., Paul, F., Raup, B., 2014. Chapter 1: introduction: global glacier monitoring – a long-term task integrating in situ observations and remote sensing. In: Kargel, J.S., Leonard, G.J., Bishop, M.P., Kääb, A., Raup, B.H. (Eds.), Global Land Ice Measurements from Space. Springer Praxis Books, Heidelberg, pp. 1–21. http://dx.doi.org/10.1007/978-3-540-79818-7_1.

Zemp, M., Frey, H., Gärtner-Roer, I., Nussbaumer, S.U., Hoelzle, M., Paul, F., Haeberli, W., Denzinger, F., Ahlstroem, A.P., Anderson, B., Bajracharya, S., Baroni, C., Braun, L.N., Caceres, B.E., Casassa, G., Cobos, G., Davila, L.R., Delgado Granados, H., Demuth, M.N., Espizua, L., Fischer, A., Fujita, K., Gadek, B., Ghazanfar, A., Hagen, J.O., Holmlund, P., Karimi, N., Li, Z., Pelto, M., Pitte, P., Popovnin, V.V., Portocarrero, C.A., Prinz, R., Sangewar, C.V., Severskiy, I., Sigurdsson, O., Soruco, A., Usubaliev, R., Vincent, C., 2015. Historically unprecedented global glacier decline in the early 21st century. Journal of Glaciology 61 (228), 745–762. http://dx.doi.org/10.3189/2015JoG15J017.

Zemp, M., Thibert, E., Huss, M., Stumm, D., Rolstad Denby, C., Nuth, C., Nussbaumer, S.U., Moholdt, G., Mercer, A., Mayer, C., Joerg, P.C., Jansson, P., Hynek, B., Fischer, A., Escher-Vetter, H., Elvehøy, H., Andreassen, L.M., 2013. Reanalysing glacier mass balance measurement series. The Cryosphere 7, 1227–1245. http://dx.doi.org/10.5194/tc-7-1227-2013.

Index

'*Note*: Page numbers followed by "f" indicate figures and "t" indicate tables.'

G

H

T

U

V

W

FSC
www.fsc.org
MIX
Papier | Fördert
gute Waldnutzung
FSC® C083411

Kolibri 360 GmbH
Ferdinand-Jühlke-Straße 7
99095 Erfurt, Deutschland
produktsicherheit@kolibri360.de